Rutile silk,
Mogok, Burma (Myanmar).

Healed fissure,
Mogok, Burma (Myanmar).

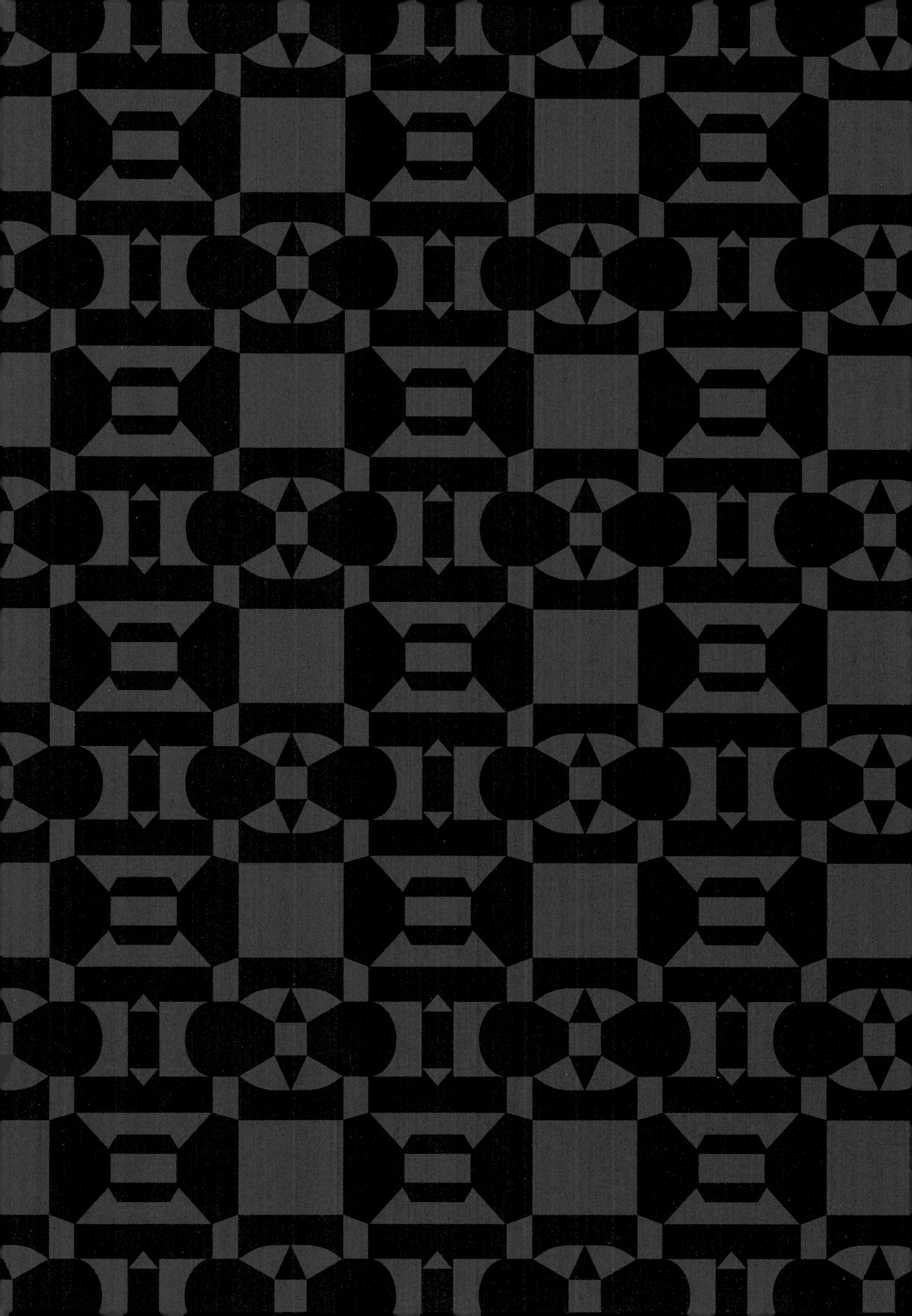

Plagioclase crystals and growth zoning,
Inverell, Australia.

Rutile silk,
Mogok, Burma (Myanmar).

Rutile crystal,
Mogok, Burma (Myanmar).

Rutile silk,
Ceylon (Sri Lanka).

Negative crystal with CO$_2$-rich fluid,
Ceylon (Sri Lanka).

Healed fissure,
Ceylon (Sri Lanka).

SAPPHIRE

JOANNA HARDY

SAPPHIRE

A
CELEBRATION
OF
COLOUR

Edited by
Robert Violette

First published in the United Kingdom in 2021 by
Thames & Hudson Ltd, 181A High Holborn, London WC1V 7QX

First published in 2021 in the United States of America by
Thames & Hudson Inc., 500 Fifth Avenue, New York, New York 10110

Published in association with Violette Editions
www.violetteeditions.com

Reprinted 2022, 2024

Edited, originated and produced by Robert Violette, Violette Editions
Designed by Studio Frith

Picture research by Joanna Hardy, Georgina Izzard and Robert Violette
Copyedited by Robert Violette and Sarah Kane
Editorial assistance and research by Georgina Izzard
Pre-press and digital proofing by Violette Editions
with retouching by Jamie Register

For their kind assistance in the origination of this book,
Robert Violette would like to extend his thanks to: its expert author,
Joanna Hardy; to the archivists at the numerous historic jewellery
houses represented in this volume; to Sean Gilbertson, Sophie Ebbetts,
Jennifer Head and Gemfields Ltd for their generous support; to Jonny
Clowes and Rachel Dewhirst at Thames & Hudson Ltd; to Frith Kerr,
Leah Byrne, Claire Köster, Hannah Mullen and Imogen Walker at
Studio Frith; and to Georgina Izzard, Sarah Kane, Jamie Register;
to Adrian C Williams at Club Type and to Giuseppe Gilardi and
all at printer Grafiche Milani.

British Library Cataloguing-in-Publication Data
A catalogue record for this book is available from the British Library

Library of Congress Control Number 2021935377

ISBN 978-0-500-024775

Printed and bound in Italy by Grafiche Milani

Frontispiece: A portrait of Queen Marie of Romania
wearing a tiara executed by Cartier Paris in 1909
for the Grand Duchess Vladimir in platinum, gold,
silver, sapphires and diamonds, published in
The Tatler, 22 June 1927.

Cover: Sapphire and diamond demi-parure,
Mellerio dits Meller, *c.* 1870; see page 111.

Back cover band: Elizabeth Taylor wearing her Tiffany Fleur de Mer brooch,
with Richard Burton and their daughters, photographed by Ron Galella in 1968;
Elizabeth Taylor's Bulgari sautoir, 1969, with a 52.72-carat Burmese sugarloaf
sapphire, gifted to her by Richard Burton in 1972 on her 40th birthday.

CONTENTS

INTRODUCTION

Above: Calibré-cut emeralds, rubies and sapphires
— the three gems of this trilogy — from the jewels
of the Duchess of Windsor are set in the plumes
of this Flamingo brooch, made by Cartier Paris in
1940 for the Duke of Windsor to gift to his wife.

Below: Fine sapphire crystals are prized for
their durability, but carving them into figures as
detailed as this Indian Buddha, which is less than
2 cm tall, takes exceptional skill. Whilst Hindus
are wary of blue sapphires, Buddhists may have
placed this sapphire in a shrine before it was
mounted as a pin, likely Victorian. This piece
is now in the collection of the Natural History
Museum, London.

Introduction

'The sapphire was like the pure sky, and mighty Nature had endowed it with so great a power that it might be called sacred and the gem of gems' — MARBOD, *Liber de lapidibus*, 1531

Mighty Nature has indeed bestowed upon us a formidable gem: sapphire. From our belief in sapphire's connection to heavenly powers, to our confidence in its capacity to protect and cure us from ailments, humans have attributed greater sacred and restorative qualities to sapphire than to any other coloured gemstone. While emerald and ruby have shared the centre stage of our gemstone fascination, sapphire has remained our constant reliable companion.

Sapphire is the final volume of this gemstone trilogy. Its companions, *Emerald* (2013) and *Ruby* (2017), also follow the finest examples of their respective prized gem through history and across the world, but neither can match sapphire's breadth of use, interpretation and, most importantly, colour. Sapphire naturally occurs in virtually every colour: blue to green and grey, yellow to orange, pink and purple. Only the red sapphire – ruby – is separated by its appearance. The most famous of sapphire's colours is blue, but there is still an infinite range of tones and hues of blue sapphire, as every stone is unique. This book's sections of wonderful full-page photographs of sapphire inclusions testify to this stone's impressive range – there truly is a sapphire for everyone.

Sapphire crystals present us with characteristics that neither emerald nor ruby can match, characteristics that make sapphire so flexible in its use. Sapphire often occurs in bigger crystals than emerald or ruby, and their scale has long been used to embolden the wearer and impress onlookers. The stones included in the chapter Showstoppers are quite incredible and extremely rare; the sheer size of some of them is astonishing, particularly that of the largest Kashmir sapphire ever to come onto the market (page 273), which was auctioned just before publication. Sapphire's large crystals, coupled with its durability, also make it a candidate for carving into cameos, intaglios and even sculptures, as seen in the chapter Profiles & Portraits. Though carving is a delicate operation with all gemstones, emerald's brittleness in particular restricts the process and possibilities, which is why emeralds have so seldom been carved as cameos or intaglios. Similarly, rubies are typically found in much smaller crystals, so they too have not lent themselves to figurative carving.

Some of the jewels in this book are as big and bold as the sapphires they hold, but others are simple, letting the sapphires speak for themselves – whether to the gods or high society. This book is therefore a celebration of sapphire: its colours, its stories and the collections it has become part of.

In the opening chapter, Early Trade, we follow sapphire's journey from the earliest mined deposits in Sri Lanka to China, into the Middle East and across to Europe as it passed through the hands of traders along the land-based Silk Road and the maritime silk routes. Intrepid travellers such as Marco Polo and Niccolò de' Conti pursued sapphire, other gemstones and spices along these routes and returned home to Europe with stories of kings harbouring great treasuries of sapphires. The blue sapphire was drilled and first traded as strings of beads, not for its monetary value, which was negligible, but for its talismanic properties.

Writing this trilogy has introduced me to a whole other world in which gemstones are appreciated and desired for their perceived healing and transcendental qualities. My first introduction to the use of sapphire for its supposed medicinal and magical properties was when holding the Middleham Jewel. This fifteenth-century gold lozenge-shaped locket (page 52) is engraved with biblical scenes and a magical incantation – a daring mix of the religious and the occult – and its talismanic power is reinforced with a sapphire to ward off disease and misfortune. I was mesmerized by this single blue sapphire bead that would have imparted reassurance in a period when death and decay were only just around the corner and good health was in no way guaranteed.

Sapphire has been endowed with cultural significance across the world. Though there are nuances in interpretation, sapphire's potency, whether thought to be good or bad, is constant. Sapphire's blue colour has linked it to the sky and therefore to the heavens and the divine. Medieval goldsmiths, many of whom were from the clergy, were kept busy making stunning reliquary jewellery and objects of devotion for the Church or for high-ranking individuals. The golden reliquary cross commissioned by Countess Gertrude of Brunswick in 1038, below, has a large sapphire at its centre, a position that highlights the importance of sapphire and connects it to the relics of St Peter that the cross was meant to contain. More stunning medieval craftsmanship can be viewed in my all-time favourite museum, the Munich Residenz, which houses three spectacular crowns that are the first jewels to greet you as you enter the Treasury vaults. I never tire of viewing the Treasury's splendid Sapphire Cup (page 85) as its craftsmanship is exquisite: the cup is delicately enamelled, and atop its lid a golden warrior holds aloft a large ring carved from a single sapphire crystal.

The crowns in the Residenz are not alone in being set with sapphires; sapphire's connection to the skies and the divine made it a useful and important addition to these highly symbolic ornaments, helping to confer divine power upon mortal rulers. Indeed, before the

In medieval Europe, blue, and therefore blue sapphire, was linked to
the heavens and divine power. Above, from left: Crown of St Wenceslas,
made 1346, part of the Bohemian crown jewels; the reliquary cross
with a large sapphire at its centre from the portable altar, 1038,
of Countess Gertrude of Brunswick (Braunschweig); and the blue
robes of the Virgin Mary and angels in the Wilton Diptych,
made c.1395–9 for King Richard II (1367–1400).

Cullinan II Diamond was set at the front of the British Imperial State Crown in 1910, it was the Stuart Sapphire that occupied the prime position. The sapphire is still set in the crown and is accompanied, right at the top of the crown's cross, by St Edward's Sapphire, the oldest gemstone in the Royal Collection. Likewise, sapphire demands all your attention when viewing the crown of Maria II, Queen of Portugal (1819–1853, r. 1826–53), pictured here, made in the mid-nineteenth century. Its five large sapphires weigh approximately 35 carats in total. These stones were chosen for their continued association with the Church, as they represent the heavens' support for the Queen's position as head of state. Sadly, her reign was short-lived as, at the age of 34, she died after complications during the birth of her 11th child.

Sapphire gained one of its firmest royal seals of approval from Queen Victoria, whose husband, Prince Albert, commissioned, designed and bestowed on her many sapphire jewels. These tokens of his love inspired many of their descendants, including several Romanovs, to gift their own sapphire-set jewels to mark engagements and weddings – especially since, in the language of gems, sapphire represents constancy.

After Queen Victoria's death at the very beginning of the twentieth century, fashions in Britain broke free from the stoicism of mourning jewellery, and sapphire, in all its colours, was the perfect gemstone to complement Art Nouveau and Art Deco designs. Discoveries of sapphires in Kashmir, Australia and Montana, USA, enabled a supply of different colours and qualities of sapphire to reach jewellers ready to appeal to all clients. The sapphire lent itself perfectly to the 1930s Art Deco period; cut in geometric shapes and paired with diamonds set in wide bracelets, blue sapphires created the dramatic monochromatic look of the moment. A bracelet made in 1937 by Van Cleef & Arpels, illustrated on the opening page of the Art of Collecting chapter (page 234), is set with baguette-cut diamonds in linear settings that frame the seven spectacular emerald-cut sapphires. It is an outstanding jewel that would be almost impossible to replicate today, as it takes years to source matching stones of such high quality.

In the chapters New Horizons and Jewels of Impact, I celebrate women and, specifically, the key roles they played in shaping the jewels of the Art Deco period. Many jewellery connoisseurs have heard of the designer René Boivin, but it was his widow, Jeanne Boivin, who really expanded the business. Jeanne continued under René's name and hired a young

Above, and right: The crown of Maria II,
Queen of Portugal (1819–1853, r. 1826–53), made *c.*1850,
with nine detachable brooch elements set, alternately,
with sapphires and diamonds; and an eighteenth-century
ring with ruby and sapphire hearts surmounted by a
diamond-set crown believed to be a gift to the young
Queen Victoria from her mother.

girl, Suzanne Belperron, to help with the designs. Belperron went on to be successful in her own right. At the same time, Jeanne Toussaint was making waves at Cartier. Toussaint was a visionary and, by all accounts, a formidable woman – she was nicknamed 'the Panther'. She had come from humble beginnings, which I think shaped her determination, but became a close confidante of Louis Cartier and later, in 1933, the artistic director of Cartier Paris. Meanwhile, at Van Cleef & Arpels, the young Renée Puissant, self-taught in design, worked closely with the *maison*'s designer René Sim Lacaze to build a formidable design legacy.

But these women were not basking in glory; like so many others, they faced considerable hardship during World War II. Their tenacity, and their passion for their work, clearly saw them through incredibly difficult times, as they managed to create some of the most iconic jewels of the period.

Bold Art Deco jewels appealed to the stars of stage and screen. In the blossoming world of Hollywood, actresses like Joan Crawford wanted their jewellery to be visible, and the right sapphire could offer them another special enticement: the bright star-like optical effect of asterism – very apt to their profession. Even though ruby can display asterism too, ruby crystals are generally much smaller than sapphire crystals. Most of the large star sapphires were found in Sri Lanka, where it was believed that these stones, each with its moving 'star', would serve as a form of protection and guard against all kinds of witchcraft. While researching for the Star Sapphire chapter I met with Salih Maricar, from a dynasty of gem dealers, and he showed me a photograph of his great-grandfather on the tarmac at Colombo airport in Ceylon (Sri Lanka) holding a briefcase that contained one of the largest star sapphires in the world. This stone, the Star of Asia, is now part of the Smithsonian's collection. In another instance, Cher, the pop icon, is pictured wearing the huge Black Star of Queensland sapphire (page 188), a stone that was used, as I found out during research for the Sapphire Discoveries chapter, as a doorstop for years by the original owner before he discovered what it was. Sapphire has many engaging stories, like these, that have a touch of humility.

Much of sapphire's story and charm comes from the collections it is part of. Collecting is a key theme of *Sapphire*, and I shine a spotlight on fabulous sapphire jewels in collections around the world. Actress Elizabeth Taylor's collection included the most incredible jewels

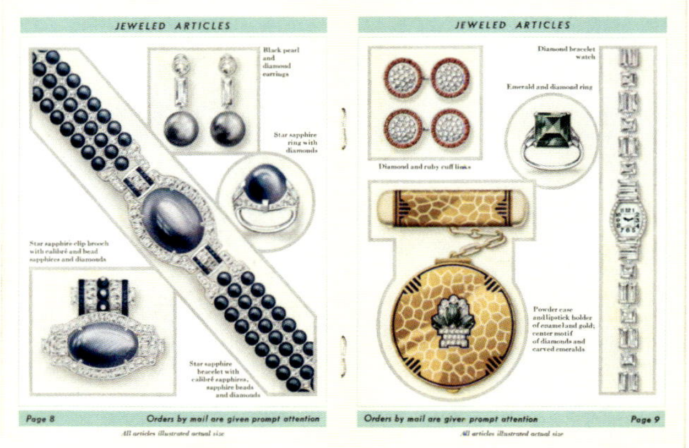

Above: A sapphire-set Thistle brooch by Verdura, 1962, with round and calibré-cut sapphires between diamond-set leaves and stem.

Left: A double page from a 1933 Cartier New York catalogue showing, on the left page, three star sapphire Art Deco jewels with diamonds and sapphire beads.

set with some of the world's best gemstones, including her Bulgari sugarloaf sapphire sautoir (page 225). Marjorie Merriweather Post, American heiress and businesswoman in her own right, with her striking sapphire and diamond Cartier necklace (page 217), and the Duchess of Windsor, who had some incredible sapphire jewels, wore their impactful sapphire jewellery with such elegance.

Part of what makes sapphire so exciting to collect is its colour – its full range of colours. Colour is very subjective – even the demarcation between pink sapphire and ruby does not have an objective measure – and a gemstone's colour can change in different light sources. But that is part of sapphire's allure: everyone will find a colour that they consider electric and that stirs an emotion. The array of unheated sapphires that are showcased by 18 leading designers in the 21st Century chapter makes clear the endless possibilities of sapphire.

When I asked a few traders about the colour they most associate with a fine blue sapphire, they said it was the colour of a cornflower, or the feathers from the neck or tail of a peacock. When I look at a peacock feather, that electric blue is breathtaking to behold – if you ever get a sapphire near to that colour, you know it is a very special stone. The unmounted 130-carat Burmese sapphire, above, from a splendid brooch is a true 'peacock' blue for me. As well as being a fabulous colour for blue sapphires, peacocks symbolize beauty, well-being, splendour and rebirth through their stunning feathers that they renew each year. Together, the peacock and the sapphire have inspired many jewellers, and representations of the peacock feather have been carefully designed, like the stunning Boucheron peacock feather Question Mark necklace on page 299.

Sapphires are also known by other colour associations, such as royal blue, velvet, indigo, twilight and padparadscha, a term used to describe a sapphire that displays a mix of pink and orange, like the colours of a lotus flower and a sunset. Many of these descriptions are becoming recognized classifications; this is a dangerous trend because it can lead to an over-reliance on gemmological reports when purchasing a significant sapphire. Pastel colours, or natural-coloured sapphires that have not been heat-treated, are becoming very popular as there is an appreciation for sapphires in their natural state. When 90 per cent of sapphires on the market are heat-treated, however, unheated stones are getting harder to find.

Above: A 130.50-carat Burmese sapphire of a fine colour and clarity unmounted from its brooch, c.1860, with an openwork diamond-set gold surround within a frame of cushion-shaped diamonds.

Left: This beautifully articulated Bulgari bib necklace, 2005, exemplifies the jewellery house's 'signature': brightly coloured gemstones, in this case 395.89 carats of blue, yellow, pink, purple and orange sapphires, taking centre stage and highlighted by small diamonds.

Stars and their sapphires. Clockwise from top: Joan Crawford (1904–1977) photographed by George Hurrell in 1935 for the film *No More Ladies* (MGM, 1935) wearing her star sapphire pendant; Elizabeth Taylor (1932–2011) shows off her sugar-loaf sapphire engagement ring from Michael Wilding before flying to her wedding in 1952; Princess Grace of Monaco (1929–1982), with Horst Buchholz at a gala to benefit the Monaco Red Cross, Casino of Monte Carlo, 1972, wearing her Marguerite clip by Van Cleef & Arpels, 1956, with brilliant-cut diamonds around faceted sapphires; and Crawford, again, wearing a sapphire necklace in her hospital bed in 1964.

The colour-change sapphire is another phenomenon that adds to sapphire's intrigue. Sometimes these sapphires are referred to as 'alexandrite' sapphires because an alexandrite – which is part of the chrysoberyl family, unlike sapphire's corundum family – changes colour from red to green when viewed in different lights. In Sapphire Discoveries, I include Shirley Nuell's account of her family's journey to becoming sapphire miners in Australia, in which she refers to the find of an 'alexandrite' sapphire that, when sold, supported her family comfortably for two years.

My own favourite sapphire 'find' occurred while I was filming as a jewellery specialist for the BBC's *Antiques Roadshow*, a television programme that has run for over 40 years and has become a 'British institution'. Each year the *Antiques Roadshow* crew and its team of specialists visit venues around the UK to meet members of the public that have brought along items that they want to find out more about – I never know what is going to turn up. During filming at Pembroke Castle in Wales, a lady brought along two rings, one of which was set with a blue stone, which she thought was glass. On inspection I could see that it was a sapphire, but the ring needed cleaning before it could be assessed properly; it was only when I left daylight to hold it in the fluorescent light of the cloakroom that, to my delight, the sapphire changed colour. I was so excited to tell the owner, who was even more surprised than I was and highly delighted.

The rare colour-change phenomenon of some sapphires reminds me to be open to understanding all of sapphire's powers. Who is to say that sapphires did not cure diseases or protect people in battle? Maybe medieval dignitaries and early Indian soothsayers were indeed aware of sapphire emitting certain energy. We look at the past through twenty-first-century eyes; we will never truly know what it felt like hundreds of years ago to believe in hope through a gem. When Shirley Nuell explained to me her 'party trick' of finding sapphires when blindfolded in the outback purely with divining sticks, she was not able to explain it. Whether there is a scientific explanation or not (it is possible that greater iron content in basaltic sapphires causes them to react to the sticks), Shirley could feel the energy from these stones, and that to me is special enough.

Writing *Sapphire* has been quite an extraordinary experience. When I left to visit sapphire deposits in Madagascar in March 2020, there was word of a virus coming our way. Talking about a pandemic reaching our shores felt surreal, but I could not pass up the opportunity to travel there. Since the discovery in 1998 of sapphires in Madagascar, the country has become the biggest producer of sapphires; it is such an important deposit that I needed to see it for myself. It was quite a memorable trip, which I relay in the Sapphire Discoveries chapter; a week after returning to the UK we were in lockdown. This has been a very special book to write, thanks in part to the pandemic isolation, strange as that may sound. I am grateful for how generous people have been with their help and knowledge, while all sitting in their various homes. I have always enjoyed travelling immensely, and suddenly, like everyone else's, my travels were curtailed. Nevertheless, there were days that by phoning Australia in the morning and the USA in the afternoon, I almost felt as if I had been changing continents.

During 2020, I was able to speak with people who are normally difficult to pin down because of their very hectic schedules; the virus had forced us all to slow down. Everyone I spoke to really wanted to talk about their passion, their collections or their stone business. One collector told me that it made him feel happy to talk about his collection. Another told me they spent time with their collection every day just to see beautiful things while the impact of the pandemic was engulfing the world. The Art of Collecting chapter is the result of six collectors generously sharing their beautiful sapphire jewels with me. Catherine Cariou, a colleague with whom I have collaborated in the past and for whom I have huge admiration, works with many international collectors that have sought her expertise. She generously introduced me to some of the private collectors she advises, for which I am incredibly grateful. Very often we see jewels at auction, or in an exhibition, or in a museum, and there is undoubtedly a feeling of disconnect between the jewels and the human touch, for jewels are made to be worn. The Art of Collecting chapter has brought spectacular jewels back into view, such as the Van Cleef & Arpels' Ballerina brooches and the spectacular sapphire bracelets. When I see a beautiful jewel at an auction, I often wonder where it will end up, so finding jewels in these collections that I have not seen for years has been a revelation; speaking to the collectors, hearing about why they bought the jewel, and knowing it is appreciated and has gone to a good home, has been a real pleasure.

Over the past 10 years I have travelled the globe researching for *Emerald*, *Ruby* and *Sapphire*, and it has been a journey I would not have missed. I see this trilogy as a global team-effort, and I count myself incredibly lucky to have met people who have shared their stories with me and to have worked with the ever-helpful world brands, auction houses, cutters, miners, goldsmiths, designers and historians who have generously shared their knowledge with me. I am often asked what my favourite stone is, and my reply has always been that I love the best of any gemstone; but, since writing this book, I think my heart has been won by the sapphire of many colours. I hope you enjoy reading *Sapphire* as much as I have enjoyed writing it.

Van Cleef & Arpels Cristaux de Saphirs clip, 2010,
with diamond-set points resembling the natural
crystal habit of the clip's pink Sri Lankan sapphires.

Rutile crystal,
Songea, Tanzania.

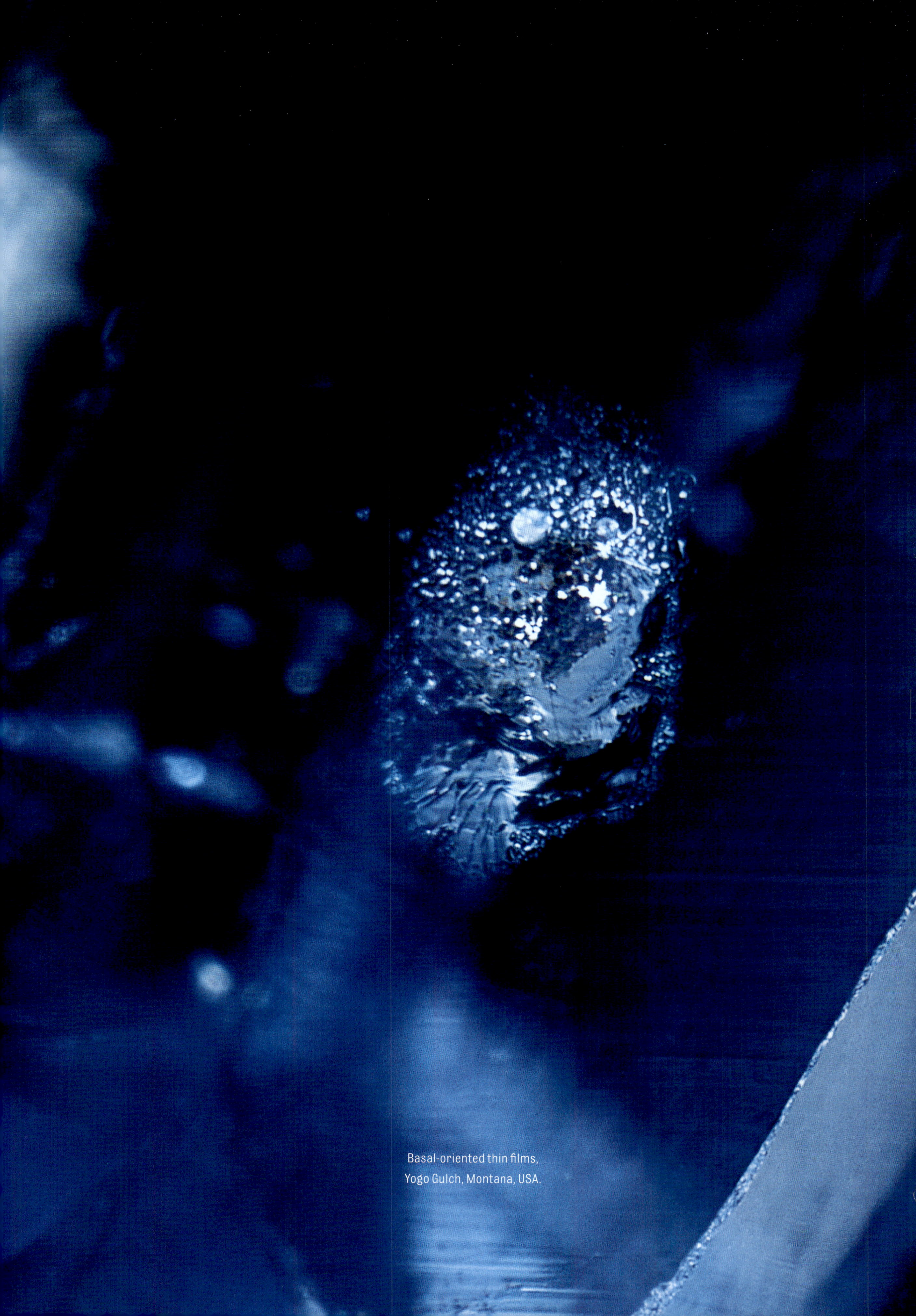

Basal-oriented thin films,
Yogo Gulch, Montana, USA.

Milky zonal clouds,
Madagascar.

Diverse inclusion patterns,
Madagascar.

Carbonate crystal with decrepitation feature,
Montepuez, Mozambique.

Diverse inclusion scene,
Garba Tula, Kenya.

Negative crystal with CO_2-rich fluid
and black graphite scale,
Burma (Myanmar).

Lattice-work of intersection tubules,
Aappaluttoq, Greenland.

EARLY TRADE

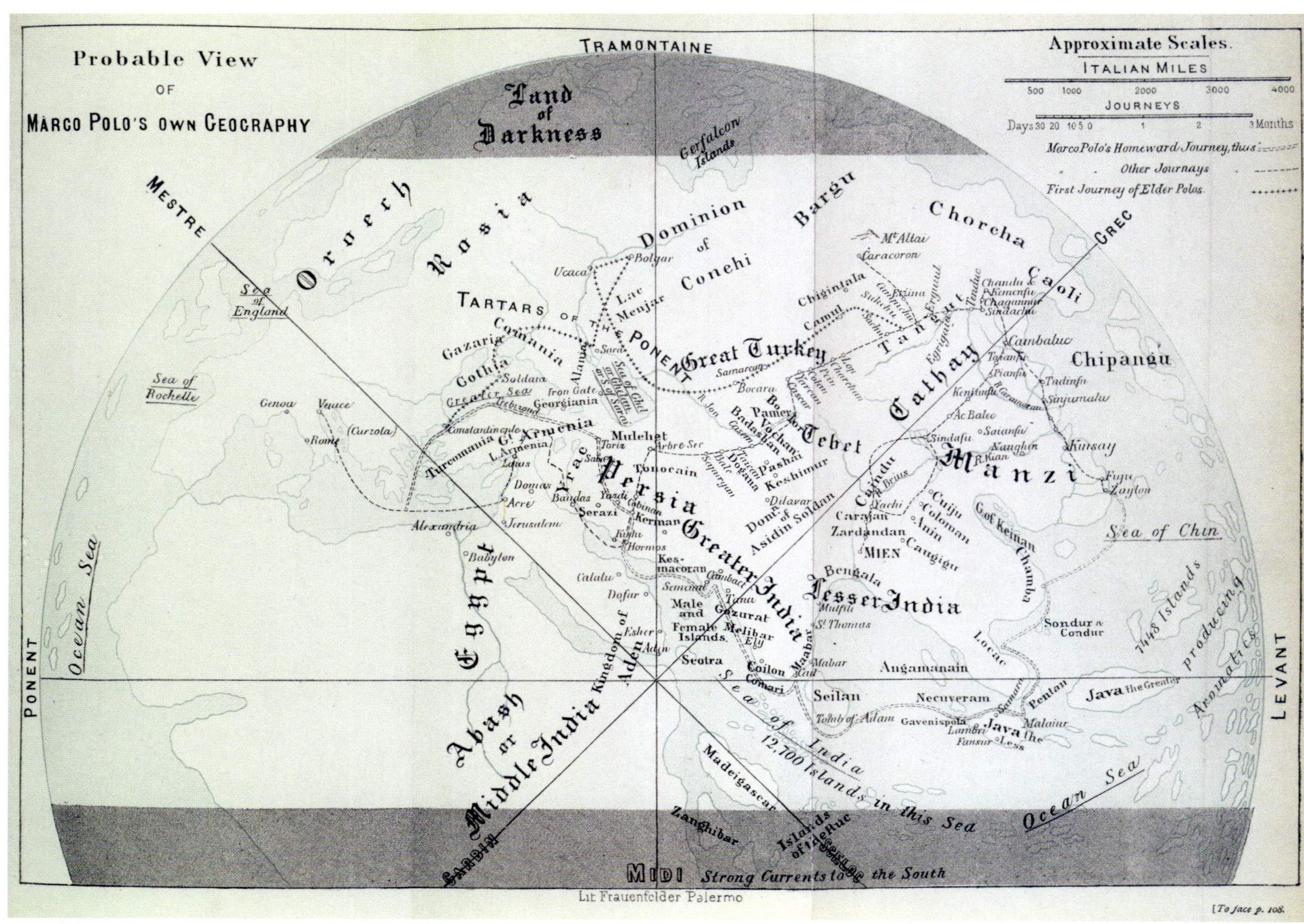

An early twentieth-century map showing the
'Probable View of Marco Polo's Own Geography',
and his travels in Asia.

Early Trade

Since antiquity, sapphires have been discovered in Sri Lanka (Ceylon), Myanmar (Burma), Cambodia, Thailand, Vietnam, Ethiopia and France, and merchants have traded gemstones between continents. Even sapphire deposits in Madagascar, which we consider as a recent discovery, may have been known to ancient miners and merchants — a possibility we should not rule out. Sapphires tended to be more accessible than many other gemstones. For more than 2,500 years, Ceylon sapphires were found in riverbeds, known as alluvial deposits, and had the appearance of waterworn crystals as they had been naturally tumbling along the riverbeds for millennia.

Egyptians used opaque stones, such as lapis lazuli, turquoise and agates, to represent blocks of colour in images; while, later, the Romans preferred transparent stones, using settings with open backs to enhance a stone's brilliance or lustre. For the rarer, hard, coloured gemstones, like sapphires, other stones often acted as substitutes: yellow sapphire or topaz would sometimes be replaced with citrine, and red garnets were substitutes for spinel and ruby. Blue sapphire, however, does not have a similar substitute, in terms of colour, and could not therefore be mimicked by less durable or less expensive stones.

Sapphire's medicinal, religious and talismanic properties, as attributed by the ancients, combined with its beauty and durability, have both driven and hindered the sapphire trade as local rulers and wealthy merchants tried to retain any locally mined stones for themselves as protection or currency.

From the first century BCE, long before ruby and spinel, sapphires begin to appear in European jewellery. Trade routes reaching as far west as Europe had been facilitated by the formidable military campaigns of Alexander the Great (356–323 BCE), who created one of the largest empires of the ancient world. Alexander's empire stretched from Greece down to Egypt and east to northwestern India and Afghanistan, spreading Greek influence through the Indian subcontinent. His campaigns, and the operations that accompanied them, secured trade routes from Sri Lanka (known to the ancient Greeks as Taprobana or Tabropane) and India across his empire, both overland and via the Indian Ocean and the Red Sea. These routes joined the Silk Road that already stretched 5,000 miles from East Asia to the Mediterranean.

Evidence of early trade's influence in goldsmithing is found in this pink sapphire ring with Indian diamonds. The ring was found in 1999 by archaeologists digging at the site of Ai-Khanoum, an ancient city established by Alexander the Great in present-day northern Afghanistan. According to Dr Jack Ogden, an eminent jewellery historian, at around 2,300 years old it is the earliest known diamond-set jewel. The ring's goldwork, with signs of enamel decoration, is very typical of Greek goldsmithing, which strongly implies that it was the work of a travelling Greek goldsmith who had set up a workshop in Afghanistan. On his travels, he would have met clients and other goldsmiths to acquire this Sri Lankan sapphire and Indian diamonds. Found along with this pink sapphire ring was another ring of Greek manufacture, less typically Greek in style but of the same age or slightly earlier, dating from 400 to 300 BCE and mounted with a Sri Lankan blue sapphire.

Until the discovery of the Ai-Khanoum rings, it was supposed that Roman jewels represented the earliest examples of set sapphires. Though the majority of sapphires set in ancient jewellery have been found to come from Sri Lanka, recent studies have shown that some sapphires have a basalt-related origin, meaning that they may have derived from the alluvial deposits of Ethiopia or the historic deposit of the Massif Central (Puy-en-Velay) in France. Sapphire deposits from Ethiopia were mentioned in manuscripts by the Roman writers Pliny the Elder (23/24–79) and Solinus (early third century), with specific reference to *hyacinthos* (or *iacinthus*), which is generally accepted as the ancient name for blue sapphire. Other basaltic sapphire sources —such as Nigeria, Rwanda, Cameroon, as well as New South Wales and Queensland in Australia – were not accessible during Roman times, and there are no historical accounts of their discovery during this period. These territories can therefore be ruled out as sources of early sapphires, but basalt-related deposits in northern Madagascar, which we consider as a recent discovery, have received little research and may have been known to ancient miners and merchants.

These particular Roman earrings, found in Turkey, which support emerald-crystal drops from Egypt, are surmounted by button-shaped cabochon sapphires that have been found to come from a basalt-related deposit, likely the Massif Central. Many of the Silk Road traders passed through Turkey with their gemstones. Sapphires arriving in Europe in medieval times from these routes – via the trading centres of Constantinople, Aleppo and Alexandria – were sometimes described as *saphirs d'orient*, presumably to distinguish them from sapphires from Puy-en-Velay in France, a well-documented source. European gem merchants did not yet understand all the differences between minerals, but the compilers of medieval inventories seem to have had a good understanding of the main gem varieties.

Above: Two rings, thought to be the oldest known sapphire rings, found at the archaeological site of Ai-Khanoum in present-day Afghanistan, set with a pink sapphire and two diamond octahedrons and a blue sapphire, respectively. The blue sapphire ring dates around 400–300 BCE, the pink likely the same or later, and both show the influence of Greek goldsmithing techniques.

Below: A pair of Roman earrings found in Turkey and set with cabochon sapphires and suspending emerald-crystal drops.

Top left: A 4.92-carat carved sapphire bead, second to first century BCE, with a motif of four leaves, front and back. The bead was reportedly found near Ai-Khanoum, like the two rings on the facing page, and also shows Greco-Indian influences. From the Derek J Content Collection.

Top right: A 23.73-carat sapphire tablet from India, likely twelfth to thirteenth century, with an octagonal girdle and angular facets meeting at the table facet. The drill hole that runs through this stone was created by drilling from both sides of the sapphire. From the Derek J Content Collection.

From the Derek J Content Collection, rings from top: A gold ring from northwestern India, Gupta period, fourth to late fifth century, set with its original deep-blue sapphire cabochon; a central Asian silver ring of stirrup shape, twelfth to thirteenth century, set with a cabochon sapphire in a gold bezel (the shoulders of the ring are inscribed with lines from the Qur'an [XXII: 56] in Arabic, reading 'Sovereignty belongs to God'); a Seljuq ring, one of a pair of gimmel rings, thirteenth century, set with a lightly faceted pink sapphire with drill hole in a setting of many wire claws arranged closely together (the second ring would fit under and against this ring); and an eighth- to ninth-century ring from Cambodia, cast in gold and then chiselled to create a wave-like design for the shoulders that support a ruby either side of the central cabochon sapphire.

LAPIS LAZULI

'Lost in translation' is a phrase that very much applies to descriptions of gemstones through ancient times. The English word 'sapphire' derives from the Old French *saphir* from the Latin *sapphirus* (also spelt *saphirus*), which in turn is derived from the Greek *sappheiros* and perhaps ultimately from the Sanskrit *sanipriya*, which means 'dear to Shani', the planet Saturn, because blue sapphire was associated with Saturn, believed to be a bringer of possible ill fortune. Modern translations of these terms may automatically assume that they refer to blue sapphire, but the majority of early Greek and Roman references would have related to the stone lapis lazuli from the ancient mines of Badakhshan in present-day Afghanistan. Indeed, Roman historian Pliny the Elder described sapphire as blue and opaque, a description that is always true for lapis lazuli, but only sometimes true for sapphire.

Lapis lazuli is an opaque, rich blue gem flecked with pyrites — 'spotted with gold dust', according to the Greek writer Theophrastus — and is one of the earliest traded gems. The stone was found in the Neolithic burials at Mehrgarh, west of the Indus Valley, Pakistan, from the second half of the fourth millennium BCE and in the Sumerian tombs of Ur from 2500 BCE. Here, the Sumerians worked it into bowls, beads and small statues. Lapis lazuli can be found in large blocks, which lend themselves to being carved into many small objects. Ancient civilizations such as the Hebrews, the Babylonians, the Akkadians and the Assyrians all favoured this gemstone for use as inlay in their jewellery and seals. In ancient Egypt, it was associated with the wisdom of the 'all-seeing' Eye of Horus and was used as make-up. Lapis lazuli is virtually non-existent within known Greek and Roman jewellery, but when introduced to Europe in the fifth century, the stone was known as *ultramarinus* (ultramarine), meaning 'beyond the sea'; crushed and ground down, with care taken to sift out the pyrite metal impurities from the blue mineral, it was used as a pigment for paints, especially in religious scenes of the thirteenth and fourteenth centuries.

The lapis lazuli pendant carved to mimic a shell, with a gold figure of Aphrodite Anadyomene wringing her hair of sea water, is filled with symbolism. Aphrodite, Greek goddess of beauty, poses in a suggestive manner with a deep blue background representing the heavens and the sea; she is enhanced by the precious materials surrounding her. This pendant is an example of exceptional jewellery craftsmanship in the early Byzantine period, a luxury jewel that only the elite could have afforded. But it is also an amulet that would have been cherished for its powers of protection, or as a charm capable of granting a secret wish.

From top: An early seventh-century Byzantine necklace with a shell-shaped pendant carved from lapis lazuli, overlain with a gold Aphrodite Anadyomene, the ancient Greek goddess of beauty, from the Dumbarton Oaks collection; and a third-century Roman ring set with a round sapphire, from the British Museum. Rich blue sapphires, like this stone, share their colour with lapis lazuli, and the two stone varieties were often known by the same name.

SRI LANKA

In jewellery from around 400 and after, larger paler sapphires become more prevalent; these sapphires are from Sri Lanka, and developments in sea trading carried them from this island to distant shores. At the time, natural pearls and sapphires were the two most important items exported from Sri Lanka to Europe. We see this combination in many ancient jewels such as these sixth- to seventh-century Byzantine gold bracelets with pearls and sapphires.

The natural wealth of Sri Lanka did not go unnoticed by the Chinese; Yi Jin, a seventh-century traveller and Buddhist monk, named the island Baozhu, meaning 'Jewel Isle'. Sri Lanka has gone by many different names over the centuries: called Taprobane, Taprobana or Tabropane by the Greeks and Romans, the island was known as Serendib by the Persians and, later, as Ceylon by the British.

Above: A pair of Byzantine bracelets, *c.*500–700, each with a rosette set with a sapphire encircled by pearls to a bracelet of amethysts, quartz, glass and pearls; gifted to the Metropolitan Museum of Art by banker J Pierpont Morgan in 1917.

Centre, left: A pair of Roman earrings, *c.*100–300, with sapphire, garnet and emerald beads, from the British Museum.

Top: A necklace, *c.*400–600, found near Asyut, Egypt, set with pearls and now-empty square bezels and suspending a fringe of sapphire drops, now in the collection of Berlin's Altes Museum.

Centre, right: A late Roman pendant, *c.*300–600, of hinged gold plaques set with three sapphires and one emerald, from the British Museum.

MARITIME
TRADE

The Maritime Silk Route, a sea thoroughfare for traders between China, South Asia, West Asia, Europe and North Africa, was a key route for international trade alongside the land-based Silk Road. Once mariners properly understood winds, tides and currents, these routes very quickly became important pathways for commodities. Trade in silk and jade artefacts spread from the South China Sea around 1000–600 BCE, and these East Asian treasures were joined by gemstones from India and Sri Lanka. Alexander the Great's troops returned home from battles by both land and sea, marking out additional viable routes that commercial traders later exploited. These routes linked camel caravan posts in the Arabian deserts with ports on the South Arabian coast. They were part of a vast trade network that brought frankincense and myrrh to the Romans (fragrant resins that were highly prized in antiquity as they could only be harvested from trees growing in southern Arabia, Ethiopia and Somalia). The path they took is referred to today as the Incense Route, but traders also used these routes to trade in other precious commodities, such as gold, ivory, pearls, textiles and gemstones. By the seventh century, Arab traders, including gem merchants, started to use the sea route to spread the teachings of Islam.

The Chinese Song Dynasty (960–1279) started to build their own fleet of ships and established a permanent standing navy in 1132. Their naval prowess helped them to control large swathes of the commodities trade, stretching as far south as Java. The wreck of a tenth-century trading vessel, found in 2003 near the port city of Cirebon off the coast of West Java, has provided evidence of Chinese influence in the trade. The shipwreck revealed a cargo of a large amount of Chinese Yue ware, as well as Indian glassware and gemstones, including 11,000 pearls from the Arabian Gulf and 400 blue sapphires. Prominent gemmologist Ken Scarratt was able to confirm that the sapphires originated from Sri Lanka.

PRINCE ZHUANG OF
LIANG AND LADY WEI

The rulers of the later Ming Dynasty (1368–1644) ensured the continuation of the Chinese maritime trade of gemstones. The mid-1400s tomb of Prince Zhuang of Liang and Lady Wei, found in 2001 at a construction site in Zhongxiang, Hubei Province, contained inside its chambers many gold and gem-set treasures, including six gold hat ornaments. Three of these hat ornaments are surmounted by sapphires, including a large, 200-carat, colourless sapphire with a drill hole running through its centre. This sapphire is one of the largest found in any archaeological dig. Each dynasty had its favoured goldsmiths who passed down their skills in filigree, inlay and goldwork techniques such as those seen in these elaborately decorated hat ornaments. The blue sapphire is mounted above a three-tiered, seven-petalled lotus flower in full bloom set with pink, blue and colourless sapphires; it also has a middle layer depicting a ten-petalled lotus flower, which is opening in the opposite direction. One of the other two hat ornaments is set with a yellow sapphire on a base studded with pink and blue sapphires and a turquoise stone.

These jewels show the extent of the gem trade six hundred years ago and reinforce sapphire's importance as a gem that travelled to all corners of the globe. Just as Sri Lankan sapphires passed through many traders' hands along these routes to reach Europe, they also passed to the Middle East, Central Asia and Southeast Asia, along with Arabian pearls, Indian diamonds and Egyptian emeralds. These hat ornaments are great examples of how trade moved in all directions between the Chinese and the Arabs, long before the Dutch, Portuguese, French and English joined.

Three hat ornaments from the tomb of Ming Dynasty ruler Prince Zhuang of Liang (*c.*1450) excavated in Zhongxiang City, China, and now in the collection of the Hubei Provincial Museum, all three of twisted gold wire that secures the blue, colourless and yellow sapphires sitting atop.

A significant hoard of early Byzantine gold jewellery was reputedly discovered in 1910 during restoration work being carried out on an old stable in the Piazza della Consolazione in Rome.

All the jewellery of this hoard has been dated to the fifth century, mainly due to the similarity in style of the pieces. The sapphires in the collection all have drill holes, which would have allowed the stones to be traded on strings. Most of the sapphires have only superficial polish and are water worn, which suggests they were mined from an alluvial deposit, most likely in Sri Lanka. The design of the jewels prioritizes the gemstones, which are set in simple gold settings in a style that is highly characteristic of the early Byzantine period. The emeralds are 'slices' of crystals, drilled lengthwise, and could have originated from Egypt, in an area much later referred to as Cleopatra's Mines. With gold, gemstones and pearls sourced from around the world, the jewels in this hoard encapsulate the global trade operating at the time of their creation.

There is conjecture, but no concrete evidence, as to who owned or buried the Piazza della Consolazione jewels. Throughout history, many hoards such as this were hastily hidden or buried due to the possibility of invasion and drastic changes in power structures. In the case of most discovered hoards – and this is no exception – detailed inventories are rarely drawn up, making it difficult to account for what was originally found, or indeed to prove that the discovery was legitimate and not fabricated for the convenience of a trader.

Jewellery hoards are very seldom kept together after their discovery; instead, items may in the past have been offered to the melting pot, sold and exchanged for financial gain, or scattered among various museums. A substantial group of this Byzantine jewellery from the Piazza della Consolazione hoard is kept at Dumbarton Oaks, Washington, DC, and a few pieces are held by the Metropolitan Museum of Art, New York, the jewels having made their way into these museum collections via dealers. In 1928, not so long after the hoard is said to have been unearthed, Mildred and Robert Woods Bliss purchased the jewels from the Paris-based antiques dealers Kalebdjian Frères, who also sold to Louis Cartier. These jewels remained in the Woods Bliss collection until 1940, when the collection became part of the Dumbarton Oaks Research Library and Collection. We may never know the full story of this jewellery and its early owners, but it does provide illuminating insights into the work of traders and goldsmiths.

Four early Byzantine gold necklaces and a pair of earrings from the early fifth century, all with sapphire, pearl and emerald beads, reputedly found as part of the Piazza della Consolazione hoard in Rome, now in the Dumbarton Oaks collection.

Marco Polo (1254–1324) was an intrepid European traveller of the thirteenth century. Both his father and uncle were merchants who travelled extensively and established trading posts in Constantinople, Sudak (Crimea), the western region of the Mongolian Empire and even in China. With their influence, it was only a matter of time before the young Marco Polo would follow in their footsteps.

In 1271, Polo left his home city of Venice to join his father and uncle on a journey to the East that would end up lasting 24 years. After spending 17 years in China, the trio made their way back to Europe, taking a sea route that crossed the South China Sea and passed through Vietnam, Malaysia and Sri Lanka. His impressions of Sri Lanka (or Seilam, as he termed it) were later recorded in the *Book of the Wonders of the World*: 'I want you to understand that the island of Seilam is, for its size, the finest island in the world, and from its streams come rubies, sapphires, topazes, amethyst and garnets.' Once round Sri Lanka, they finally landed in India and travelled overland back to Venice, arriving in 1295.

Marco Polo's detailed account of his epic travels became akin to the *Lonely Planet Guide* for future travellers, explorers, missionaries and merchants. But he did not write his journal whilst travelling; instead, it wasn't until he returned and was imprisoned by the Genoese, who were enemies of the Venetians, that he met a fellow prisoner, Rustichello da Pisa, who was able to transcribe Polo's travelogue. The original manuscript is lost, but its copies tell his story of journeying through extraordinary lands to which few Europeans had ever ventured before and how he ended up in the court of the fifth Great Khan, Kublai, at his summer palace at Shangdu (Xanadu). It is incredible to think of the many hazards and challenges the group must have faced along the way; no wonder, then, that their story became a bestseller, even before the first printing presses came into action around 1440.

Accounts like those of Marco Polo inspired many travellers to explore the riches of the East. Though few reliable documents remain to tell us of these voyages, extant accounts inform us of the increasing importance of global trade networks and the place of gems and jewellery within these systems. In 1435, over 150 years after Marco Polo's travels, another Italian explorer set off on his own trip. Niccolò de' Conti (*c*.1395–1469), a merchant and intrepid traveller from northern Italy, immersed himself in the countries he travelled through: he learnt Arabic when he stayed in Damascus, Syria, and Persian in Oman, and he respected the religions and practices he encountered. Travelling incognito allowed him to venture extensively across Asia into India and over to Burma (Myanmar), where he discovered rubies, sapphires and other gemstones. We do not see many Burmese sapphires before the late 1400s; instead, spinel seems to have been the most-traded stone. Across Europe and Asia, sapphire was considered the finest of gemstones up until the late thirteenth century – 'fit for the fingers of kings', according to an ancient lapidary – until ruby started to take centre stage in the fourteenth century.

Top: A late medieval pendant, 1400–64, in the form of a roundel set with a sapphire and with white enamel beads; part of the Fishpool Hoard of high-status jewels found in Nottinghamshire, England, and held in the British Museum.

Inset: An engraving depicting Marco Polo, included in *Kristofer Columbus, Hans Lefnad och Resor*, the 1893 Swedish translation of Washington Irving's book on Christopher Columbus, who was supposedly inspired by stories of Marco Polo's travels.

In the history of maritime trade, it was not until relatively late that the Dutch, Portuguese, Spanish and English started to travel and trade directly with Southeast Asia, instead of receiving commodities via intermediaries. In Europe, this period of new trade in the sixteenth century was known as the Age of Exploration.

Christopher Columbus (1451–1506), Italian by birth, went to sea at a young age and, after travelling extensively, persuaded Queen Isabella I of Castile and King Ferdinand II of Aragon to sponsor a journey west. In 1492, he landed in the Bahamas. A mere two years later, and with the Pope's approval, Spain and Portugal divided this 'New World' between them. A line was drawn longitudinally between the Americas and Africa, with Spain taking North and South America, except Brazil, which came under Portugal's control. Portugal established trading posts along the west coast of Africa, monopolizing the extremely lucrative spice trade from the 'Spice Islands' of Macao, India, Ceylon, Sumatra and Java. Spain's invasion and conquest of the Aztec and Inca empires included much pillaging of these empires' land and seas for pearls, gold and gemstones, especially emeralds. The large quantities of goods that Spain shipped to Europe transformed commerce from around 1545.

When Elizabeth I succeeded to the English throne in 1558, her challenges were not insignificant: not only was she to convince her court that her gender would not hinder her ability or authority to govern, but she also needed to find ways of countering Spain's and Portugal's control of important trading routes. As Elizabeth did not yet have the finances to fund a worldwide colonizing campaign, she at first looked to subsidize foreign trading ventures, so long as she shared in the profit. During the 1570s, the explorer Francis Drake (c.1540–1596) was a close acquaintance of the Queen; he gained her 'unofficial' support to raid Spanish treasure ships on their return from their South American conquests, which brought Spanish riches to English shores. In 1580, Drake became the first Englishman to circumnavigate the globe, a journey he undertook in a tiny vessel only 70 feet long and 19 feet wide. This achievement brought great wealth to the trip's investors, which included the Queen; to reward him, she knighted him in April 1581. At this time, the English sometimes joined with the Dutch to attack Spanish and Portuguese outposts in the eastern empires in a bid to capture the very profitable spice trade. Trade within Europe during this period was immense and intense: the array of goods, including spices, gemstones, Chinese silks and lapis lazuli from Afghanistan, were exchanged with wool and woollen cloth from English merchants.

From the sixteenth century, Sri Lanka's coastal areas were controlled principally by the Portuguese, but also briefly by the Danish, the Dutch and the British; however, the central area, home to the sapphire mines in the Kingdom of Kandy, was stringently ruled by the Sinhalese. They ensured that this central area remained independent from 1597 to 1815; they expelled the Portuguese forces from the island in 1656, and the Dutch had control only of the major towns. The kings of Kandy retained much power over the island's sapphire mines and forbade the mining and trading of any sapphires; all sapphires had to remain on the island, with the threat of dire consequences for anyone caught engaging in the sapphire trade. Consequently, hardly any sapphires reached Europe, with the result that very few European jewels made from 1600 to the early 1800s are set with blue sapphires. Kandy maintained its independence until 1815, when the British, who had already established the British Ceylon colony in 1796, took control.

A 1683 map by Alain Mallet of Taprobane (or Taprobana), the ancient Greek name for an island described by Roman geographer and mathematician Ptolemy and thought to be the island known today as Sri Lanka.

Corundum, the gemmological family of sapphire and ruby, is a very hard gemstone, and amongst gemstones second only in hardness to diamond. Only a diamond is able to cut a sapphire or ruby.

Due to sapphire's hardness, gemstone cutters were not able to facet sapphires until much later than they were other gemstones. Holes have been drilled through gemstones since the fourth millennium BCE, but it was not until 2500 BCE that drilling with corundum became possible. From 500 BCE, cutters used hand-held diamond drills to drill through sapphires. With their drill holes, the sapphires could be strung and worn for safe-keeping or easily traded in larger numbers as currency, and the drill holes allowed for more secure wire settings through the sapphires in jewellery. The holes would be drilled at both ends, with the hope of meeting in the middle, so as not to split the crystal; however, often the holes would not quite meet in the middle, which is why in some transparent stones you can see a 'dog-leg' attempt to marry up the holes. Sometimes the drill hole would be aligned through the stone to eradicate an unsightly inclusion or flaw. The edges of the drill holes would be rough and so, to avoid the sapphires cutting through their string, the holes would be smoothed by threading the sapphires along an oiled cord impregnated with diamond particles. Sapphires from Sri Lanka were sometimes surface polished and drilled at source, but many stones found in the island's alluvial deposits had been naturally weathered and were worn and smooth, so they only needed to be drilled.

It is interesting to note in this collection of 77 ancient, drilled beads that some have been drilled twice. The second drill hole suggests that the stone may have been reused, either as a bead or as a pendant, and the jeweller wanted the stone to hang in a different orientation. The collection encompasses an array of different sapphire colours, shapes and sizes. Although the beads were collected from sites across the Eastern Mediterranean and the Indian subcontinent, most of these stones are likely to have travelled from Sri Lankan deposits – the major sapphire source since antiquity.

Drilled sapphires are very difficult to date. Though drill holes and the 'polished' pebble or cabochon shape are common in stones predating the development of faceting techniques around 1400, we should not assume that sapphires with these features were fashioned and traded before 1400, as even today you can find sapphires recently fashioned in simple cuts.

The only reliable ways to date sapphires are when they are found at an archaeological site that is datable and when they are set in jewellery. Gem-quality sapphires do not seem to have been used to any extent in Europe and the Mediterranean before the Roman period. Blue sapphires are uncommon in Indian jewellery: not being auspicious gems for Hindus, they were not revered and were therefore used only as currency and beads.

In Europe, wearers of drilled-sapphire jewellery believed that the drill hole increased the sapphires' power, or 'virtue'. This was a convenient belief for traders, as most sapphires in the medieval period had drill holes from being traded and transported on strings. A few gem-quality sapphires were used for carving intaglios, as seen in the Profiles & Portraits chapter of this book, but during the Roman period sapphires were much admired as beads and unengraved stones for their hardness and colour. Sometimes one can see asterism – the star-like optical effect created by a stone's inclusions – in the sapphire beads, but this would have been more by accident than by design; although the ancient Roman author Pliny the Elder (23/24–79) does indeed describe star sapphires, few other Romans would have known about the effect.

A collection of 77 ancient sapphire beads of different hues and sizes, from 21.4 to 5.2 mm in diameter, in the Derek J Content Collection. These sapphires were found in assemblages from the Eastern Mediterranean to the Indian subcontinent, but their believed age and drill holes suggest they travelled as part of the Sri Lankan gem trade.

Inscribing gems is a very skilled task, and the lapidaries capable of making inscriptions would generally be found in larger metropolitan centres, where practitioners of different crafts gathered. If a patron could not secure the services of a skilled lapidary, they might have their ring mount inscribed rather than the stone itself. Caution is needed when trying to date a sapphire in a ring mount because you must ascertain whether the stone is the original sapphire. Sometimes, studying the inscription and text on engraved stones can also help with the dating, even if this reading can present its own challenges.

A form of early faceting is evident in a handful of known sapphires today. These gems have escaped later recutting and show us a grooved 'proto-faceting' style that is completely unique to each stone. Medieval Arab writers recorded that lapidaries created these concave cuts to remove unsightly surface-reaching imperfections in the sapphire, and additional cuts may have been added in some stones to create a pattern before polishing. This technique probably only existed because sapphire was too hard to facet in a way that more drastically changed the shape of the individual stone and because cutters did not want to lose too much weight from each stone. A few sapphires found at archaeological sites and now in the incredible collections of gem historians Derek Content and Riaz Barbar display these concave cuts. A blue sapphire bead, dating between the tenth and twelfth centuries, bears on one side a Kufic inscription surrounded by cuts that appear to mark the removal of imperfections, while the other side has polished dimples arranged in a more stylistic design of concentric circles. The combination of cuts for these two different aesthetic reasons points to the development of this sapphire faceting style.

On some of these sapphires the grooving is very delicate, to the point that you cannot always see the dimpled effect unless you look very closely. An extreme example of this fine dimpling is evident on the Buddha Blue sapphire (following page), which is covered in tiny concave indentations. The play of light created by the dimples on its surface has a magical quality; it may have been cut that way for maximum decorative display.

By the fourteenth and fifteenth centuries, medieval lapidaries could further facet stones. They would grind the stone on a flat surface with an abrasive paste, which is quite a limiting technique for cutting small stones as it is difficult to make the fine angle changes to cut the facet angles (although I did recently see a cutter in Burma using this technique to cut the main facets — preform a crystal — without a loupe or electricity to an amazing degree of accuracy). In India and Sri Lanka, cutters used the bow drill/lathe, which included a disc impregnated with an abrasive paste that rotated when the bow connected to it was moved back and forth; the cutter held the stone next to the disc as it moved, which slowly ground and filed the stone. This technology was mainly used for preforming, not faceting, stones because the cutting accuracy was also limited. Technological limitatons, principally a lack of diamond powder, may explain why early faceting is not seen in stones coming from India and Sri Lanka. This bow technique is still used by some cutters in India today.

The drilled or preformed stones arrived in Europe through the trade routes and were then faceted. In the 1400s an important technological advancement in Europe revolutionized cutting. The invention of the continuous rotary motion wheel, which may have been inspired from silk spinning, made it possible to cut small stones very accurately as they could be clamped to arms (dops) to ensure they were still while being cut and the cutter did not have to readjust their position to keep the wheel spinning. Small bits of gems that were left over from the cutting were ground up and used for medicinal purposes.

By the 1600s, faceting had advanced, and lapidaries sought new and exciting ways of using their faceting equipment. In the Cheapside Hoard, an important collection of Elizabethan and Jacobean jewellery introduced in the At Court chapter, there is a rectangular sapphire set in a ring that, at first glance, appears to have a smooth, flat table facet, but that actually has an amazing pattern of intersecting facets. Jewellery historian Dr Jack Ogden believes this stone may have been cut in Portugal. When I looked at it, I did not see the 'faceting' on the table, so subtle is the cutting.

A nineteenth-century Indian watercolour of a lapidary cutting and polishing stones. He uses the bow lathe, still in use today, to rotate a disc coated with abrasive paste, against which he could grind the surface of gemstones.

Examples of early faceting styles, particularly an unusual dimpling with grooves that were likely intended to add brilliance and, in some instances, to remove surface flaws.

Above: A late medieval French ring with gold dragon wings meeting at a cabochon sapphire with five 'dimples' to its surface, from the collection of the British Museum and recorded as having been found in 1829 at the 1165 tomb of Thierry, Bishop of Verdun.

Left: Three views of a gold ring, eighth to ninth century, from the Kingdom of Champa, now Vietnam, from the Derek J Content Collection; rounded shoulders sit either side of a sapphire pebble with neat rows of 'dimples'.

Top and middle right: Two views of a tenth-century 4.35-carat sapphire, likely intended to be set in a ring, cut with eight angular facets meeting in a high dome with a flat base. From the Derek J Content Collection.

Bottom right: The 15.73-carat Buddha Blue Ceylon sapphire, extensively analysed and researched by Dr Çigdem Lüle, has many repeated grooves cut into its free-form cabochon shape. Comparison with other patterned sapphires suggests a cutting date between the thirteenth and fifteenth centuries.

Below: A late thirteenth- to early fourteenth-century ring set with a hexagon-shaped sapphire with prominent surface grooves; part of the Erfurt Treasure discovered in 1998 at a site once belonging to Kalman von Wiehe, a Jewish merchant who is thought to have buried the treasure before the pogrom of 1349 that destroyed the community around the synagogue in Erfurt, Germany.

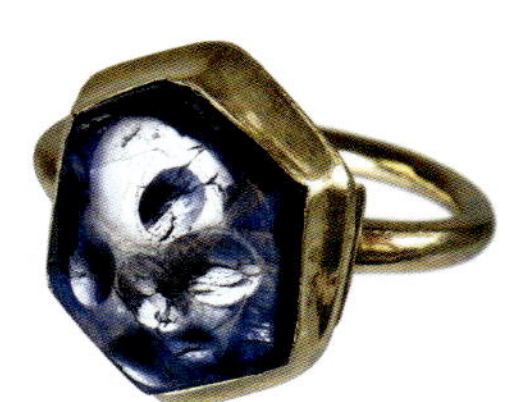

MEDICINE & MAGIC

A nineteenth-century drawing depicting the front of the Talisman of Charlemagne. Written in ink around the drawing is a transcription of the 1804 description of this reliquary jewel by Berdolet, Bishop of Aachen, as well as an inscription by Prince Louis Napoleon, who later became Napoleon III.

Over millennia, cultures around the world have shared the belief that sapphire possesses a great, mystical potency. The healing and restorative powers that have been attributed to it are influenced by its colour and durability. Sapphire's blue colour linked the stone to the heavens, the sea and the endless horizon; owning a sapphire therefore brought you closer to the infinite. Still today, blue is linked to fidelity and purity.

This chapter opens with Charlemagne, Charles the Great, the most powerful and famous ruler of the early medieval period, who wore an amulet pendant known as the Talisman of Charlemagne. This jewel was originally set with two very large, transparent, cabochon sapphires protecting hair from the Virgin Mary said to be held within the amulet. The sapphires symbolized a mighty power that would protect Charlemagne in this life and the next. During this period, a gemstone's colour influenced its identification and meanings. The translucent blue tones of sapphire, stones that had travelled around the world from the distant shores of Ceylon, were the obvious candidate to associate with the ecclesiastical world because of this colour's resemblance to the skies and heavens. Sapphire's durability also linked it to the everlasting. When worn, sapphire offered protection and a dialogue with ethereal spirits. During medieval times, life was hard and, in many cases, short, even for the wealthy; the power that money brought was only useful in this mortal world, whereas the power that sapphire bestowed on its owner when worn gave reassurance of protection in all worlds.

In Asia, blue and also yellow sapphires held a revered position. In India, the powers of the sapphire and of the planets have always been closely aligned. Indeed, for Hindus certain astrological alignments meant that wearing or gifting a blue sapphire was required, or alternatively to be avoided. In the superstitions and beliefs of cultures across the globe, sapphire continues to hold a universal importance.

A fifteenth-century love ring in the British Museum, likely made in England or France, with a sapphire bead set in a circular bezel. Its broad gold band bears an inscription on the exterior, surrounded by flowers and leaves; inside, engravings of a woman and squirrel are accompanied by the inscription 'S(I?)S AMOUR EST INFINITI(V)E GE VEU ESTE SON RELATIFF' ('My love is an infinitive which wants to be in the relative').

THE TALISMAN OF CHARLEMAGNE

Charlemagne (*c.*748–814) was the most powerful and famous ruler of Europe in the early medieval period. King of the Franks and then the Lombards (Italy), 'Charles the Great' was crowned Emperor of the Romans in 800 by Pope Leo III, politically uniting Western and Central Europe, after he had saved the Pope and his position from enemies in northern Italy. A thousand years later, such was the Emperor's reputation, Napoleon Bonaparte compared himself to Charlemagne to endorse his own leadership.

Charlemagne died in the imperial capital of Aachen, known as Aix-la-Chapelle in French, and was buried in the cathedral he had commissioned there. His body was exhumed a number of times, but it was not until around 1200 that a manuscript listing relics found in Aachen first mentions a reliquary connected to Charlemagne. Although this jewel has not always been attributed to Charlemagne, nineteenth-century documents allude to the Emperor wearing the reliquary in battle and to the fact that he was buried with it. In the 1200 manuscript, and in a later document produced in the early 1500s, this jewel is described as containing the hair of the Virgin Mary. Such was the importance of reliquaries at the time, the religious contents would have been of more interest to the scribe than the jewel itself. During this period, hair and nails were considered superfluous for the immortal body and so were often used in reliquaries. By the early nineteenth century, the jewel was reported to contain a small cross made of wood from the True Cross.

The form of the Talisman of Charlemagne is derived from that of a pilgrim's ampulla, a small round flask for holy water or oil, which would traditionally have been made of terracotta and later lead. It consists of two circular parts with a broad edge, all made in a high-carat gold with a filigree and repoussé decoration. The front is encrusted with pearls, garnets and emeralds, centrally set with a later blue glass cabochon displaying a visible relic of the True Cross inside. The reverse is set with a large sapphire of approximately 190 carats, believed to be one of the largest sapphires used in a jewel before the seventeenth century. The sapphire is accompanied by pearls, garnets, emeralds and an amethyst. The edge is set with a similar array of gemstones.

The jewel has recently been studied closely by a team of experts, who determined that the sapphire originated in Sri Lanka, the emeralds in Egypt (bar one from Austria), and the garnets in southern India or Sri Lanka. The jewel would originally have been set with two large sapphires, instead of the current blue glass cabochon. Further research of detailed nineteenth-century engravings suggests that one of the stones was replaced in the eighteenth century and before 1843.

Reliquary jewels containing 'holy' relics would be worn by the clergy during ceremonies. The large size of the sapphire dominating this jewel suggests its importance and subsequently it may have been worn by a bishop. Sapphires were chosen both for their celestial connotations and for their supposed ability to cool the body and purify the spirit. Worn next to the body, as opposed to other relics kept in cases, a reliquary jewel would have had a very intimate relationship with the wearer. Given the significant size of this jewel, its presence would be impossible to forget. The jewel would often be held in the palm of the hand, and a constant inspection would have fuelled a greater belief in its magical powers.

In 1802, after the French Revolution, Aachen acquired French status, becoming the administrative headquarters of the Roer department. In 1804, Napoleon crowned himself Emperor and, considering himself a descendant of Charlemagne, went to visit his tomb. On this visit, he was offered the reliquary jewel by Marc-Antoine Berdolet, Bishop of Aachen. The Empress Joséphine became the custodian of the reliquary; since it was her personal possession, she was free to pass it on to her daughter, Napoleon's step-daughter, Hortense de Beauharnais, upon her death in 1814. From this time, the jewel was the subject of elaborate storytelling about its important role both in Charlemagne's battles and in those of Napoleon, who declared that he had worn it on his breast during battle just as Charlemagne had 900 years before.

The talisman was passed from Hortense to her son, Napoleon III (1808–1873), who apparently became quite reliant on it; he kept the jewel in his bedroom until his death, having had a casket made for it by the goldsmith François-Désiré Froment-Meurice. Not having an heir, his widow, Empress Eugénie, decided to gift the jewel. She was very saddened by the news of a fire that broke out in 1914 at Reims Cathedral, where many French coronations had taken place, so she declared her wish for the reliquary to have a final resting place at the cathedral. In 1962, the Talisman of Charlemagne was classified as a historic monument and is on permanent display in the treasury of the Palace of Tau in Reims.

Above: The front blue glass cabochon of the Talisman of Charlemagne showing the cross of wood believed to be a relic of the True Cross of Jesus; an 1834 portrait by Félix Cottreau of Hortense de Beauharnais (1783–1837), shown wearing the talisman as a cloak clasp. It is unlikely that the talisman was ever really worn in this manner.

Facing page: The Talisman of Charlemagne, measuring 73 x 65 x 35 mm; even with its 190-carat sapphire, its size does not compare to its formidable potency.

Front and back views of the
Middleham Jewel, a gold pendant locket made
around 1450–75, inscribed and engraved with
scenes from the Nativity and Crucifixion
and set with a sapphire bead.

Unearthing a golden object that has not seen the light of day for centuries must be an exhilarating experience. In 1985, Ted Seaton, an experienced metal-detectorist, was searching the land around the ruins of Middleham Castle, with prior permission of the land-owner, the land's tenant farmer and two colleagues. Together, they shared an understanding that, if a discovery was made, the proceeds would be divided equally between them.

Middleham Castle, in North Yorkshire, England, is best known as one of the childhood homes of Richard III (1452–1485). Between 1260 and 1471, it was the seat of the Neville family, who were one of the most powerful families in England. Richard Neville, 16th Earl of Warwick, 6th Earl of Salisbury, was referred to as the 'Kingmaker' due to his role as arbiter of royal power during the Wars of the Roses (1455–85), a bloody and bitter feud between the houses of Lancaster and York for the English Crown. Following Warwick's

death in 1471, Richard III acquired the Nevilles' lands, including Middleham Castle, and, to cement his position, he married Warwick's daughter, Anne Neville.

By the mid-fifteenth century, Middleham had been converted into a palatial residence. This was also a period when superstition and reliance on magical intervention were prevalent; residents and visitors to the castle would have carried with them various talismans to alleviate fear and worries.

One morning in September, Seaton found a few artefacts with his metal detector near an ancient pathway between the two abbeys of Jervaulx and Coverham. Having initially thought he had uncovered a powder compact, it was not until he got home and was cleaning away the centuries of dirt that he realized he had found something very special indeed. It was a magnificent jewel, crafted by a medieval goldsmith around 1450–75, made of gold and set with a large blue sapphire – the gem of the Middle Ages, which was thought to be able to cure ailments such as ulcers, poor eyesight, headaches and stammers, and even to cleanse the soul. Medieval lapidaries recorded that a sapphire 'comforted the soul, could reveal fraud or malevolent magic and connected the wearer to the divine'.

The rhombic-shaped pendant locket is an incredible survival, and its reappearance transports us back in time. The wearer of the Middleham Jewel would have appeared to others adorned with multiple codes, and finding it allows us to interpret this jewel's role in Middleham society. Its height alone (64 millimetres) underscores its importance.

The front s skilfully engraved with the Trinity, showing the Crucifixion of Jesus, surmounted by a sapphire bead. The sapphire emphasizes the importance of prayer and is associated with the Virgin Mary, the colour blue generally being used to depict the colour of her robe in medieval images. The sapphire bead has a hole drilled through it, which indicates that the stone had been strung together with others while being traded en route from its origin, Ceylon. The sapphire is set within a gold circle of reeded decoration.

Along the outer rim, two 'magical' words are engraved. One is 'Tetragrammaton', a reference to the four letters, read from right to left, of the Latinized Hebrew spelling of the unutterable name of God, which was recited by Christians as a magical mantra in the Middle Ages. The other is 'Ananizapta', which is believed to have been an invocation intended to protect people from epilepsy, or 'the falling sickness', which was attributed to demonic possession. The word is also frequently interpreted as being formed from the initial letters of the prayer 'Antidotum Nazareni auferat necem; intoxicationis santificet alimenta poculaque Trinitas alma', which was recited to protect against intoxications, not only alcohol, but also poisons that could have been hidden in food and drink at the time. Additionally, a Latin extract from the Mass frames the engraved Crucifixion: 'Behold the Lamb of God, that takest away the sins of the world. Have mercy on us.'

Along the edge of the pendant are holes suggesting that there may have been a further gold frame around the jewel, plus one surviving gold loop; the frame and loops may have been decorated with pearls, which were popular at the time. There would also once have been a key to lock the sliding panel.

Very small traces of enamel suggest that the front and back of the jewel were decorated, but nearly all the enamel has since chipped away.

The reverse slides open to reveal a compartment containing four fragments of gold embroidery silk; they may have been believed to be from a cloth owned by one of the saints depicted around the edge, or the fragments may not be original. Regardless of their age, they add to the mystery of this pendant. Jewelled amulets and objects that contained a holy relic from a saint or from their clothing, known as reliquary jewels, were very popular throughout the Middle Ages.

The reverse is meticulously engraved with the Nativity scene; with this side worn to the chest, a suggestion of intimacy is created. Below the Nativity scene is an image of the Lamb of God within a border engraved with 15 saints, some of whom have been identified: St George, Catherine of Alexandria, St Peter, St Barbara, St Anne, Dorothea of Caesarea and St Margaret of Antioch. The depiction of the Virgin Mary, the particular combination of female saints and the associations with childbirth have led historians to suggest that the jewel might have been owned by a royal or noble lady keen to protect herself from the risks of childbirth, one of the most perilous ordeals faced by all medieval women. The jewel may have been owned by Richard III's wife, Anne Neville, or by his mother. The former British Museum curator John Cherry has suggested that Anne Beauchamp (1426–1492), widow of the Earl of Warwick and mother-in-law of Richard III, may have owned the jewel since she is known to have frequently assisted at childbirths. She was at Middleham Castle around the time of the birth of Prince Edward, in 1473, the only legitimate son to Richard, who died at the age of 10.

The Middleham Jewel shows how close medieval faith and religion stood to the precarious belief in magic and its curative powers. Every possible meaning, association and symbolic potential has been carefully thought through by the goldsmith.

I once had the opportunity to handle this marvellous jewel, and I remember it feeling very heavy, not only because of its gold weight but also because of the weight of history in my hand. It was an extraordinary moment.

The discovery and legal reporting of an artefact like the Middleham Jewel follow a strict protocol. On realizing the importance of his find, Ted Seaton would have taken it first to a local 'finds liaison officer', who records and photographs the object to prepare a report for the local coroner. For items that are more than 300 years old and are more than 10 per cent gold or silver, this stage must legally be completed within 14 days. The report is then passed to an expert at the British Museum, who will advise the coroner whether the item can be classed as 'treasure'.

For an item to be deemed treasure under the Treasure Act 1996, the team must decide whether the treasure was hidden with the intention of being recovered, or whether it was simply lost. If its burial was intentional, it is classed as treasure and becomes the property of the Crown. A committee of antiquities experts, known as the Treasure Valuation Committee, then ascertains its monetary value, before it is offered to a museum and the finder is compensated. The museum local to the find is usually given first refusal to acquire the item, and the British Museum will declare any interest should the local institution decline. If no museum shows any interest, then the item will return to the finder and the landowner. This process often applies to hoards.

If it is not deemed to be treasure, the item is returned to the finder or the landowner. If an item is considered lost or abandoned, it belongs to the owner of the land on which it was lost and they can sell or keep the item; the proceeds from any sale are expected to be split between the original parties.

Single objects are usually classified as lost, since it is difficult to lose a hoard accidentally. The Middleham Jewel, as a single item, was deemed lost and so went up for sale at auction in 1986. It sold at Sotheby's to a private collector for £1.3 million. An export licence was temporarily refused to allow for funds to be raised to keep the jewel in the UK. Eventually, the sum of £2.5 million was put up so that the jewel could be acquired by the Yorkshire Museum in York, where it can be admired today.

The ruins of Middleham Castle,
built in the twelfth century in North Yorkshire,
England. Richard III (1452–1485, r. 1483–5)
spent much of his late childhood at Middleham
Castle, where he trained as a knight under the
supervision of his cousin Richard Neville,
16th Earl of Warwick. This statue of Richard III,
made in 1996 by artist Linda Thompson,
stands in the bailey of the castle.

Displayed in the Victoria and Albert Museum is a gold pendant set with three stones: a peridot, a hessonite garnet and a blue sapphire drop. It is a delightfully colourful jewel, made in England *c.* 1540–60, but, when turned over, reveals a darker story.

Jewels of this period in their original condition are extremely rare, as many would have been melted down, lost or remodelled according to changes in fashion or purpose. Surviving jewels are invaluable in giving us an insight into medieval people's beliefs about the metaphysical, which was no less a part of their reality than the physical world in which they lived.

The elaborate gold pendant of stylized acanthus leaves frames an open-backed peridot and hessonite garnet, an unusual setting for the period, as many jewels had gems set in closed foil-backed settings. The open setting of these stones meant that they could touch the skin of the wearer and release their attributed powers. The sapphire, which would have already been drilled from previous use, is suspended below. Both front and back have traces of translucent blue enamel decoration, which has since been damaged.

The reverse reveals black enamel inscriptions: around the peridot are the words 'ANNANISAPTA • DEI' and around the garnet 'DETRA-GRAMMATA • IHS • MARIA'. The Latin inscriptions IHS, MARIA and DEI refer to Jesus Christ, the Virgin Mary and God, respectively. Like the Middleham Jewel, the pendant uses the word *tetragram-maton* (here given in the variant form *detragrammata*) to invoke

the Hebrew name for God and the word *Annanisapta* as a magical formula to ward off the demons that were believed to present themselves in epilepsy sufferers.

The combination of stones was also important. Peridot was believed to dispel nightmares. Garnets were thought to cheer the heart and drive away sorrow, while protecting the wearer from pestilent diseases. Sapphires had a prophylactic function, bestowing on the wearer protection against a multitude of ailments, as well as promoting peace, constancy and freedom from fear or envy. Together, the energy from these stones and the potent words touching the skin made this jewel a formidable talisman.

In the Middle Ages, life was very harsh. The plethora of diseases, the constant fear of sudden death, and the mistrust of life's twists and turns only confirmed the need to supplement one's religious faith. Believing in combinations of words and materials from the natural world, like those found within this pendant, provided a sense of safety and a moment's relief from a very unpredictable world.

Front and reverse views of
a sixteenth-century English pendant in
gold with enamel. The hessonite garnet and
peridot are mounted in open-backed settings
and the sapphire as a drop, so that the
stones could touch the wearer's skin
and impart their mystical powers.

Blue sapphire was the most used stone in medieval times, believed to possess the 'highest virtues' as well as divine powers, which could protect the wearer against deceit and treacherous plots. It was also a symbol of friendship. In the thirteenth century, the colour blue was associated with the heavens and the robe of the Virgin Mary. It is unsurprising, then, that blue sapphire became the favourite stone of kings and bishops.

Ecclesiastical, devotional or bishop's rings – or, as they are sometimes called, pontifical rings because of the status of the Pope as the Bishop of Rome – must be made of gold and set with a gemstone that is not engraved. In the thirteenth century, the pontifical compiled by Durandus (William Durand) for the diocese of Mende, France, stated that the ring of gold symbolized perfection and, being gem-set, would signify the splendours of the gifts of the Holy Spirit, 'which Christ has received without measure'. Bishops received their rings at their investiture ceremony, as they vowed to serve the Church faithfully. The sapphire represented chastity in marriage and symbolized the heavens, so it was the natural choice of stone for an ecclesiastical ring. A bishop wore his ring as a symbol of office; it was like a wedding ring, since a bishop was deemed to be married to the Church and would 'lay down his life for her, like Christ'.

Sapphire was deeply associated with the clergy; in the known graves of English medieval bishops, 20 rings have been recovered, of which 12 are set with sapphires. The graves of both Walter de Gray, Archbishop of York (d. 1255), and William Wytlesey, Archbishop of Canterbury (d. 1374), have yielded rings set with a sapphire.

A fine ring shown here, above left, likely to be a bishop's ring, features a water-worn sapphire pebble that has been slightly polished to remove surface-reaching blemishes, which has created a dimpled effect. It is set in a four-claw gold mount, and the pierced openwork decoration around the bezel depicts winged beasts. Two of these beasts have wide-open mouths, almost like lions. The shank is incredibly detailed and engraved on both sides in Latin with the words 'AVE MARIA GRATIA PLENA DMI' ('Hail Mary, full of the grace of the Lord'), an inscription frequently used in medieval jewellery.

Three rings, likely for bishops or abbots.
From left: A thirteenth-century gold ring with
engraved shank and bezel holding a sapphire
pebble; a chased gold ring inscribed with the
name William Wytlesey, Bishop of Rochester
and later Archbishop of Canterbury, and set with
an octagonal sapphire bead (this ring dates to
his time as bishop, between 1362 and 1374);
and a thirteenth-century gold ring set with an
oval sapphire cabochon, the shank inscribed
with a Latin prayer in Lombardic lettering.

The Massif Central, France, pictured above,
was a notable source of sapphires in the
medieval period. This group of sapphires,
below, weighing between 1.92 and 0.04 carats,
is a rare recent find from river gravels near
Issoire, Puy-de-Dôme.

In the medieval period, one of the known sources for sapphires, garnets and zircons was the Massif Central in the Auvergne region of France. The area was affected by volcanic activity, which resulted in alkaline basaltic flows that provided the conditions for sapphire formation. Sapphires were found in riverbeds, and miners exploited these sites until the nineteenth century — few are found today. The natural colours of these sapphires ranged from dark blue and black, to greenish yellow and greenish blue, to saturated blue. It would be interesting for gemmological laboratories to conduct tests on sapphires set in medieval jewels to see if some of them were from France instead of Sri Lanka, the typical assumed origin for sapphires in this period. During the Roman period it is very likely that some blue sapphires set in jewellery came from France, and it is even possible that some came from Ethiopia.

MEDIEVAL GOLDSMITHS

During the medieval period, 500–1450, high demand for the use of gold and silver objects in religious services meant that monasteries were the main source of patronage for goldsmiths. A few large monasteries in Europe even had their own goldsmithing schools, such was the importance of their contribution to teaching. Many of the monks became goldsmiths, which saved on the expense of having to employ outsiders.

However, in the later centuries of this period, the number of monk-goldsmiths declined as more secular goldsmiths set up their own workshops. Some of these goldsmiths were employed on commissions for aristocratic courts, emperors, kings and noblemen, whilst others worked as independent goldsmiths. The goldsmithing trade increased during the fifteenth century; at this time, the largest concentration of goldsmiths was in Cologne, Germany. Trade was booming across Europe during this period, and prolific commissions from royal patrons kept goldsmiths and silversmiths very busy.

As well as the result of institutional commissions, jewels, and especially rings, were also given as gifts and thus played a significant role in medieval society: consolidating friendships, cementing business contracts, reinforcing strategic relationships and serving as tokens of love and affection.

Very little gold or silver was found in Western and Eastern Europe, which meant that these precious metals tended to be recycled from other jewels, or gold coins that had travelled along the Silk Road, looted from Egyptian tombs or melted down from Roman and Byzantine jewels and objects. Originally, the goldsmith would also cut stones, but soon lapidary (the art of cutting gems) became a skill in its own right. Lapidary started to develop in Europe around 1200, though cut styles were still basic and many gemstones had already been roughly cut when imported. In the fourteenth century, gem cutting was still in its infancy and something of a novelty – indeed, the cutting of crystals was still viewed with caution and conducted in secret on account of the widespread belief that cutting gemstones, which were seen as having great potency, could bestow ill fortune on the cutter.

Medieval treatises recorded gemstones and their magical, spiritual and talismanic properties. These treatises – or lapidaries, as they were known – helped goldsmiths understand the symbolism of gems, which they drew on for their designs. Some concentrated on stones' association with the signs of the zodiac, others on their medicinal qualities. One thing was common to all approaches: you did not want to make jewels set with stones that did not complement either each other since that might hinder their curative powers.

One such tract was the *Liber de lapidibus* (Book of Stones), a compendium written in Latin verse by Marbod (*c.*1035–1123), Bishop of Rennes, in which he documented 60 stones. He described sapphire as an appropriate gemstone for a king's fingers because of its cooling qualities, which would aid against headaches, ulcers and profuse perspiration, along with being good for the eyes and speech. Sapphire not only generally protected the body, it also had spiritual uses: it brought peace and reconciliation, encouraged chastity and paved the way for prayers to be heard favourably by God. The meaning of stones was up for interpretation: other writers suggested that sapphire could expel envy, cure snake bites and detect fraud and witchcraft.

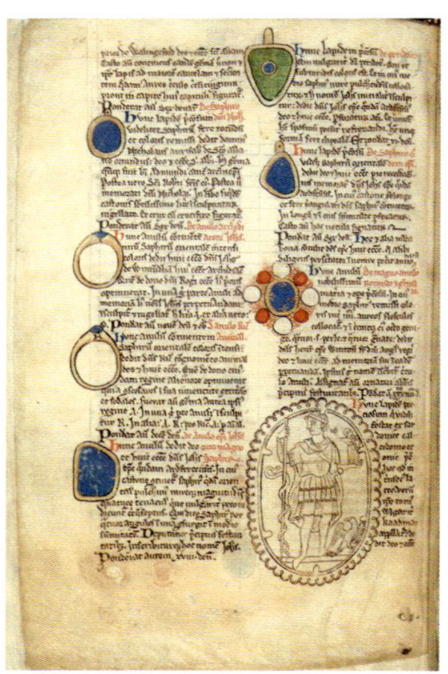

Top: A fourteenth-century amulet ring with a thin, inscribed shank holding a bezel set with a sapphire.

Above: A folio from the *Liber additamentorum*, a collection of treatises and documents assembled by Matthew Paris between 1250 and 1259, describing sapphire-set rings and other jewels in the treasury of St Albans, England.

Mortality rates were high in the seventeenth and eighteenth centuries due to epidemics and wars. The departure of a family member in the eighteenth century was followed by a respectful period of mourning. This was a time not just to lament the loss of a beloved or admired individual, but also to pay homage by following a series of protocols: widows would wear black for a year, mirrors were covered and elaborate tombstones were commissioned.

Jewellery, worn by the privileged few, would also adhere to a strict code. Mourning rings in the seventeenth century often contained a plaited lock of hair from the deceased, or bore an inscription stating the date of death in white or black enamel in the gold mount. In the eighteenth century, mourning rings were often set with gemstones, with or without hair mounted underneath, with the name and date of death enamelled in white, for the unmarried, or black for the married. The significance of using a blue sapphire in a mourning ring is to connect the deceased with heaven.

Four English sapphire-set mourning rings in gold,
all bearing inscriptions on black or white enamel
grounds with dates between 1731 and 1742.

BLUE SAPPHIRE IN
HINDU ASTROLOGY

Although European cultures believed the sapphire to have protective powers and bring good fortune, in ancient Hindu astrology blue sapphires were associated with the dark and turbulent planet Shani (Saturn), and were treated with great caution. Still today, many people in India are hesitant about blue sapphires.

Ancient Indian mythology held that gems originated from the slaying of the demon Bala. Every dismembered part of his body created a gem: diamonds were from his bones, pearls from his teeth, rubies from his blood, emeralds from his bile and sapphires from his eyes. Varahamihira, author of the encyclopaedic text *Brihat-samhita*, written around 500, associated gemstones with nature because they came from below the earth. He believed they were intrinsically imbued with magical powers that could influence destiny and cure disease, while counteracting strong planetary forces. In India, the combination of gemstones and gold – both products of nature – has been considered particularly potent. Jewels became important tools in combatting the unknown, controlling fate and protecting the wearer from ill health.

Sapphire was considered the most powerful gem. *Sanipriya* (from *sani*, Saturn, and *priya*, dear, therefore 'dear to Saturn') is Sanskrit for 'blue sapphire'. Sometimes the terms *indranila* or *neelam* are also used to mean sapphire. It is believed that anyone worshipping Saturn while wearing a blue sapphire will receive positive blessings, as the stone has been associated with love and purity, but it is a stone that Hindus, in particular, will not wear unless under specific instruction from their astrologer.

Saturn returns every seven years in an individual's astrological chart. Their astrologer may then advise them to wear a sapphire during the period of Saturn's return, but this advice is not given lightly; all planetary configurations have to be taken into account and, crucially, the sapphire must be of a certain weight and quality for its magical powers not to be hindered. I often wear a blue sapphire ring and I notice that it always attracts a sideways second glance in India. Blue sapphires, if worn, are normally set in a ring; it is important for the stone to be in an open setting, so that the gem touches the wearer's skin. Light can travel through the stone, transmitting positive energies to the body to protect the wearer from negative influences.

Not all sapphires, however, are viewed with suspicion; the yellow sapphire promises intelligence and good health. I remember, during one of my many visits to India, an astrologer reading my astrological chart and promptly pulling open a draw to show me a selection of yellow sapphires that I could purchase to aid my good fortune.

Hindu astrology identifies seven planets and the ascending and descending nodes of the moon. Each planet is assigned a specific deity, who works closely with the universe and mortals. Individual gems are deemed potent in their own right – and all the more so when placed together with other gems. Jewels have long been seen as a tangible link between the human and the divine. A jewel was therefore designed to combine all these powerful astrological influences: the *navaratna* nine-gem jewel represents the symbiotic relationship between the planets, their deities and man.

The nine gems are usually set in a square in a specific formation; ruby represents the sun and is set in the centre so that its powers can flow into the eight other gems. The placement of the stones must follow accordingly:

In the centre	Sun	Ruby
To the east	Venus	Diamond
To the southeast	Moon	Pearl
To the south	Mars	Coral
To the southwest	ascending node	Jacinth (orange/ red zircon)
To the west	Saturn	Blue Sapphire
To the northwest	Jupiter	Yellow Sapphire
To the north	descending node	Cat's Eye
To the northeast	Mercury	Emerald

The quality of the gems is equally important, for Hindus believe that the beneficial powers of every gem depend upon its perfection – a poor-quality gem would risk attracting unhappiness and misfortune.

Despite the mystical caution that surrounds blue sapphire, Usha Balakrishnan, a highly respected and knowledgeable historian of Indian jewellery, has seen blue stones set in jewels given as offerings. The presence of these blue sapphire jewels implies that good fortune had been bestowed on the worshipper while Saturn was in his astrological chart and, wanting to thank the gods for their good fortune or good health, they had made an offering of a sapphire. In the South Indian Srirangam Temple, the processional figure of Sri Ranganathaswamy has been gifted five very large cabochon sapphires set in a pendant. Queen Victoria reputedly heard about a deity's pendant set with ten cabochon sapphires in the Meenakshi Temple in Madurai and had it shipped to London for her inspection.

Traditional Indian doctors advocated that disease could be cured by first wearing precious stones, then by the power of prayer; only if these two remedies had not worked would medicine be tried. Ayurveda, a system of traditional medicine originating thousands of years ago, believes in the healing powers of metals and precious stones. The powers of particular gemstones are released when placed on the afflicted area of the body in such a way that light passes through them into the body. Sometimes, gems would even be crushed up and consumed for their immediate effect.

Above, left and right: A pair of matching gold
and green enamel *navaratna* armlets, shown on
their reverse and front, respectively, made in
Delhi, India, *c.*1850. Each armlet bears the nine
auspicious stones of the *navaratna*.

Centre: A Hindu processional figure in
Sri Ranganathaswamy Temple in Srirangam,
South India, bedecked with flowers and
jewels, including large sapphires.

An early twentieth-century, Indian, fan-shaped
gold pendant known as a *pankhi*, set with
two sapphires cut in the shape of feet, called
vishnupada or *pagalia*, and nine triangular,
flat-faceted, or *parab*-cut, diamonds.

Jewellery has adorned not only men and women; the importance of bedecking Indian deities with jewellery has long been understood. Presenting jewellery to images or figures of deities in their temples of worship has been integral to Hindu rituals. In the *Bhagavata Purana*, Krishna is cited as saying, 'Whatever is best and most valued in this world and that which is most dear to you should be offered to me and it will be received back in immense and endless quantity'. Even today, jewellery is first presented to the deity in the family shrine at home before it is worn.

The daily prayer (*puja*) in a Hindu temple involves giving 16 offerings to a deity; gems and jewellery are the ninth offering. As gems and precious metals come from nature, and therefore from the gods, it is appropriate to offer these riches back to them.

Over the centuries, temples throughout India have received offerings of gems and jewellery from devotees. These jewels are referred to as 'temple jewellery', retained in the temple never to be released or even made available for viewing. The size of these temple collections is never disclosed for fear of theft or unwanted publicity, or, indeed, so as not to alert government officials who may find a way of claiming taxes from the temples. Much to the frustration of the jewellery historian, these temples must hold invaluable collections of classical Indian jewellery that have survived the ravages of changing fashions yet remain hidden from view. The only opportunity to see these jewels is to watch devotees in the act of offering their precious treasures to deity figures already dripping in sumptuous jewels.

There is a charming fable, recited in the ancient Sanskrit scriptures, which highlights the astrological powers of blue sapphire. King Dushyanta was hunting in a forest when he came across a young girl called Shakuntala, who was the daughter of a powerful sage, a yogi who lived in the wilderness, and instantly fell in love

with her. The King gave her a blue sapphire ring to signify his intentions of marriage and for her to come to his palace when she was ready to marry him. One morning a few months later, Shakuntala was drawing water from the river for her father when the sapphire ring fell off her finger into the river and was promptly swallowed by a fish. She later went to the palace to declare her consent to be married but the King did not recognize her or recall his proposal, as he had been cursed, and so she was dismissed. A few months later, a fisherman caught the same fish that had swallowed the sapphire and, finding this beautiful ring, took it to the King, who immediately remembered the girl and his proposal of marriage. Shakuntala was sent for and they were married and lived happily ever after, believing their destiny had been orchestrated by the potency of the blue sapphire.

Above: A gold ring with a carved sapphire bird,
with almandine garnet beak and ruby eyes, set
atop two outstretched gold hands, potentially
inspired by the clasped hands of European *fede*
rings, made in India, likely in Calcutta, *c.*1850;
a late nineteenth-century painting of the Hindu
god Krishna with blue skin in his form known as
Srinathji, in watercolour and gold on
paper from Nathadwara, India.

Facing page: An impressive sapphire carving from
the eighteenth century depicting a Hindu saint.
This sapphire weighs 37.82 carats
and is just 29 mm tall.

Simply suspended from a chain-link necklace is this solitary faceted sapphire, known as a *taveez* cut, a very traditional Indian cut. It has been hand cut, and its visible horizontal unpolished groove suggests that a lapidary cut away a surface imperfection.

With blue sapphire's negative connotations, this simple necklace is a powerful statement of allegiance to and confidence in the sapphire – its owner, the Maharajah of Indore, must have trusted his astrologer implicitly.

In the early twentieth century, the Maharajah and Maharani of Indore were India's most glamorous and stylish couple. Known for their impeccable taste in fashion and style, they combined Indian and European cultures and became patrons of many contemporary European and Indian artists and designers. Young and madly in love, they created an incredible modernist palace in the heart of India on the Malwa Plateau. It was the perfect setting for their huge art collection. From their visits to the cities of Europe, they would bring back to India design ideas and treasures. They loved the new music genre of jazz, and to accommodate their love of dancing built a huge ballroom in the palace. Sadly, the Maharani died in 1937 at the tender age of 22, which devasted the young Maharajah, being only a few years older.

Left: A 1927 Man Ray photograph of the Maharjah of Indore, Yeshwant Rao Holkar II (1908–1961), with his wife, Sanyogita Devi of Indore (1914–1937). The Maharajah wears the Indore Sapphire, above, an eighteenth-century sapphire *taveez* bead of 23.20 carats, later mounted on a simple white gold chain by Cartier.

Members of the Mughal court, as well as the Islamic sultans of the Deccan, felt very differently to Hindus about the blue sapphire. Instead of being feared, sapphire was highly prized, often worn and regularly given as part of a special gift.

The wearing of sumptuous jewels to demonstrate wealth and regal magnificence amongst the ruling Mughal emperors was vastly increased with Jahangir (1569–1627, r. 1605–27) and Shah Jahan (1592–1666, r. 1628–58). One of the most prominent jewelled statements was the all-important turban ornament or *sarpech*. It signified power at court because it was only worn by emperors, male members of regal families, their favoured relatives, ambassadors, foreign rulers and noblemen. This esteemed jewelled ornament found particular favour in the eighteenth century. Coupled with the development of the decorative arts and the art of enamelling, the turban ornament was a popular vehicle for showcasing each region's specialized enamel motifs and colour combinations.

This splendid turban ornament, above, was probably made in Murshidabad, the capital of Bengal, around 1755 and would have been worn at court. It is centrally set with a large water-worn sapphire that has been lightly polished, surrounded by lasque-cut diamonds, cabochon rubies and emeralds. The design of the ornament, which terminates with an emerald bead drop, confirms that it was made in Bengal and especially in Murshidabad, where similar jewels are depicted in contemporary paintings.

The other jewel depicted here – a *sarpatti* – would have been worn at the same time as the turban ornament. This jewel is set with an emerald with two cabochon rubies either side, lasque-cut diamonds and a natural pearl drop. The reverse is enamelled in the traditional colours of green and red over a white background.

In Bengal, the British felt that the Muslim ruler Siraj ad-Daula, the Nawab of Bengal, could not be relied upon to maintain trading interests with the British East India Company, so they decided to install another ruler who would be more amenable. This plot resulted in the Battle of Plassey, for which Robert Clive led the British army and Admiral Watson commanded the fleet. With British victory, they installed Nawab Mir Ja'far, and he presented these jewels to Admiral Charles Watson on 26 July 1757. The jewels stayed in the Watson family until 1982, when they were eventually sold at auction and acquired by the Victoria and Albert Museum.

Top: The jewelled front and enamelled reverse of two Mughal turban jewels, to be worn together, presented to Admiral Charles Watson on 26 July 1757 by the Nawab of Bengal. The larger jewel is set with a large sapphire, surrounded by diamonds, rubies and emeralds.

Left and right: A Mughal-dynasty ring of the eighteenth century, richly chased and set with a cross of rubies and a central sapphire; an eighteenth-century Mughal gold ring with flower motifs, each with ruby petals and set with a sapphire at their centre.

Both these pendants could be worn individually, or they may have been part of a much larger collection of necklaces that were worn together to create a sumptuous display by wealthy men, women or children.

The blue sapphire pendant, *c.* 1850–1900, is interestingly set with a sapphire originating from Australian deposits discovered in the 1850s, though the pendant's gold and enamelwork could be earlier. The fact that the centre stone is a sapphire suggests that it had been owned by a Muslim, which is in keeping with other sapphire-set jewels belonging to the Nizams of Hyderabad. The reverse is decorated with green and red enamel, depicting a

poppy, on a white background, which is characteristic of Deccan enamelwork. The enamelling colour combination was inspired by Shah Jahan's marble monuments, which were inlaid with red carnelian and green jade. Over time, the range of colours used in enamelling widened, and cobalt-blue enamel is found on nineteenth-century jewels.

The pendant featuring the octagonal yellow sapphire from Sri Lanka, *c.* 1650–1700, is unusual for Deccan jewels of this period because it has an open frame; typically, jewels had stones in closed-back settings. The sapphire is surmounted with three table-cut diamonds supporting a pearl drop.

Two antique, Indian, sapphire pendants: the
pendant on the left dates to the mid to late
nineteenth century and is set with a hollow-back
cabochon sapphire from Australia, surrounded by
rubies, emeralds and diamonds, with a pearl drop.
The pendant on the right, from the seventeenth
century, is set with an octagon-shaped yellow
Ceylon sapphire, surmounted by three table-cut
diamonds and with a pearl drop.

PROFILES & PORTRAITS

CARVED SAPPHIRE
DRESS PIN

The carved sapphire in this dress pin in the collection of the British Museum is thought to date from early in the second century, while the mount is certainly later. There is no drill hole through this gem, as is usually found in sapphires of this period coming from Sri Lanka, so its Roman date is open to debate. With a height of only 18 millimetres, including the setting, this is a fine little jewel. The figurine is reputed to depict Plotina, who was a Roman empress and wife of the Roman emperor Trajan. It's an intriguing jewel; when you turn the miniature bust around there is the face of a child carved at the back of the head. Plotina and Trajan were unable to have children, so maybe this jewel represented the child they never had. Conversely, the two faces may depict Plotina at two stages of her life or, indeed, may reference the past while looking to the future. The carver's intended meaning is an enigma we may never be able to unravel.

Profiles and Portraits

Small, well-executed details in a jewel are beautiful but, in the case of all the jewels in this chapter, small details are also a window onto a much bigger picture. The intaglio, the reverse intaglio (a carving that has been executed from the reverse of the gem) and the cameo offer an intimate look into human cultural history and how it has developed over the centuries to accommodate changing beliefs. The carved image was a way of communicating wealth, status, power and learning. Across history, many people could not read or write and so the world was explained by means of myths, legends and gods, whose powers were often represented through the carved image. The cameo and intaglio are both types of glyptic art (from the Greek word *glyptos*, meaning carved), and were revered art forms.

Cameos and intaglios were also more than just art: they would serve as an amulet, a talisman, a way of displaying political allegiance or sealing friendships and relationships. They were not worn just for adornment but for a purpose. For example, a wax seal bearing the impression of an intaglio ring might have been used to identify and authorize documents, to mark property or assert ownership and discourage people from tampering. The tool used for cutting and engraving the cameos and intaglios was made of iron and fitted with a diamond crystal, the sharp octahedral points of which were used to chisel out fine details. Moreover, all kinds of materials were used for seals, such as wood, ivory and stone, but to have a seal, intaglio or cameo in carved sapphire was special as the gemstone was considered to be metaphysically potent. Communities across the world believed that sapphires had the power to dispel envy, detect fraud and prevent witchcraft, while also being associated with heaven — which is why they tended to be reserved for dignitaries, emperors and rulers.

Particular details in the carved and engraved gemstones can help us to date the works. Specific hairstyles, for instance, point to particular Greek and Roman periods. However, since sixteenth-century carvers would copy and replicate classical Greek and Roman subjects, other characteristics besides the hairstyles should also be considered. Ancient Egyptian craftsmen were governed by their priests and had to follow a strict style of human representation, which meant there was no freedom of interpretation or expression. It was therefore left to the Phoenicians, Greeks, Etruscans and Romans to teach gem engraving to future generations and to expand and lift the craft so that functional objects became desirable works of art.

Recently, gemmological studies have helped to reveal the origin of ancient sapphires, from which we can learn about ancient trade routes and whether those routes were maritime or across land along the Central Asian Silk Road. Furthermore, these historic carvings can help us piece together many other details about the lives of sapphires. Not all jewels, however, are suitable or available for this level of study: some will forever remain mysterious.

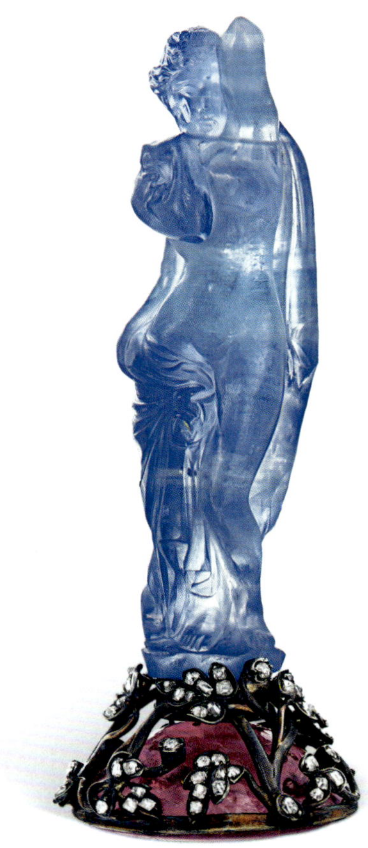

An unusual and fairly erotic mid-nineteenth-century carving, this sapphire jewel makes quite an impact considering it is only 10 centimetres in height. The figure of Venus stands on a carved intaglio spinel, weighing 116 carats; the intaglio depicts the profile of Medusa, in the style of the British Museum's Strozzi Medusa chalcedony intaglio carved *c.* 50 BCE, and is encased in a foliate rose-cut diamond-set cage.

This carved sapphire was once owned by Prince Felix Yusupov (1887–1967), who is best remembered for his role in the assassination of Rasputin, the Russian mystic and self-proclaimed holy man who had a great influence on Tsar Nicolas II and his family, especially his wife Alexandra Feodorovna, because of his alleged healing abilities. The Yusupovs were incredibly wealthy and owned many opulent residences, including a palace on the banks of the Moyka River in St Petersburg as well as an Empire-style chateau at Arkhangelskoye. In these palaces, the family protected their vast holdings of paintings and objets d'art, as well as the significant jewellery collection of Felix's mother Princess Zinaida, the last in the Yusupov line. Princess Zinaida's collection of jewels was considered second only to the one held in the imperial vaults.

During the 1917 Revolution, these jewels and objects were hidden by Prince Felix but, unfortunately, many were later found by the Bolsheviks. Felix himself managed to escape to Paris with a large number of jewels that he later sold to various dealers to fund his new life in exile. Amongst the jewels was this sapphire statuette of the goddess of beauty and love. When exactly the Blue Venus came into the Yusupov family's possession is unclear.

In Greek mythology, Medusa was the daughter of the sea gods Phorcys and Ceto; unlike her siblings, Medusa was not born a monster but, because of an ill-fated love affair with Poseidon, she was turned into a vicious creature with snakes for hair and met her fate in an encounter with Perseus, who beheaded her. Anyone who looked at the head of Medusa would be turned to stone. For the Blue Venus to be standing on the head of Medusa is a clear demonstration of the triumph of love and beauty, as well as constancy (represented by sapphire) over lust.

AUGUSTAN CAMEO

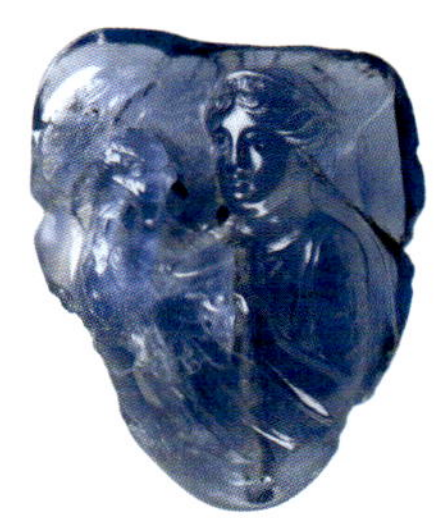

This large (35 x 30 millimetres) sapphire bead has been carved into a cameo depicting a three-quarter profile of a youthful woman. Curators at the Fitzwilliam Museum, the art and antiquities museum of the University of Cambridge, where this stone is held, believe the cameo may depict the Greek goddess of love and beauty, Aphrodite, later recognized as the Roman goddess Venus, because she is being led by an eagle to Hermes. However, Aphrodite is sometimes linked to Hebe, the Greek goddess of youth. Hebe is more often depicted with eagles, particularly seen giving an eagle, who represents her father, Zeus, a drink from a cup. The woman in this cameo has her long curls tied up in a loose bun, decorated with a head ornament, and holds in her left hand a cup for the eagle to drink, its swept-back wings representing imperial power. Due to their connection to Zeus, eagles in mythology were signs of immortality and, like the phoenix, were able to renew themselves and in so doing return to their youthful state. The sapphire originated from Ceylon as a bead, as the drill hole can be seen running vertically through the stone, and was carved later, possibly in Rome. According to the Fitzwilliam, this early cameo may date from the reign of Augustus, the first Roman emperor (r. 27 BCE–14). The execution of the carving is quite superb. It is only fitting that sapphire – the stone that represents the heavens – should have been chosen to depict Hebe's eternal youth.

POMPEIIAN HIPPOCAMP

This sapphire, housed in the National Archaeological Museum in Naples, is a glyptic (stone-carving) masterpiece beautifully engraved with a winged hippocamp, a sea-horse possessing the upper body of a horse and the tail of a fish. To the ancients, this beast symbolized Poseidon/Neptune and was frequently depicted with Tritons (mermen). Hippocamp engravings are often found on 'sea-coloured' gemstones because of the creature's association with the sea.

Weighing 11.62 carats, measuring 21.78 x 12.27 millimetres and having a pale, milky, bluish grey colour, this oval cabochon sapphire is a Roman reverse intaglio found in 1986 in Pompeii. Many ancient intaglio sapphires were carved from the front, but this sapphire is unusual in having been carved from the reverse so that the image is seen through the smooth domed front. The clarity of the sapphire enhances the readability of the engraving. The fact that this sapphire is of basaltic origin rules out the possibility of it coming from Sri Lanka, where most Roman sapphires were sourced. The team of experts that carried out a recent gemmological analysis of this stone suggests that it could possibly have come from Ethiopia instead, though as yet there has been no concrete evidence of early mining in Ethiopia to confirm this. New research on mining locations could shed light not only on the ancient gem trade, but also on broader trade and diplomatic relations.

PTOLEMAIC REVERSE INTAGLIO

This gem, just 18 millimetres in diameter, is another example of a reverse intaglio sapphire depicting a Ptolemaic queen or princess from the Greek family that ruled Egypt from 323 to 30 BCE. We see her in profile, facing left, and wearing a head ornament; her bust is three-quarter turned to face the front. The wings emerging from her back, level with her right breast, connect her to Nike, the Greek winged goddess of victory. The curls surrounding her face and the plumage of the wings are rendered by small, engraved circles, all incised into the reverse of this gem. The stone is from Ceylon and was likely carved in the second century BCE. This intaglio is part of the Coins, Medals and Antiques Department of the National Library of France.

ANCIENT GREEK INTAGLIO

This carved sapphire, probably from Ceylon, is much earlier in date than the gold ring mount it sits in. The stone is likely to have been carved around the first century BCE in Alexandria (present-day Egypt), then the centre of the Greek world. Like the reverse intaglio on the previous page, the sapphire was probably originally designed to depict a Ptolemaic princess. However, the gem would likely have been recontextualized within a later Christian framework as an intaglio of the Virgin Mary, as she too was often depicted with a veil over her face.

The gold medieval signet ring, measuring 25 millimetres in diameter, was possibly made in Europe around 1275–1325. In his 2006 analysis of the jewel, British archaeologist Reverend Professor Martin Henig of Wolfson College, Oxford, found evidence that the stone had been cut down in size, likely to make it suitable for setting in this ring. The bezel surrounding the sapphire is engraved in Lombardic lettering 'TECTA LEGE, LECTA TEGE', which translates as 'Read what is written, hide what is read'. This cryptic reminder suggests that the ring was for personal use. The sapphire is set in an open-backed mount so that the stone can touch the skin and impart its supposed medicinal and talismanic qualities.

This ring was found in a well in Hereford in 1824 and documented in the *Gentleman's Magazine* in 1830. Edmund Waterton (1830–1887), one of the foremost ring and engraved-gem collectors of the nineteenth century, exhibited it to the Society of Antiquaries in 1855. His collection, of which this jewel was part, amounted to 760 items. Waterton was known to have an extravagant lifestyle; when he ran into financial difficulties, he resorted to pawning his collection with Robert Phillips, a renowned London jeweller of the Victorian period. In 1871, when Waterton could not repay the loan, Phillips sold the collection to the Victoria and Albert Museum. In 1899, the V&A also acquired a small group of rings that Waterton had kept back.

INTAGLIO OF JULIA DOMNA

This is a second- to third-century Roman sapphire intaglio in a later setting, 20 x 13 millimetres, bearing the portrait of Julia Domna, which resides in the Hermitage, St Petersburg. The hairstyle shown is typical of depictions of Julia Domna (*c.*160–217), a Syrian-born Roman empress consort during the reign of her husband, Libyan-born Roman emperor Septimius Severus (145–211). After the Emperor's death, she became more involved in imperial politics, as well as being instrumental in advancing philosophy amongst the imperial court.

LATE ROMAN RING

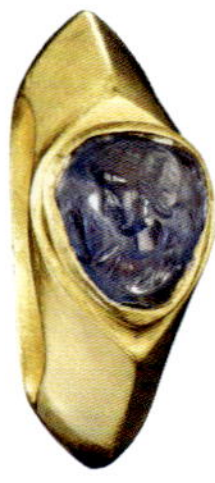

This ring is set with a wonderful sapphire intaglio bearing a portrait bust of a Roman imperial woman, dating to the early fourth century. The detail of this engraving is amazing for something so small, only 12 millimetres in length. The woman's hair is arranged in what is called a *Helmfrisur,* a pulled-back, smooth style with a small bun at the back. Underneath the intaglio are three inscribed letters, which could read *SND.* This sapphire is from Sri Lanka, and most Ceylon stones reached the Mediterranean as traded beads with drill holes that allowed them to be strung for ease of transport. This sapphire's lack of a drill hole may be a useful pointer for dating the ring, as the sapphire was never used as a bead – its first use may have been in this ring. The ring is possibly Constantinian in date and the portrait may be of Empress Fausta (Flavia Maxima Fausta, 289–326), who was the daughter of the Roman emperor Maximian (*c.*250–310) and in 307 became the wife of Constantine I (*c.*272–337) as part of a political alliance between Constantine and Maximian to gain control of the empire. The style of the facial features and the hairstyle can be compared to coins of the period that bear Empress Fausta's portrait.

This ring was acquired by collector of carved gems Giorgio Sangiorgi (1886–1965) and passed down through his family until it was sold at Christie's New York in 2019. The high prices realized by the items in this sale of his collection – the highest achieving over US$2 million – testify to the recognition of the fine craftsmanship of these carved gems.

This is a splendid sapphire intaglio ring, now part of the Derek J Content Collection, complete with its original setting, which is very rare as so often intaglios are set in later mounts. The ring is solid in manufacture, rather than being constructed from multiple parts. It dates to the fifth or sixth century and was made somewhere within the borders of the Sasanian Empire, the longest-lived Persian dynasty (224–651). The sapphire intaglio depicts a peacock with open tail that is facing the front, which is an unusual angle in images of the bird. The peacock was viewed in Persian culture as a guardian to royalty and as a symbol of splendour and power. Around the world, peacocks have been associated with rebirth because they shed and regrow their feathers to their full grandeur every time. With these connections, this ring may have been worn by a member of the royal entourage.

An unusual, pale-coloured sapphire cameo, this stone has been carved to depict a half-length figure of Christ with a cruciferous nimbus (or halo) around his head. Just 28.2 x 22.4 x 8.9 millimetres, he holds a closed book in his left hand and raises his right hand in benediction. The Greek inscription, 'IC XC', stands for 'Jesus Christ'. Sapphire's association with the heavens, because of its blue colour, may have been a factor in its choice as the material for this cameo. The cameo dates to the Middle Byzantine period (eleventh–twelfth century). It was bought by the British Museum in 1869 as part of a collection of Byzantine cameos.

In the shape of a star, this brooch has been made by linking together a series of individual gold cells to create an architectural form in 3D relief. This structure is a distinctive characteristic of Ottonian goldsmith's work, produced during the second half of the tenth century in territory that is now part of modern Germany. The brooch was inspired by Carolingian and Byzantine architecture and is centrally set with an earlier star-sapphire intaglio dated 337–50 and depicting a Roman emperor. The intaglio is surrounded by pearls (some missing). The Metropolitan Museum of Art team, which has displayed the brooch at The Met Cloisters, New York, suggests that the four exterior mounts at the compass points would have been set with amethysts, a stone that represents devotion.

On the left, a sapphire intaglio in the British Museum depicting a figure, possibly the Duke of Berry (1340–1416), the third son of King John II of France and Bonne of Luxembourg, seated in a grand high chair with decoration to the arms and back and wearing a hat and a long robe. The early fifteenth-century sapphire is mounted in a more recent gold ring, 13.7 millimetres in length.

On the right, this early fifteenth-century gold ring, made in the Netherlands and now in the collection of the State Hermitage Museum, St Petersburg, is set with a sapphire intaglio depicting a front view of the tilted head of an angel. The hair and facial features are very detailed. This sapphire is encased in an octagonal bezel setting measuring 20 x 17 millimetres. The sapphire's angular shape implies that the cutter has tried to follow the natural shape of the crystal. Sapphires crystallize in the trigonal subdivision of the hexagonal crystal system.

FREDERICK III'S SEAL RING

This intaglio sapphire (19 x 16 millimetres), set in a closed-back ring, bears the portrait bust of Frederick III the Peaceful (1415–1493) surrounded by the inscription 'REX/FRIDRIC' and the motto 'AEIOV' (AEIOU). The carving is attributed to the stone-cutter Heinrich Reweling, who is known to have worked for Frederick III. The gold ring mount is in the shape of two interlaced tree trunks. This distinctive design suggests it was made in Nuremberg, where pieces in a similar style have been produced. Frederick III was the first Holy Roman Emperor of the House of Habsburg; he was elected King of the Romans in 1440 and crowned Emperor in 1452. We can date the ring to this period due to its inscription, which not only states Frederick's position as King of the Romans, but also gives his enigmatic motto 'AEIOV', a motto in use by Frederick since at least 1437 and subsequently adopted by the Habsburg family, the exact meaning of which has never been unravelled.

Princes and emperors from Roman antiquity chose to have themselves depicted in sapphire cameos and intaglios because they had become a highly regarded representation of their status. However, this ring is the only known surviving medieval or Renaissance ring to feature an imperial portrait in sapphire.

This large ovoid sapphire intaglio also from the Hermitage collection is set within an octagonal gold mount measuring 31 x 29 millimetres, made in the sixteenth century in Italy. The detailed intaglio depicts the profile of Trajan, Roman emperor from 98 to 117 and considered to be a successful leader for expanding the empire further east, notably into Dacia, Arabia, Armenia and Mesopotamia. This carving is executed to an incredibly high standard, with his hair and the decoration on his tunic very finely engraved. Trajan's wife, Plotina, is depicted in the first sapphire sculpture in this chapter.

THE KNYVETT SEAL

During the Renaissance period, developments in jewellery design saw jewellers offset gemstones, pearls and gold with coloured enamels to create lavish decoration. This large sapphire seal has been meticulously engraved with the Knyvett arms and set in an elaborate gold pendant mount with enamel decoration to the surmount and bezel, front and back. The pendant fits into a decorated, black-and-white enamelled, protective gold case, both made in England around 1580. The seal and its case are on display in the jewellery gallery at the Victoria and Albert Museum.

This jewel may have been the personal seal of a member of the Knyvett family, but it is difficult to narrow down which family member it first belonged to since many Knyvetts held powerful roles in English society and would have been in a position to commission and use a seal of this extravagance. Sir Thomas Knyvett the elder (1485–1512), Master of the Horse to Henry VIII, had three grandsons that were knighted in the late sixteenth century, all of whom were therefore entitled to use the Knyvett arms. Sir Thomas Knyvett the younger (1539–1618) held the position of High Sheriff of Norfolk and was knighted in 1579/80; he was married to Muriel, daughter of Sir Thomas Parr, Treasurer of the Household to Queen Elizabeth I, so had important connections. Alternatively, another grandson, Sir Henry Knyvett (1537–1598), was knighted in 1574, five years before his cousin. Sir Henry's younger brother, also named Sir Thomas Knyvett (1545/1546–1622), was knighted in 1601. This Sir Thomas was Keeper of Westminster Palace, Groom of the Privy Chamber, a Member of Parliament and, from 1607, a peer. His claim to historical fame is that he arrested Guy Fawkes, who in 1605 infamously attempted to blow up James I and the House of Lords in Westminster.

The detail in this seventeenth-century sapphire seal brooch, measuring 40 x 26 millimetres, is superb. It is centrally set with an octagonal sapphire engraved with the coat of arms of Johann Hugo von Orsbeck (1634–1711), who was appointed Archbishop Prince Elector of Trier on 13 July 1676 and Bishop of Speyer just three days later. Johann Hugo also later added Abbot of Prüm and Provost of Weissenburg to his list of titles.

This armorial sapphire brooch was not used as a seal because the engraving is not in reverse; instead, the inscription was intended to be read as seen and the seal had a ceremonial role. The letters around the seal, 'I.H.D.G. ARC. TREV. PR. EL. EP. SP.', are the abbreviation of 'Iohann Hugo Deo Gratia Archiepiscopus Trevirensis Princeps Elector Episcopus Spirensis', a list of his titles that helps us date the jewel to the period in which he was both archbishop and an elector of the Holy Roman Empire. This combination of positions was unusual as few people were officially powerful in both the Church and the Empire; Johann Hugo united both spiritual and political authority.

In its ceremonial role, this armorial gem was set in the Archbishop Elector's archiepiscopal mitre; having his coat of arms on the mitre meant that the mitre represented him and he would have used it as an extension of himself during ceremonies. An account book in the State Archive of Rhineland-Palatinate (Rheinland-Pfalz) details a payment made in 1691 to the gem engraver Johann Helfrich Ries (1656–1725) for cutting Johann Hugo's arms onto a large sapphire for the mitre, so it is likely that Ries engraved this sapphire. The stone was removed from the mitre in 1822. Since then, it has been set in a brooch mount of scrolled foliate design in gold, silver and diamonds and was likely worn by a woman. Over the years, this sapphire has been separated from its original owner and meanings.

The Cheapside Hoard is an invaluable collection, mainly housed at the Museum of London, of over 500 pieces of Elizabethan and Jacobean jewellery and gemstones discovered by builders in 1912 while they were demolishing a house in Cheapside, London. Buried beneath the foundations, this collection had not been touched for over 300 years. Cheapside was the goldsmithing street of London in the 1500s and 1600s. Many independent workshops and traders occupied premises in the area, so there is speculation that the hoard had been a jeweller's 'stock in trade', or it may have been hidden for safe-keeping while the owners were called to fight in the English Civil War (1642–1651). Evidently, no one ever returned to collect the jewels.

Amongst the treasures were many sapphires, including this fine Byzantine white sapphire cameo pendant depicting the New Testament story of the Incredulity of St Thomas. The carving, measuring 33 x 22 millimetres, possibly dates to the tenth or eleventh century and the setting, decorated with enamel and surmounted with a natural pearl, to the sixteenth or early seventeenth century. The sapphire is foiled on the reverse to add colour to the stone. The cameo shows the moment Jesus asks Thomas to put his fingers into his side to prove that he has indeed risen from the dead.

At the beginning of the first century, sapphires and natural pearls were the most important export from Sri Lanka to the Western world. Pearls symbolized purity and were associated both with the moon and with the planet Venus, named after the goddess of love who came from the sea. The combination of the sapphire and the pearl, together with the depiction of St Thomas and Christ, makes this jewel extremely important as it conveys a strong message of unity and faith.

Instead of being mounted in a closed-back setting to enhance the colour of the stone, this pale blue sapphire intaglio is set in an open ring mount that allows the sapphire to sit against the skin, thereby strengthening its talismanic properties. This ring, in the British Museum collection, dates to the eighteenth century and was made in the Balkans during the long period of Ottoman rule (1299–1922). The 14 millimetre-tall intaglio depicts a mounted horseman holding a hawk. The horse was the most important means of travel (and battle) among the peoples of Central Asia. The terrain was harsh and the horses were bred for their endurance, capable of sustaining a constant trot over many miles. This sapphire intaglio could have been made for an Ottoman sultan, whose imperial stables would have contained as many as 6,000 horses. The imperial sport of falconry was very popular but was reserved for the elite ruling class. Falconry skills and traditions are centuries old, and the knowledge would have passed down the generations.

CASTELLANI

Fortunato Pio Castellani (1794–1865) founded his jewellery business in 1814 in Rome. During the nineteenth century, fascination with the ancient world was growing amongst the wider population of Europe as more archaeological sites were discovered. Excavations of Roman, Egyptian, Hellenistic and Etruscan sites inspired an archaeological revival in the arts. Castellani saw the opportunities this fascination with antiquity brought for the jeweller, and over time his workshop became known for jewels inspired by ancient classical styles. Rome was one of the most important destinations for Victorians when they were on the 'Grand Tour', and a visit to the Castellani showroom became a fixture on the itinerary. Castellani's two sons joined the business, and from the 1850s they solely produced jewels that were archaeologically inspired.

Castellani's workshop went to great lengths to replicate the original goldsmithing techniques used in ancient artefacts. When Etruscan burial sites yielded ancient gold jewellery pieces, Castellani looked to recreate their characteristic granulation and filigree decorations, creating jewels that connected Italy with its heritage. Shell cameo carvings were very popular in Italy during the nineteenth century, but Castellani turned to gemstones instead (even though the process involved was more time-consuming) and commissioned gem carvers to recreate ancient hardstone intaglios and cameos, which were then set in their archaeologically inspired jewels.

The sapphire intaglio set in the Castellani pearl brooch, which is on display at the British Museum, depicts the Greek mythological figure Medusa. The brooch was made around 1870 with the intaglio set in a ropework surround that is bordered with natural pearls, each pinned with a tiny cabochon ruby. It measures 33 millimetres in height.

This large brooch, measuring 72 x 45 x 13 millimetres, with pendant sapphire in the Smithsonian's Cooper Hewitt collection is a perfect example of Castellani's work, being centrally set with an engraved sapphire depicting the personification of Italy in the form of a young seated woman of ancient Rome, either Cybele or Isis, holding a globe, a symbol that would have appealed to supporters of Italy's unification during the late nineteenth century; its setting uses the house's characteristic style of gold ropework decoration. The brooch is also set with four sugarloaf sapphires and circular-cut diamonds and supports a large, faceted sapphire drop, weighing approximately 13.5 carats. It is decorated with small emeralds and rubies. The reverse of the brooch has the two entwined Cs of the Castellani monogram and is dated *c.*1898.

REGAL TREASURE

A nineteenth-century lithograph depicting the
votive crowns and crosses of the Guarrazar
Treasure, a hoard of gold chalices, crosses and
20 crowns believed to date to the seventh century.

Regal Treasure

Royal jewels intended to impress and intimidate with their spectacle of riches are important examples of craftsmanship, and for hundreds of years goldsmiths were kept busy as fashions in regalia changed. During the ninth century, European monarchies did not represent sovereignty with just a single crown: new crowns would often be made for each successive coronation, and sometimes three crowns were required for one ceremony alone. When Charlemagne crowned his son Louis, in 813, he wore a crown himself, while placing a crown on the altar for his son's coronation ceremony; a third was sent by the Pope for the new Emperor. All three were of equal importance: there was no dispute about who was the sovereign, so there was never a need to have one individual superior crown. Later, in the fourteenth century, inventories seem to suggest again that the monarchs of France and England owned multiple crowns of various designs.

Common to all was the use of gemstones. Of the gemstones most frequently used — notably rubies, emeralds, sapphires, pearls and diamonds – it is the blue sapphire that we see most often in royal jewels because of its association with heaven and celestial powers. The majority of the sapphires used in historic regalia are assumed to have originated in Ceylon (Sri Lanka), because of their size, hue and tone. Ceylon mines yielded very large sapphire crystals, which has helped to confirm their status and importance. The lapidary took advantage of this, cutting them down carefully to produce maximum yield. But it is not only about size: the two sapphires in the British Imperial State Crown, for example, are very different in size and yet are equally revered for what they represent.

Many of the crowns in this chapter are no longer used for ceremonies and are kept behind glass in museums. Although it is rarely possible to carry out a detailed analysis of these jewels, a team of historians and gemmological experts has recently managed to study the Royal Crown of Blanche of Lancaster in the Munich Residenz. Their investigations have revealed valuable information about the gems that were used. Today, questions may be asked about the necessity of regalia, but these crowns are historic examples of craftsmanship and genuine feats of engineering – if we only ever look at them through glass cabinets we miss an opportunity to uncover their true value in the study of jewels and gemstones.

Disbelief and confusion must have reigned on 25 August 1858 when, as the story goes, Francisco Morales and María Pérez, a farming couple who lived in the village of Guarrazar, Spain, made a great discovery. After a night of heavy rain, Morales and Pérez were tilling virgin land in a deserted area 15 kilometres from Toledo when they stumbled across something glistening in the mud. After further digging near a masonry well, they discovered treasure that had been exposed by the rains: a hoard of gold chalices, gem-set crosses and over 20 now-famous votive crowns.

Word of their discovery prompted other treasure hunters to scour their land and, a few days later, local gardener Domingo de la Cruz also found treasure in his own garden, which he hid for more than two years.

The crowns of the Guarrazar Treasure were not worn by humans; as votive crowns, they were hung on chains over altars to adorn the sacred host during mass. It is not known when this tradition began, but it was clearly practised during the reign of the western Goths (Visigoths) in the seventh century.

Crowns from the treasure of Guarrazar have a hanging cross at their centre, a symbol of Christ as King of the Christians. Some are set with large sapphire beads representing chastity, faithfulness and the heavens. It is believed that these gold crowns were offered to the Roman Catholic Church in seventh-century Spain by Visigoth kings as a demonstration of their faith and submission to the ecclesiastical hierarchy. Most of the votive crowns have an inscription of the name of the donor, which makes their offerings clear to other devotees.

The two most important votive crowns in this treasure are those believed to have been gifted by kings Recceswinth and Suinthila, respectively, though King Suintila's crown was stolen in 1921.

Both crowns are made of gold and studded with gems, including large blue Ceylon sapphires of typical tumbled form. Tumbled stones have been naturally polished by being rubbed against each other in riverbeds; this process gives them a smooth, pebble-like finish that is now recreated by simple machines. King Recceswinth's crown is a spectacular homage to the sapphire: the blue stones are set around the crown (see facing page), as well as suspended from chains of gold, interspersed with pearls and decorated with filigree. Filigree is typical of Iberian goldsmithing techniques and has been a tradition for hundreds of years. The letters hanging from the base form the Latin phrase 'RECCESVINTHVS REX OFFERET' – 'King Recceswinth offered it'. The choice of sapphire for these crowns highlights the importance of the gem during the Middle Ages and the stone's global connections.

It is estimated that 23 crowns were originally discovered, though only 10 are known to have survived. Inevitably, pieces have been sold or stolen, and the whereabouts of many are now unknown. The remaining treasure has found a resting place in various museums, including the Museum of Santa Cruz in Toledo, the National Museum of the Middle Ages (Musée de Cluny), Paris, and in Madrid at both the Royal Palace and the National Archaeological Museum, which notably houses King Recceswinth's crown.

Six votive crowns and two crosses of the seventh-century Guarrazar Treasure in the collection of the National Archaeological Museum, Madrid. The gold crowns are too big to be worn and are likely votive gifts to the Church from Visigoth kings, including the crown of King Recceswinth, above left and facing page, which is set with sapphire cabochons, ruby beads and pearls.

THE CROWN OF
THE HOLY ROMAN EMPIRE

Charlemagne's coronation as Emperor on Christmas Day in 800 marked the establishment of the Holy Roman Empire.

In the early medieval period, the sacredness of the empire was conveyed through symbolism and insignia. Christian doctrine combined with secular emblems of power to enforce the strict hierarchy of the divinely ordained. The Pope made the Emperor the head of the empire by investing him with the crown, orb, sceptre and imperial sword.

This iteration of the Crown of the Holy Roman Emperor was created, in its basic form, in the late tenth or early eleventh century, possibly for the coronation of Otto I in 962. Making a new crown echoed a decree issued by Charlemagne that called for *renovatio Romanorum imperii*: the renewal of the Roman Empire. The crown became its most important symbol.

Forged from high-carat gold in an unusual shape, this crown consists of eight hinged panels — representing the eight walls of the Holy City of Jerusalem — that together form an octagon rather than a circle. The front panel is surmounted by a cross, from which a single arch curves to the back. The arch is encrusted with gemstones and spells out the name of the Emperor Conrad II (r. 1027–39 as Emperor) in seed pearls set into fine goldwork. The entire crown could be dismantled for convenience when being transported.

Three panels beautifully depict in enamel three kings from the Old Testament: Solomon, David and Ezekias. The prophet Isaiah bears a scroll inscribed with quotations from the coronation liturgy that promote the royal virtues of justice, wisdom and faith. The fourth panel depicts in enamel Christ as Pantocrator ('the Almighty') seated on a throne flanked by two angels. Each plaque has a surround of large sapphire beads, set in traditional medieval filigree and granulation decoration, with pearls in between. The two other side panels are completely encrusted with gemstones, each set with trefoil claws to add further decoration.

The front and back plaques are both set with 12 large gemstones, including blue sapphires, pink sapphires or even pink spinels (as gemmological tests have not been undertaken), emeralds, amethysts and pearls. The 12 stones symbolize the 12 apostles and the 12 tribes of Israel. The arrangement of the stones echoes the design of the breastplate that, according to the account in Exodus, was worn by the Jewish high priest in biblical times. The stones used in the breastplate were supposed to represent the colours of the 12 tribes, so different gemstones were used depending on availability; therefore, ancient texts differ on their assessment of the gems that should be used. One text lists the gemstones set in the breastplate as carnelian, peridot, emerald, ruby, lapis lazuli, onyx, sapphire or jacinth (zircon), banded agate, amethyst, topaz, beryl and green jasper or jade.

The crown today resides in the Imperial Treasury at the Hofburg Palace in Vienna, Austria.

A portrait by Albrecht Dürer (1471–1528), of Emperor Charlemagne (742–814), *c.*1512, depicted wearing the Crown of the Holy Roman Empire.

The Crown of the Holy Roman Empire, likely
created in the late tenth or early eleventh
century, with eight gem-set panels, including blue
and purplish sapphires and an arch decorated
with seed pearls to spell the name of Emperor
Conrad II (r. 1027–39). The four enamelled panels
depict the kings Solomon, David and Ezekias and
Christ on a throne with two angels (right); all in
surrounds of large sapphire beads and pearls.

RESIDENZ TREASURY
AND MUSEUM, MUNICH

Munich was the seat of the court of Emperor Louis IV 'the Bavarian' (r. 1328–47 as Emperor), who transformed the ancestral castle and his father's residence into the imperial seat and who continued to expand the palace and gardens. From the seventeenth century, the so-called Residenz was home to the rulers of the House of Wittelsbach, who governed Bavaria as dukes until 1623, as electors until 1806, and then as kings until 1918.

It was Louis's later successor Albert V (Duke of Bavaria, 1550–79) whose vision it was to establish a *Kunst- und Wunderkammer*, literally an 'art and wonder chamber', to show off his art collection and treasures. At his death, he instructed that his precious collection never be sold, a wish that led subsequent generations to honour and preserve his legacy by enriching its holdings. In the late eighteenth century, treasure from the Palatine Wittelsbachs was transferred to the Residenz. The location of this palace in Munich led to a growth in the collection for a very different reason. In 1803, Bavarian abbeys and monasteries were secularized due to the turmoil of the French Revolutionaries' advances; many of the abbeys' and monasteries' outstanding medieval works of art were confiscated as booty for the Bavarian princes after they acquiesced to negotiations with Napoleon. These works found new homes in what was now the *Schatzkammer* (Treasury) of the Munich Residenz.

Opened briefly to the public before World War I, the Treasury survived the fall of the Bavarian monarchy in 1918. Unfortunately, during World War II, the palace and grounds were very badly damaged and required gradual reconstruction from 1945 until they reopened to the public in 1958.

Every time I find myself in Munich, I always make time to visit the Treasury. The objects on display are mesmerizing and I never tire of seeing them. As soon as you walk through the double doors to the Treasury, you immediately enter a room displaying three impressive medieval crowns standing tall in their cabinets – it is almost too much heritage and craft to digest in one view.

Top: The Antiquarium, Duke Albert V's room for his collection of antique sculptures in the Munich Residenz. Built 1568–71, it is the oldest room in the current Residenz complex and was used by Albert's successors as a banquet hall.

Above: A portrait, *c.*1556, of Duke Albert (Albrecht) V of Bavaria (1528–1579) by Hans Mielich (1516–1573).

Facing page: One of the finest pieces in the Munich Residenz is the impressive gold standing cup, known as the Sapphire Cup, designed by the court painter Hans Mielich and attributed to the Munich goldsmith Hans Reimer (1555–1604), who made this cup in 1553 for Duke Albert V of Bavaria. The white enamel is lavishly inlaid with leaves and tendrils of gold leaf and decorated with gold swags and frames that encircle 36 large blue Ceylon sapphires – white and blue are the colours of the Bavarian coat of arms. The classical warrior atop the cup raises a ring of carved sapphire, which must have been expertly cut from a large crystal. The cup is 48.6 cm tall and weighs 5.5 kg.

This crown (1010–20), reputedly owned by Empress Kunigunde of Luxembourg (*c.*975–1040), consists of a gold-coloured band with gemstones and filigree decoration. The condition is quite superb, particularly that of the pearls, which, considering the age of the crown and the fact that pearls are very susceptible to environmental effects, have barely deteriorated at all.

The front of the band is decorated with a circular filigree motif, set with a large, tumbled pale blue sapphire that has a drill hole running horizontally through it. This drill hole tells us that the sapphire was once part of a trading string of sapphires. Observing the crown through glass limits the identification of the stones, but surrounding the sapphire are four pale pink cabochon-cut stones, which could possibly be rose quartz or pale pink sapphires or spinels, with two natural pearls top and bottom. All the stones are set in individual collets decorated with fine granulation.

Either side of the central circular motif are two cartouche-shaped designs – each, again, set with a large, tumbled pale blue sapphire, this time vertically set. These sapphires also have drill holes, running vertically. One sapphire is quite badly damaged, so it is hard to tell if it has a drill hole, although it probably does.

The rest of the band is set with a variety of stones in a three-row formation. These stones are mainly cabochon cuts, with a few sugarloaf cuts, and they are a mix of coloured glass and gemstones. Interspersed between these stones and glass are natural pearls, which are likely a mixture of pearls from freshwater oysters in European rivers and saltwater oysters in the ocean. The smaller pearls display their drill holes. A few stones seem to have been replaced, as some colours are out of keeping with the rest of the colour palette. Sapphire is the predominant stone; its presence aligns with the medieval reverence for blue sapphire and confirms the idea that it was considered the most appropriate stone to be set in a crown.

Kunigunde of Luxembourg was crowned Queen and Holy Roman Empress in 1014, together with her husband, Emperor Henry II. Following his death, she founded the monastery of Kaufungen, where she ended her days as a nun. Like her husband, she was later canonized, in 1200.

An early eleventh-century gold band set with
large, tumbled pale blue sapphires surrounded
by other gemstones; purportedly the crown of
Kunigunde of Luxembourg, now in the
collection of the Munich Residenz.

Near to Kunigunde's crown is the crown of her husband, Henry II. Born in 973 and son of the Duke of Bavaria, Henry II succeeded his cousin as King of the Romans in 1002 and, two years later, also became King of Italy. In 1014, Henry II was crowned Holy Roman Emperor by Pope Benedict VIII. A great supporter of the teachings of the Church, Henry II was admired for his decisive political leadership. He and Kunigunde were deeply devoted to each other and to the Church. Although they did not have children, Henry's deep love for Kunigunde and his profound respect for the sacrament of marriage meant that they remained faithfully together.

This huge medieval crown was formerly part of the treasury of Bamberg Cathedral. Though it is known as 'Henry's Crown', it is too large for a human head — instead, it was used to decorate the head reliquary of Emperor Henry II, who was canonized in 1146 by Pope Eugene III.

Made of silver gilt, this crown consists of six fleur-de-lys panels joined with hinges. Each hinge is fixed with pins that terminate in angels standing on acanthus leaves, a motif that represents enduring life, common in funerary floral displays of the time. All the panels are encrusted with substantial gemstones, including various agates, moonstones, natural pearls, amethysts, drilled pale blue sapphires and two shell cameos, all surrounded by embossed acanthus-leaf motifs applied throughout.

It is comforting that both Henry's and Kunigunde's crowns now reside in the same location, so that visitors can be reminded of their commitment to each other.

A crown of six silver-gilt panels, shaped as fleurs-de-lys and encrusted with gemstones including pale blue sapphires, flanked by golden angels, from the collection of the Munich Residenz. This large crown adorned the head reliquary of Henry II (973–1024), Duke of Bavaria and later Holy Roman Emperor, when he was canonized in 1146.

The impressive crown of Blanche of Lancaster
(1392–1409), part of her dowry when she married
Louis III, Elector Palatine of the Rhine from the
House of Wittelsbach of Bavaria, and supposedly
made *c.*1380 for Anne of Bohemia (1366–1394),
Queen of England. Its tall gold fleurons are set
with blue and pink sapphires, plus pearls and
other smaller gemstones.

THE ROYAL CROWN OF
BLANCHE OF LANCASTER

A beautiful and elegant crown meets the eye as you enter the Munich Treasury. Every time I see it, I am struck by its magnificence; then I try to imagine what it must have been like to witness it being worn. Displayed between Henry's Crown and Kunigunde's, it is one of the most elegant crowns I have ever seen. Its towering fleurons – the tall, elongated spikes that are attached to the circlet – stand at 18 centimetres in height, with the six smaller fleurons in between at 14.5 centimetres.

There are many theories about when this crown was created and for whom, but the most probable is that it was made in the 1380s for Anne of Bohemia (1366–1394) after her marriage to Richard II. Anne was the eldest daughter of Charles IV, Holy Roman Emperor and King of Bohemia, who was the most powerful monarch in Europe at the time. She became Queen of England when she married Richard II in London in 1382, when they were both only 15 years old. They were married for 12 years before Anne was struck by the plague and died in 1394, aged only 28.

Such was the turmoil among the ruling families in England and the Holy Roman Empire that in 1399 Richard II was forced to abdicate, succeeded on the throne by his cousin Henry IV. A marriage that would be beneficial to both Germany and England was being arranged by the families of Henry IV and Rupert III, Elector Palatine. It was decided that Henry IV's daughter, Blanche of Lancaster (1392–1409), also known as Blanche of England, would marry Rupert III's son Louis III, Elec-

tor Palatine, in 1402, bringing with her a substantial dowry that included this crown. Repaired in March–April 1402 by London goldsmith Thomas Lamport, the crown travelled with Blanche to Cologne, where she was married to the future Louis III in July of that same year.

Blanche died young and her crown remained with her husband, apart from its brief time at the monastery of Maulbronn, after Louis needed to pawn it in return for 3,000 guilders. The transaction records from 1421 do not detail why Louis needed the money, but it may have helped finance his significant library of religious manuscripts that later became an important part of the Bibliotheca Palatina, now split between Heidelberg and the Vatican. The loan was later paid back, and the crown was returned to Louis.

For the next four centuries, the crown was moved to various locations in Germany for safe-keeping as wars and unrest developed. In 1720, Mannheim took over from Heidelberg as the capital of the Palatinate and the crown was listed in the Palatine inventories and resided in the treasury. In 1782, the treasury was transferred to Munich, after Charles Theodore of the Palatine branch of the House of Wittelsbach became Elector of Bavaria by succession. An inventory was made in 1783 of the jewels owned by the Wittelsbach family that includes the crown but, in 1818, the new constitution declared that the treasury belonged to the State.

It is remarkable that this crown has survived with only a few repairs and stone replacements. A team of experts have recently published an extensive report on the crown, in which they reveal the results of their in-depth study both of the goldsmithing techniques employed and of the gemstones used. The team were able to completely open the crown and lay it flat by removing a pin from one of the hinges; its now open form allowed further examinations to be made with gemmological equipment.

According to the researchers' findings, there are 47 blue sapphires of assumed Ceylon origin, with nine that display asterism (star sapphires), and three dark blue sapphires that may have come from the Le Puy-en-Velay sapphire deposit in France, plus one blue imitation jewel. Other stones include 10 pink sapphires, 22 octahedral diamond crystals, 53 spinels, five garnets and nine emeralds. The settings of this huge selection of gemstones are decorated with pearls and with a few imitation stones. The majority of the stones are mounted in foil-backed, closed settings, which is a typical style of setting for the medieval period. Many of the gems were also found to have drill holes, which is a common feature of gemstones from this period.

This study has been invaluable in helping us to understand and appreciate the techniques used by medieval goldsmiths; it also brings to life the skill of the lapidary and the arduous journey these gemstones took to end up in such an important jewel. These jewelled objects are a window into a moment of history and into a world that technological advancements are able to reveal to us.

An illuminated page from the *Liber Regalis*, a 1390s manuscript
in the collection of Westminster Abbey, thought to have been
written for use during the coronation of Queen Anne, depicted
here with her husband Richard II and believed to be the original
owner of Blanche of Lancaster's crown.

With all the splendid gems in the Imperial State Crown, there is one gem that stands out: St Edward's Sapphire, set in the Maltese cross above the diamond-set orb.

Edward the Confessor (1002/1005–1066) was known for his generosity and the giving of alms. Legend has it that one day, when he was walking to Westminster Abbey (which he had founded), he was accosted by a beggar. Having already parted with all his coins to previous people in need, he had no more money to give. Not wanting to pass the person by without offering something, he took his ring, which was set with a sapphire, and gave it to the beggar, not thinking any more of it. Later, in Syria, the same beggar came across two pilgrims who had lost their way in a storm and were looking for shelter, so he guided them to an inn. In the morning, they went to thank the old man and mentioned they were English and that their king was Edward. Hearing this news, the beggar gave them the sapphire ring and asked that they give it back to the King with the message that he was, in fact, St John the Evangelist and that the King would join him in heaven.

The pilgrims returned to England and managed to meet the King to return the ring; the King recognized the sapphire ring he had given to the beggar many years before. Upon his death in 1066, he was buried with the ring, along with other regalia. In 1161, he was canonized by Pope Alexander III and became St Edward.

Later, King Henry III built a splendid tomb for St Edward and so, after 200 years, his coffin was opened so that his body could be transferred to the new tomb. It was decided that all the jewels and regalia that had been buried with him, including the sapphire ring, should be removed. As emblems of divine authority, these jewels were revered by all subsequent monarchs and became part of the Crown Jewels of England.

At the end of the coronation ceremony, the monarch exchanges the heavy St Edward's Crown for the lighter Imperial State Crown. This practice remembers the pre-Civil War tradition of the coronation crown residing only in Westminster Abbey. An imperial state crown has been in existence in various forms since the fifteenth century. In 1838, a new crown was made for Queen Victoria's coronation. Its design includes closed arches, which symbolize that Britain is not subject to any other earthly powers. Interestingly, many medieval crowns were designed with an open top and no arches. By the time of George VI's coronation in 1937, the Imperial State Crown was over 100 years old and a bit worse for wear. Its supposed fragility is captured in a story from George V's funeral. The crown was travelling on top of George V's coffin when the Maltese cross, containing St Edward's Sapphire, toppled off the crown; the procession carried on but, luckily, the company sergeant major of the Grenadier Guards managed to rescue the cross from the tarmac and placed it in his pocket. A new crown was commissioned and was a virtual replica of Queen Victoria's, except it was made 10 ounces lighter, with St Edward's Sapphire reinstated.

The particular importance accorded to St Edward's Sapphire, believed to hold divine authority and representing the heavens, explains why it has always been set high on the Imperial State Crown.

A detail from the Wilton Diptych, *c.*1395–9, by an unknown artist, in the National Gallery, London. St Edward is depicted holding his sapphire ring while supporting a kneeling King Richard II as he is presented to the Virgin and Child.

Another sapphire of great importance set in the Imperial State Crown is the Stuart Sapphire, named after the House of Stuart. How the drilled Ceylon sapphire, weighing approximately 104 carats, made its way to England is a mystery. It has been suggested, though, that James II of England (VII of Scotland, 1633–1701) in fact smuggled the sapphire out of England into France in 1688 when he was escaping the Glorious Revolution, which brought an end to the absolute monarchy and established Parliament as the ruling power of England.

The sapphire had already been passed down through a succession of monarchs, when, on his death in 1701, James II left it to his son James Edward, 'The Old Pretender' (1688–1766). The sapphire continued to descend in the royal family until it eventually came into the possession of George IV (1762–1830) through an Italian dealer. After his death, the sapphire took centre stage and was set, along with the Black Prince's Ruby, in the band at the front of Queen Victoria's newly made Imperial State Crown. The Stuart Sapphire stayed in its prominent position until the Cullinan Diamond, found in 1905, was cut and one of its two principal stones, Cullinan II, was chosen for the crown. The Stuart Sapphire was replaced in 1909, but only relegated to the back of the crown, where it resides today.

The British Imperial State Crown,
made in 1937 to the same design and with
the same stones as the crown Queen Victoria
commissioned 100 years before. On the reverse,
pictured here, is the Stuart Sapphire, a Ceylon
sapphire of approximately 104 carats.

The Crown of the Absolute Monarch, set with two
large table-cut sapphires, commissioned from
jeweller Paul Kurtz by King Christian V of Denmark
(1646–1699) for his coronation in 1671.

The Crown of King Christian V of Denmark, also known as the Crown of the Absolute Monarch, is spectacular. Its two stunningly large sapphires, set in the band of the crown, strike you first. More than likely from Ceylon, the largest of the two sapphires can be traced to the Duke of Milan in 1474. One sapphire has obviously been cut from a very large crystal as it is set very proud of the band, showing the depth, or chunkiness, of the stone. Neither sapphire has been compromised by being cut down or by having extra facets added; instead, the lapidary has chosen the simple table cut to enhance the importance of the gems themselves. The sapphires are matched by two other table-cut gems, both large garnets, with similar cut diamonds set throughout. The cross of the crown's orb is topped with an unusual corundum cabochon: a blue sapphire with a streak of pinkish ruby. The contrasting colours of the yellow gold, red garnets, blue sapphires and white diamonds make a strong visual impact, which would have undoubtedly been the intention of the goldsmith.

Paul Kurtz, a jeweller, came to Denmark from Germany in 1655 and, by 1659, was referred to as 'the King's goldsmith'. In 1670–71, he was commissioned to make Christian V's crown. The closed design was inspired by the crown of Louis XIV of France but, instead of using lily-shaped arches, Kurtz designed the crown with palmettes adorning the sides. Golden acanthus leaves represent immortality; they are set throughout with table-cut diamonds.

The crown was first used for the coronation of Christian V on 7 June 1671. When the Danish monarchy was abolished in 1849, the coronation ceremony was discontinued; since then, the crown has only been used at funerals, placed on the coffins of monarchs.

The crown commissioned from jeweller
Humbert in 1889 by Wilhelm II (1859–1941), Emperor of
Germany, for future German emperors. A large sapphire
orb sits atop the diamond-set crown.

The Franco-Prussian War of 1870–71 meant that the briefly occupied France saw the proclamation of the head of the Hohenzollern dynasty, King Wilhelm I of Prussia, as the first Emperor of Germany. Wilhelm I had already been made King of Prussia on 10 October 1861, an event for which he had needed to commission a new crown, predominantly set with diamonds, because the previous Prussian crowns were all too big: they had been made to accommodate the fashion for wearing large wigs.

When Wilhelm II, the last prince to be made a German emperor, took the oath as Emperor of Germany in 1888, he decided that the crown from his grandfather was not grand enough. In 1889, he commissioned Berlin jeweller Humbert to make a better model. He was very particular about the gemstones to be used and approved a large sapphire sphere that surmounted the pearl- and diamond-decorated arches of the crown. When the crown was completed, it joined the rest of the regalia of the Kingdom of Prussia, but it was never worn. The monarchy relinquished power in 1918 and the crown remains part of the Hohenzollern family collection, despite supposedly having been hidden behind a church wall for its protection during World War II.

THE ARCHDUCAL CORONET OF AUSTRIA

Made in 1616, the Archducal Coronet of Austria
has a large sapphire bead set at the intersection
of two pearl- and diamond-set arches.

This incarnation of the Archducal Coronet of Austria was made in 1616 and given by Archduke Maximilian III (1558–1618, r. 1612–18), Grand Master of the Teutonic Order, to Klosterneuburg Monastery. The monastery is home to the relic of the skull of St Leopold, the patron saint of Austria; by giving the holy crown of the land to rest alongside the relic, the Archduke intended for the divine powers of the saint to be transferred to the 'crowned ruler'.

This coronet is unusual for its time in having only two intersecting arches, but these arches are lavishly decorated with pearls, diamonds and enamel and centrally support a large blue sapphire bead, surmounted by a gem-set cross. The base of the crown is wrapped in ermine. It made its last public appearance in an official capacity at the funeral of Zita of Bourbon-Parma, the last Empress of Austria, in 1989.

THE RUSSIAN IMPERIAL ORB

A sovereign's orb is an important article in coronation regalia. Just as the cross that surmounts it denotes Christ's dominion over the world, so the orb represents the power of the monarch as God's representative on earth – a belief that dates back to the Christian Roman emperors.

The Russian Imperial Orb, which is 48 centimetres high and 24 centimetres across, was made in 1762 for Catherine the Great (1729–1796), likely by court jeweller Georg Friedrich Eckart and possibly with goldsmith Jérémie Pauzié. It is made of gold, which has a reddish tint due to the percentage of copper Russian goldsmiths used in alloying gold, and is wrapped with silver ribbons pavé-set with 1,370 diamonds. The orb is surmounted by a Ceylon sapphire weighing approximately 200 carats, according to an inventory entry made in 1898. The front is adorned with a pear-shaped diamond weighing 46.92 carats, while the diamond-set cross was added later, during the reign of Paul I (1796–1801).

In 1900, the jeweller Fabergé replicated the imperial Russian regalia for the Paris Exposition Universelle after the organizers' request for the original coronation regalia had been refused. The miniature replicas were made with gold and real gemstones. The imperial crown replica was set with 1,328 diamonds, the orb with 719 diamonds, the smaller Empress's crown with 1,384 diamonds, and the sceptre with 125 rose-cut diamonds and one large stone, maybe of rock crystal, representing the historic Orlov diamond. Fabergé's workshop, led by August Wilhelm Holmström who made the replicas, was recognized for their skills when the house of Fabergé was awarded the gold medal at the Paris fair, despite not officially competing, and subsequently Peter Carl Fabergé was awarded the Légion d'honneur. Tsar Nicholas II was so impressed with the replicas that he bought and displayed them in the Jewellery Gallery at the State Hermitage Museum in St Petersburg.

Top: Miniature replicas of the imperial Russian
regalia made in 1900 by Fabergé, including
the orb set with a sapphire.

Above: A portrait of Tsar Nicholas II of Russia
(1868–1918) on his coronation day holding
the Russian Imperial Orb, surmounted with
a 200-carat Ceylon sapphire, published in the
French newspaper *Le Petit Journal*, 24 May 1896.

AT COURT

The early nineteenth-century parure of Stéphanie de Beauharnais (1789–1860), Grand Duchess of Baden, niece and later adopted daughter of Joséphine de Beauharnais and her husband, Emperor Napoleon I. This parure is thought to have been given to Stéphanie by her cousin and adoptive sister Hortense de Beauharnais and later inherited by Stéphanie's daughter Josephine, Princess of Hohenzollern Sigmaringen. The bandeau tiara was likely adapted by Josephine from a belt, a popular jewelled accessory within Napoleon's court. The tiara, necklace, bracelet, earrings, brooches, rings and pendants are all set with sapphires (some confirmed as from Sri Lanka) and rose-cut and cushion-shaped diamonds.

At Court

Throughout history, in kingdoms and empires around the world, the courts of sovereigns and aristocratic families served their rulers. Court culture developed through the need to consolidate political, intellectual and artistic patronage and to strengthen regional bureaucracy and influence; it led to displays of great pageantry and expectations of courtly etiquette in Europe, Russia, India and Ceylon (now Sri Lanka). In addition to the royal courts, with monarchs at their centre, there were also provincial courts, with their dukes and counts.

Men and women alike wore jewels to signify power and rank. Large jewels were useful for disarming or distracting one's rivals. During the sixteenth and seventeenth centuries especially, jewels set with precious stones were unequivocal evidence of a kingdom's international trading prowess.

Later, in the eighteenth and nineteenth centuries, court etiquette required the wearing of tiaras and other regalia as standard. Gas lighting illuminated jewels in the evening salons, enhancing their allure and impact at court and state events. As a consequence, diamonds and sapphires found particular favour among jewellers and their clients; since sapphires are found in sizeable crystals, they could be cut into large shapes to make an instant impression when set in a head ornament or in a *devant de corsage*.

The natural world had a strong influence on jewellery design, with tiaras of jewelled flower sprays and ears of corn. Big and bold jewels could be found in all courts, but the most opulent of all were surely those of the Romanovs, who acquired and had set some of the world's largest sapphires.

By 1900, breakthroughs in oxyacetylene torch technology allowed platinum to be worked by the goldsmith, since its high melting point could now be reached. The use of platinum for jewels meant that sapphires and other gemstones could now be set in less metal but the volume previously achieved with silver and gold still be retained. Platinum is much stronger than silver, so fine lines can be created without compromising the strength of the jewel. The craftsmen at Chaumet were the masters of tiara design; all their creations from the 1900s show a lightness in style made possible only by platinum.

The engagement ring of Napoleon Bonaparte
(1769–1821) to Joséphine de Beauharnais
(1763–1814) is set with a pear-shaped sapphire
and a pear-shaped diamond in the *Toi et moi* –
You and Me – design, symbolizing two souls
becoming one. The sapphire represented
'constancy' and the diamond meant 'for ever'.

In Marco Polo's accounts of his travels from Europe to Asia along the Silk Road in the thirteenth century, recorded by writer Rustichello da Pisa, he introduces the 'Indies' as rich in gold, silver, pearls, jewels and spices — enough to dwarf the splendours of the courts of Europe. Three hundred years later, in 1583, English merchant Ralph Fitch leaves for India and returns with similar reports. When Fitch tried to gain access to Emperor Akbar's court, he was apprehended by the Portuguese, who at that time policed the export of Indian gems, and was remanded in their colonies in India. Fitch and his travelling companions managed to escape and arrived at the Mughal courts of Agra and Fatehpur Sikri in 1585 to behold the wonders of Akbar's court. Emperor Akbar was a keen patron of the arts and culture. According to his reports, the vast wealth Fitch witnessed was far greater than that of any European court. With news of these riches reaching Europe, it became even more important for European rulers to be seen as equal rivals in gemstone and jewellery collecting and displays. The courts of Burgundy and Berry in France, the wealthy families of Italy — such as the Este, Gonzaga, Medici, Montefeltro, Sforza and Malatesta — and the royal families of England and Spain were almost in competition with each other to display their magnificence and importance through jewellery.

Queen Elizabeth I (1533–1603) was the first English monarch to send an emissary to Mughal India, a world power in international commerce; this emissary, Thomas Roe, was tasked with finding new allies for England so that the country could try to rival Spain and Portugal's dominance in trade with the East. The English court's contact with the Mughal Empire's opulent jewelled culture had an enormous impact on jewellery patronage in England. Jewellery and adornment were important parts of social identity in European and Mughal courts. Miniature portraits of Elizabeth wearing sumptuous jewels became diplomatic gifts at home and abroad. Mughal emperors institutionalized the practice of wearing lavish displays of jewellery by using jewels as regalia or as insignia of a rank of authority.

These cultural exchanges drove demand for jewellery and gemstones, which, in turn, promoted advances in both lapidary and jewellery design. Ideas and techniques flowed in both directions. Mughal jewels, portraiture and gems greatly influenced and inspired jewellery designs back in English goldsmithing workshops, while European jewellers worked in Emperor Akbar's workshops. Mughal craftspeople further developed techniques popular in European works of art, such as *pietre dure*, the hardstone inlays seen in Italian works. They took the practice one stage further by replacing the hardstones with precious gemstones — setting rubies in jadeite, for example — which were much more difficult to cut. The extensive use of coloured enamels in Mughal jewellery may have stemmed from European influence, but the major centre for enamelling remains Jaipur to this day.

Despite the obvious exchanges of techniques, styles and raw materials, the distinct identities of Indian and European jewellery in the sixteenth and seventeenth centuries were retained. Indian goldsmiths set gems in a beaten gold setting called *kundan* work; in this technique, the metal wraps round the irregular shape of the gem so that stones could be set securely without cutting in order to maintain the size and weight of the stone. In Europe, jewellers preferred cut and polished gemstones so the flat-faceted surfaces could reflect the candlelight.

Elizabeth I was not the only royal that knew the importance of jewels. Anne of Denmark, who was Queen Consort of Scotland, England and Ireland by marriage to King James VI and I, adored jewellery. In 1597, she appointed an Edinburgh jeweller, George Heriot, as her personal goldsmith for life; he made jewels for her that totalled more than £40,000, equivalent to £12 million today. James and Anne recognized two other goldsmiths for their services to the court: Sir John Spilman, who was knighted by James I in 1605, and William Herrick.

Sapphires are less well represented in European Renaissance jewellery of the sixteenth and seventeenth centuries than they were in medieval jewels. Although they did not seem to regain their earlier prominence until the late nineteenth century, this drop in use is not necessarily an indication that sapphires were any less popular. Instead, it may have been due to difficulties in the stone's availability. In the seventeenth century, most sapphires came from Ceylon (now Sri Lanka), which was ruled by the Portuguese alongside local kings and later by the Dutch and local kings. These kingdoms often battled each other for control of the island, which included the sapphire deposits. A captured British sailor, Robert Knox, witnessed the effect of these different rulings during the 20 years he spent imprisoned there from 1659. In *An Historical Relation of the Island Ceylon* (1681) Knox wrote:

> In this island are several sorts of Precious Stones, which the King of his part has enough of, and so careth not to have more discovery made. For in certain places where they [gemstones] are known to be, are sharp Poles set up fixed in the ground, signifying, that none upon pain of being struck and impaled upon those Poles, presume so much as to go that way.

Knox's description of the measures kings took to protect their sapphire deposits points to the difficulty faced by merchants that wanted to channel a supply of sapphires outside of Ceylon. As sapphires were difficult to procure, when they did reach northern Europe they often became the centrepiece of jewels.

Paintings and portraiture give us an insight into how jewels were worn in the European courts during the sixteenth and seventeenth centuries. Both men and women wore jewels either over their clothes for display; or under their clothes, so that the stones — if believed to have medicinal or talismanic qualities — could touch the skin and thus be effective and provide immediate reassurance; or sewn onto their outer garments as ensigns or badges to indicate rank and status. Brooches were worn on hats or used to hold folds of fabric together. Rings were worn on every finger — every joint and thumb on each hand — by both sexes. Rings could sit either over gloves or under gloves and sometimes with gloves slashed to reveal the stones underneath. Wearing so many rings apparently sometimes resulted in 'finger fatigue'. Men wore earrings more than women, as women's ears and hair were usually covered. It had been thought unclean for a married woman to show her hair, but Elizabeth I was proud of her 'virgin status' and exposed her auburn hair, an act that encouraged ladies of the European courts to leave their heads uncovered and even emphasize their new, free style by wearing jewels in their hair.

A turban button of quartz inlaid with a central octagon-shaped rose-cut sapphire and flower motifs of rubies and emeralds in *kundan*-style gold settings.

This turban button has a central rose-cut sapphire set in quartz using the *kundan*-style setting typical of Indian goldsmithing. The goldsmith wrapped the sapphire in gold to hold it securely in the button. The button is decorated with flowers of rubies and emeralds. The jewel is part of the vast collection that Sir Hans Sloane left in his will to the British nation and that formed the basis of the holdings of the British Museum, British Library and the Natural History Museum.

Hans Sloane was born in 1660 in County Down, Ireland, and lived to be 93 years old; his long life may have been due in part to his careful control of his genetic illness by strict monitoring of his diet and alcohol consumption, a rare action for someone living in the seventeenth and eighteenth centuries when diets were not varied and alcohol was safer to drink than water. He moved to London aged 19 to study medicine and botany; at 24 years old he was elected to the Royal Society of London for Improving Natural Knowledge and then to the Royal Society of Physicians. His first appointment overseas, and the catalyst for his collecting habit, was his appointment in 1687 as personal doctor to Christopher Monck, 2nd Duke of Albemarle, who was then serving as governor of Jamaica. Jamaica had been under Spanish rule but was now a British colony, and the country was immersed in a plantation system that survived on the appalling institution of slavery and the work of enslaved people. Sloane's lifetime passion was collecting countless medicinal plants, specimens, books and curios, and, while he was in Jamaica, he searched for new plants that might be introduced as food. He is known for introducing milk chocolate to England. He also collected wild animals, but they never survived the passage back to England. Enslaved people helped Sloane acquire his collection, both physically, by being forced to collect samples for him, and financially, in that Sloane's involvement with the slave trade allowed him to indulge and prosper in his collecting career. By the time Sloane died, he had amassed an enormous collection.

A miniature watercolour painting on ivory, *c.*1790 by an unknown painter, of Caroline (1753–1829), daughter of the Duke of Saxe-Coburg-Saalfeld, who in 1795 became Deaconess of Gandersheim Abbey, a secular community of unmarried noblewomen in Germany. Caroline is depicted wearing sapphire earrings and sapphires in her hair.

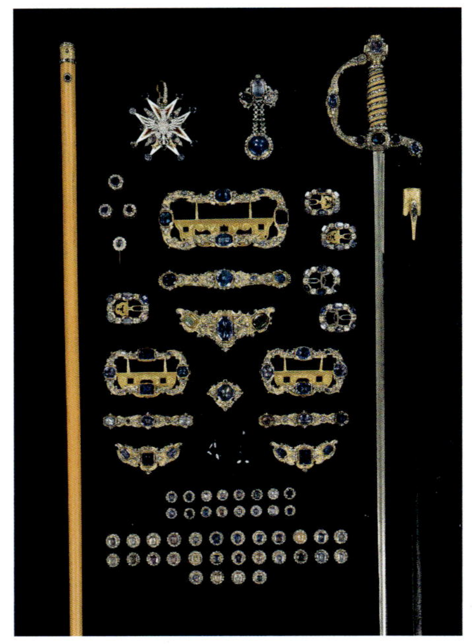

Sapphire dress jewels of
Augustus II (1670–1733), Elector of
Saxony, King of Poland and Grand Duke of
Lithuania; one of his 10 sets of ceremonial jewellery
(above) made in the workshop of his court jeweller
Johann Melchior Dinglinger (1664–1731), which are
now housed in Dresden's Green Vault.

Top: One of nine jewelled pieces connected to
a ribbon sash to suspend a sword; this piece, made
around 1710, is set with a large octagon-shaped
sapphire and 27 rose-cut diamonds, added in 1721.

Centre: An agraffe – a decorated clasp – from around
1722, to wear on a brimmed hat, set with three large
sapphires from which hang two rows of diamonds and
a lynx sapphire, the eighteenth-century name for this
blue cordierite that was hollowed and lined with silver
foil to make it appear transparent, like a fine sapphire.
The cordierite was sourced in 1701 by Saxon Grand
Chancellor Wolf Dietrich Graf von Beichlingen (1665–1725)
and recorded as the 'extra-large sapphire in the
form of a small round mountain'. Augustus's
son wore the lynx sapphire atop his coronation
crown in Krakow in 1734 as he became
King Augustus III of Poland.

Bottom: The sapphire-tipped cross, with sapphire-set
flames between its points, of the Jewel of the Polish
Order of the White Eagle, was likely made in 1713 for
Augustus after he renewed the medieval order in 1705
to strengthen ties between Saxon and Polish-Lithuanian
nobility. As Grand Master of the order, the King's jewel
reads in enamel: 'Pro Fide, Lege et Grege' – 'For loyalty,
law and people' – whilst the knights' motto was
'For loyalty, law and king'.

A chatelaine, made in Britain c.1720, set with large sapphires and faceted white glass and suspending a similarly set watch made in 1724 by Charles Cabrier I, whose work established the Huguenot Cabrier family as a celebrated lineage of watch- and clockmakers in London; now in the collection of the State Hermitage Museum, St Petersburg.

CHATELAINES

During the eighteenth century, chatelaines were widely popular, being worn by both men and women as day jewellery, but in particular by women. The chatelaine consisted of useful accessories, such as tweezers, pencils, an ivory tablet, folding scissors and, later, a pocket watch, that were attached and suspended from chains that would hang from a long clip to be fastened onto a belt. Chatelaines were made in many metals including gold, silver, gilt metal or cut steel and were often elaborately decorated with chasing and enamel and set with gems. Their popularity continued into the nineteenth century.

Above, right: A necklace and sapphire drop pendant with highly detailed enamel of the popular white, turquoise and black combination to its design of bows and knots. Made in Europe c.1660, the central bow has been added to a later nineteenth-century chain and pendant, now on display at the Victoria and Albert Museum.

KNOTS AND BOWS

Some jewellery designs were inspired by ever-popular motifs, such as knots, which carried a religious meaning. In the sixteenth century, a friar's rope belt would consist of three knots symbolizing both the Holy Trinity and the religious vows of poverty, chastity and obedience. Developing from the religious knot came the bow, used for more sentimental and less sombre purposes – 'the greater the bond the tighter the knot'. Fashion adopted the bow, which came in many guises: from double bows to graduating bows and jewelled bows, enamelled and gem-set. These bows were either pinned to a dress, ruff or sleeve. On the Continent, a seventeenth-century French goldsmith, Jean Toutin (1578–1644) of Châteaudun, is renowned for having first developed a colour combination of turquoise, black and white opaque enamel, which he painted onto gold jewels. This colour combination became very popular and was often seen decorating the front and/or the reverse of jewels in the seventeenth century – a period that also began to witness the rise in prominence of the rich bourgeoisie in European countries, who wanted to emulate 'old-monied' families by owning magnificent jewels.

THE
CHEAPSIDE
HOARD

If the Cheapside Hoard – a collection of over 400 pieces of Elizabethan and Jacobean jewellery hidden below the foundations of
a London workshop in the 1620s – had not been accidentally found
by workmen in 1912, we would have had to rely on representations
in portraits and paintings for much of our knowledge about jewellery of this period. The Cheapside Hoard is a spectacular collection
of post-medieval jewellery. The gemstones in the jewels show the
range of stones entering northern Europe during the fifteenth and
sixteenth centuries: diamonds from Borneo and India, emeralds
from Brazil and later Colombia, rubies from Burma and sapphires
from Ceylon. The sheer number and quality of the jewels found in
the hoard are a reflection of jewellery's importance as an asset
in this period.

Such was the desire of princes and despots to show their wealth
and power that jewellery was in great demand and saw a greater
increase in its production. Queens Elizabeth I and Anne of Denmark
knew how to wear jewels for maximum impact and effect, and they
cultivated their image to achieve respect and status. However,
in the case of Elizabeth, she bedecked herself in jewels as she
grew older to distract people's gaze from her aging skin and
yellowing teeth. Francis Bacon (1561–1626), philosopher and Lord
Chancellor, commented a few years after her death in 1603 that
Elizabeth I 'imagined that the people who are much influenced
by externals, would be diverted by the glitter of her jewels from
noticing the decay of her personal attractions'. Impressive jewellery played a large part in presenting a monarchical image to
the public.

Sapphires come in a variety of colours, and this variety is well
represented in the hoard, with stones ranging in tone from deep
blue to pale pink and even to light greyish, almost colourless. Many
sapphires set in closed settings have foil behind the stones to
enhance and even out the appearance of their natural colour and
to prevent the eye from seeing any obvious inclusions. This setting
style is very typical of rings from this period, and rings like those
in the hoard were later found in shipwrecks. Many of the rings
in the hoard have sapphires set in uncomplicated gold mounts,
with no evidence of enamel work; instead, attention is focused
on the stones and their cuts. Starting in the fifteenth century, the
increased availability of gems from Ceylon and Burma, together
with the introduction of grinding and polishing wheels, promoted
rapid changes in stone-cutting technology, which replaced the
up-and-down motions of earlier bow, treadle and pole lathes.

Sapphire-set Elizabethan and Jacobean jewels
from the Cheapside Hoard, Museum of London.

From top: Two earrings or pendants, both with
a black-and-white enamelled reverse, one set
with three step-cut pink sapphires between small
round emeralds and suspending a pink spinel
briolette, and the other with two triangle-shaped
sapphires suspending a kite-shaped sapphire,
all three of pale blue colour and backed by foil,
decorated with six small opals; a pendant of rose-
cut pink sapphires and diamonds set in the form
of a cross with enamelled reverse; and two gold
rings set with a rose-cut sapphire in a hexagonal
bezel and a foiled rose-cut sapphire in
an oval bezel, respectively.

This pendant, also part of the Cheapside Hoard, is set with two blue Ceylon sapphires that both display exceptional clarity. Their clarity is emphasized by the precision of their step cuts, which would have been made by European lapidaries as most stones from Ceylon were only polished, not cut, at source in this period. The spinel drop would have been cut at source as it has been superficially polished – a method for giving a smooth finish to the rough crystal. The spinel has also been drilled, which was a typical processing technique for early spinel beads. The holes were drilled in a five-stage process, from top and bottom, and show slight misalignment between the two entry points. According to Hazel Forsyth, Cheapside Hoard curator at the Museum of London, there is evidence of one drill hole having a stepped profile, which would have guided the diamond drill bit so it did not wander, reducing the possibility of it splitting the spinel bead. Though understanding of mineralogy was not formalized until the 1790s, earlier traders and lapidaries would have recognized the difference between ruby and spinel, distinguished as 'balas ruby'.

The spinel would probably have come from Tajikistan, and the jewel itself would have been assembled in Europe. There is evidence of white enamel around the delicate settings of the sapphires, and a natural pearl (its companion is missing) decorates the central step-cut sapphire. It is a significant jewel for its size and quality because sapphires of this period were usually cut to preserve weight and were not therefore as precise as these stones. The clarity and size of these stones indicate that they may have been cut from very high-grade rough crystals – indeed, they do not show the concave 'dents' of other sapphires where a lapidary has cut out an imperfection.

The Cheapside Hoard includes a number of jewelled items that feature a suspension loop at one end and a flared open socket at the other. These sockets suggest that the items were fan handles used to hold plumes of feathers. Richly decorated with gemstones and enamel, these fan handles are set with rose-cut foil-backed sapphires, probably from Ceylon (the left also with amethysts), and with foliate motifs decorated with white enamel – a striking colour combination. The Cheapside Hoard handles are the only known examples of this type of handle.

BLACKWORK
ENAMEL RING

Though not part of the Cheapside Hoard, this sixteenth-century ring from the Victoria and Albert Museum has intricate enamel work, like items in the hoard, and is therefore a useful comparison. An elaborately decorated Renaissance ring, this jewel is centrally set with an irregular-shaped foiled pink sapphire. Its bezel setting is surrounded by a quatrefoil-shaped petalled mount. The stone may be a later replacement as it does not fit perfectly in the setting. The bezel is decorated with black-and-white enamel, known as 'blackwork' enamel, which was a popular decorative style from around 1585 until the 1620s. Other colourful enamels were also popular for opulent jewellery mounts. This ring is the culmination of work by a combination of craftspeople, its creation requiring the skills of a chaser, engraver, enameller and lapidary.

The jewelled emblem of the Order of the
Golden Fleece, bestowed upon a member of
the Bavarian ducal family in the 1760s, set with
diamonds and two large, step-cut sapphires.

One of the most splendid courts in Europe was the court of Philip the Good (1396–1467), Duke of Burgundy. His court influenced court life in all of France and Europe, but it is today principally known for its founding of the Order of the Golden Fleece. The Duke of Burgundy established the order in 1430 by selecting esteemed members of his court and allies to become knights of the order. These knights were expected to defend Roman Catholicism, uphold chivalry and maintain the right to trial and charge their fellow knights if found guilty of treason or rebellion. Philip's descendants maintained the Order of the Golden Fleece, but a disagreement about royal lineage resulted in the Habsburg emperors of Austria and the kings of Spain agreeing that they could both confer knighthoods of the order. Once knighted, its members could proudly wear the jewelled emblem of the order.

One of these jewels is held in the Munich Residenz, which served as the seat of government and residence of the Bavarian dukes, electors and kings from 1508 to 1918. The Residenz houses many spectacular jewels and objets d'art including the highest award of chivalry, the Order of the Golden Fleece. The Bavarian rulers generally all belonged to the order, and this particular jewel was bestowed on a member of the family in the 1760s. It is made of silver and gold articulating pendants set with two large octagonal mixed-cut Ceylon sapphires. Sapphires were regarded as the jewel to represent heaven and constancy, two connections that marry well with the order's intentions. The sapphires represent the order's motif of a flint being struck by steel surrounded by flames set with cushion-shaped diamonds. The two sapphire pendants support a golden ram fleece from the Greek myth of Jason retrieving the fleece of a golden ram. The design originated from the Duke of Burgundy's Latin motto, 'Ante ferit quam flamme micet' – 'It strikes before the flame flickers'.

COUNTESS WALEWSKA'S PIN

The early nineteenth-century pin given by
Napoleon I to Countess Maria Walewska,
set with a 45.79-carat Ceylon sapphire, diamonds
and seed pearls and with emerald-set canon,
eagle and sword motifs.

This rare, early nineteenth-century sapphire, emerald, ruby and diamond pin was given to Countess Walewska (sometimes spelled Waleska) (1786–1817) by Napoleon I. It is embellished with symbolic motifs that represent the Roman Empire because Napoleon always wished to associate himself with the power of that ancient empire. The jewel is set with a cushion-shaped sapphire, weighing 45.79 carats, surmounted by a Roman helmet motif with calibré-cut emeralds and flanked either side with a flag and two Roman lances, a monogram and a shield set with seed pearls. Below the sapphire sit a sword, laurels, hatchet and crossed canons with a drum in the centre. At the top of the jewel rests an eagle set with emeralds and a ruby, supported by a later-added segment of pavé-set circular-cut diamonds. The jewel was later cased by Cartier with a plaque inscribed 'Presented by Emperor Napoleon 1st to Countess Waleska'. This is a very personal jewel as the eagle, which would normally be depicted standing up, is in a recumbent position, implying that Napoleon wished to protect the Countess during their love affair that led to the birth of their son, Alexander, in 1810.

During the middle of the nineteenth century, jewels were worn in abundance at court. Many goldsmiths took inspiration from the world of nature. In society generally, there was a keen interest in plant collecting, which introduced new species into Europe. Jewellers took advantage of this curiosity about flora by creating jewels that were realistic in their depiction of bouquets of exotic flowers and wildflowers, twining branches and leaves. Goldsmiths such as François-Désiré Froment-Meurice (1802–1855), from Paris, and Jules Wièse (1818–1890), originally from Germany but who worked in Paris, combined these ideas with sculptural, Gothic human figures and angels, elements derived from the Romantic movement, as well as heraldic and architectural motifs, to create elaborate jewels. This bracelet, a collaboration between both Froment-Meurice and Wièse, is a perfect example of Romanticism as its design draws on the symbolism of flowers and gemstones, both of which had their own popular symbolic languages. This bracelet is a visual stepping stone to designs of jewels made for the court.

Wearing a full parure was a prerequisite for court appearances. A parure typically included a tiara, pair of bracelets, a ring and a necklace that all matched and were meant to be worn together. Sometimes, in royal circles, the parure would also include aigrettes, buttons and stomachers. In the 1840s earrings were hardly worn, as hairstyles covered the ears, and so very few parures included earrings. Earrings regained popularity from the 1850s, when hairstyles started to reveal the ears. Tiaras were a highly popular jewel of the early nineteenth century. Designs for tiaras followed naturalistic themes, depicting sprays of flower buds, leaves and berries with trailing vines and entwined cascading tendril motifs, which were set mainly with diamonds. These trailing jewels were known as pampilles or aiguillettes.

The marriage between Napoleon III and Eugénie de Montijo (1826–1920), Countess of Teba, in January 1853 resulted in a period of material prosperity and extraordinary luxury in France. The French crown jewels were dismantled and the stones repurposed in new parures, the work entrusted to the workshops of Lemonnier, Beaugrand, Mellerio dits Meller, Kramer, Ouizille-Lemoine and Viette et Fester. These new parures followed the romantic theme and were modelled on Louis XV-style jewellery. They were set with coloured gemstones, including the regal stone, blue sapphire. Tiaras were often designed so that they could be dismantled and converted into brooches. Sadly, this transformability has meant that there are so few surviving complete examples, which makes the parures in this chapter extremely rare.

The 1840s sapphire and diamond demi-parure from the Royal House of Württemberg (illustrated overleaf) is a stunning example of naturalism in jewellery, the tiara and brooch designed as a spray of forget-me-nots and anemones within a foliate and floral natural arrangement. The sapphire and diamond necklace is a later addition, made around 1860. The importance of this demi-parure and other parures is highlighted by the use of the blue sapphires, the gemstone of royal or aristocratic status.

An elaborate gold, silver and enamel bracelet,
1850–5, made in the Romantic style by goldsmiths
François-Désiré Froment-Meurice and Jules Wièse,
with a cabochon sapphire, diamonds and rubies
to the lid of a small casket.

Above: A parure of diamond-set flowers and branches linking oval- and cushion-shaped sapphires. The tiara, necklace, brooch and ear pendants, made around 1850, were once owned by the Italian Barberini family.

Facing page: A demi-parure of sapphire- and diamond-set flowers, clusters and drops. The spray brooch/hair ornament dates around 1840, the tiara slightly later, and the necklace around 1860. Formerly in the collection of the Royal House of Württemberg, it was last worn by Princess Marguerite (1901–1975, also known as Margarete) of Urach, Countess of Württemberg.

Created around 1850 and of similar age to the Württemberg demi-parure on the facing page, the Barberini parure also features sapphire and diamond flowers. The tiara has seven graduated flower heads, each with a sapphire at its centre, on a rose-cut diamond branch with additional sapphires set in collets along the branch. It is accompanied by matching ear pendants, a necklace and a brooch. The parure once belonged to the Italian Barberini family, who resided in a palazzo designed in 1633 by the celebrated founder of baroque sculpture Gian Lorenzo Bernini (1598–1680). The palazzo now houses the National Gallery of Ancient Art. The parure was sold at Christie's in 1971 and then again in 2009.

Court jewels were constantly altered over the years, particularly by new owners of inherited jewels who wanted to reuse the same gemstones but in new designs. This practice is true of the sapphire parures that belonged to Marie-Amélie (1782–1866), Queen of France through her marriage to Louis Philippe I (1773–1850). Her husband, who also had a personal fortune, was the last King of the French from 1830 to 1848. He wanted initially to marry Princess Elizabeth, daughter of King George III of the United Kingdom of Great Britain and Ireland, but was refused due to his Catholic religion, so a year later, in 1809, he married Princess Maria Amalia of Naples and Sicily, the niece of Marie Antoinette. This second choice of marriage also very nearly did not take place, since Louis-Philippe's father, Louis-Philippe II, Duke of Orléans, had played a part in the execution of the French royal family, which included the death of Marie Antoinette.

In an 1836 painting attributed to Louis Hersent (1777–1860), Marie-Amélie is seen wearing a combination of sapphire, pearl and diamond jewels from two parures. The painting shows how Marie-Amélie wore the jewels: many sapphire, pearl and diamond clusters are sown onto her dress and the articulated tiara is worn across the rim of her hat. The flexibility of the tiara is testament to the superb craftsmanship of the crown jeweller, Bapst, who likely made the tiara. This parure, shown below in its fitted case from Bapst, includes three brooches, a belt buckle, a pair of earrings, which were now becoming fashionable, and the small tiara. The parure was subsequently owned by direct descendants, the counts of Paris.

Queen Marie-Amélie's second sapphire and diamond parure can now be seen in the Louvre. From this parure come the necklace, earrings and brooches she is wearing in the painting. The necklace is rumoured to have once belonged to her aunt, Marie Antoinette. Other sapphires used in the two parures may have come from the brooches of Queen Hortense of Holland (1783–1837), daughter of Empress Joséphine and mother of Emperor Napoleon III. It is believed that the Duke of Orléans, later becoming Louis-Philippe I, purchased the sapphires from Queen Hortense in 1821. All the sapphires in both parures are from Sri Lanka.

Queen Marie-Amélie subsequently divided the jewellery amongst her grandchildren, and the parures were passed down until they were both sold; one set was bought by the Louvre in 1985, where it can be admired today, and the other was sold through Sotheby's to a private client. None of the jewels is signed or bears any marks that might allow us to identify the workshops that made them; only the fitted case of the parure sold at auction is labelled.

Above: Depicted here in a portrait from 1836 attributed to Louis Hersent, Queen Marie-Amélie is wearing the tiara, below, the necklace, facing page, and many sapphire cluster brooches.

Below and facing page: The two sapphire- and diamond-set parures of Marie-Amélie, Queen of France, elements of which were sold in their fitted cases created by Bapst, below.

Above: A portrait, 1806, by Henri-François Riesener of Empress Joséphine of France wearing sapphire-set jewels, possibly including the sapphires her daughter, Queen Hortense of Holland, is believed to have sold to the Duke of Orléans and that later became part of his wife, Queen Marie-Amélie's two sapphire parures (previous page).

Left: The sapphire pendant/brooch, 1865, of Sophia of Nassau (1836–1913), Queen of Sweden and Norway, with a diamond-set double surround. This French jewel was recorded by Sophia's descendants as a gift from Emperor Napoleon III (1808–1873) to the future Queen when she visited Paris.

Facing page (and the cover of this book): A demi-parure, c.1870, by Mellerio dits Meller, purchased by Eugène, 8th Prince of Ligne. The chain of sapphire clusters suspends three sapphire and diamond cluster pendants with bow motifs, shown here with matching earrings.

CHAUMET'S
SAPPHIRE
JEWELS

One of the most prominent jewellers of the First Empire was Marie-Étienne Nitot (1750–1809), who was called upon to make the important suites of jewellery – using the stones from the old crown jewels – for Napoleon Bonaparte's coronation as Emperor of France in 1804. Bonaparte collected together the crown jewels in preparation; some of the missing French crown jewels, stolen in 1792, had been found, and he had secured other stones that had been pledged against foreign loans. He also added new stones to the coronation collection and again for his later marriage to Marie-Louise.

Nitot started out as a modest jeweller and clock maker in Paris in 1780 and was later joined in business by his son, François-Regnault Nitot (1779–1853). In two decades, Nitot et Fils had established its reputation to the extent that it received an order to create the first piece of regalia, a sword, for Bonaparte's new empire. But for Nitot to receive the highest honour, jeweller to the Emperor, it took a more personal meeting. Supposedly, after a horse bolted in front of the Nitot premises, the person Nitot helped to his feet was Napoleon himself, who was eternally grateful. However, a definite meeting was recorded in May 1805, when Nitot personally delivered the papal tiara commissioned by Napoleon for Pope Pius VII; Empress Joséphine was so impressed she had Nitot made jeweller to the Empress and a royal warrant followed.

Nitot et Fils made jewels for both Empresses Joséphine and Marie-Louise, as well as Napoleon's coronation crown. The firm was later sold to Jean-Baptiste Fossin (1786–1848), who had worked for Nitot, and he was joined by his son, Jules Fossin, who set up in London with Jean-Valentin Morel (1794–1860) and his son, Prosper. They exhibited in London at the 1851 Great Exhibition to great acclaim and were given a royal warrant by Queen Victoria. When they returned to Paris, Prosper Morel's daughter met, fell in love with, and married a young goldsmith called Joseph Chaumet (1852–1928), who took over the business and grew the *maison*, to which he gave his own name. Having been a jeweller himself, Chaumet had a real eye for detail and oversaw all the production; consequently, the *maison* received great

recognition at major exhibitions. Naturalistic jewels had been very popular in England since the 1840s and '50s, and Chaumet was inspired by flowers, insects and birds. He was known for his extraordinary enthusiasm and energy, which really drove the business and he quickly set up benches alongside his goldsmiths for lapidary and enamelling and workshops for box makers, as well as a gemstone laboratory so he could study and identify different gemstones. His understanding of gemstones and craftsmanship came together in his stunning creations.

Chaumet has continued to create jewels of high craftsmanship, and its archive houses the 66,000 design drawings that it has presented to clients over the years. The drawings in this chapter show the delicacy and the elegance of the French Belle Époque jewels of 1890–1914. Scrolls, feathers, stars, palmettes, tassels, trefoils and bows were crafted in platinum, the 'new metal' of choice. Platinum's strength allowed designers to create jewels that were light and delicate in design and that accentuated the gemstones, in this case yellow and blue sapphires.

The *maison* also holds over 500 *maillechort* replicas in nickel silver of the tiaras that it has made since the early 1900s, each one unique. These tiara replicas are displayed in Chaumet's museum on Place Vendôme, Paris. In total, the *maison* has made over 3,500 actual tiaras since 1780. Each tiara takes approximately six to twelve months to make. Tiaras and aigrettes were social symbols and continued to be an important fashion accessory during the early twentieth century. Jewelled head ornaments became a Chaumet speciality. A fine example is the early twentieth-century sapphire and diamond headband, which can be transformed into a *collier de chien*. It is set with eight beautifully matched graduating cabochon Ceylon sapphires, which would have taken months, if not years, to collect, but Chaumet was discerning and patient when it came to rare finds. The headband was made in 1921 and is stamped with the maker's mark of Michel Ballada for Chaumet. Its design is important in that it exemplifies how styles were starting to change quite quickly from the delicate Belle Époque taste to the bold geometric style of Art Deco.

Facing page: A transformable headband/
collier de chien, *c*.1921, by maker Michel Ballada
for Chaumet, and set with seven oval cabochon
Ceylon sapphires, graduating in size – the largest
weighing 47.34 carats – along three diamond-set
lines and within crescent-shaped surrounds. An
8.20-carat sugarloaf sapphire decorates the clasp.

Top: Chaumet's Nœud Écossais (Scottish Bowknot) brooch, 1907, set with sapphires, diamonds and other gems in a pattern reminiscent of tartan. The bowknot motif features in graphite, gouache and ink studies from Joseph Chaumet's drawing studio for other sapphire-set Chaumet jewels from the early twentieth century, including the brooches and *devants de corsage*, or stomachers, all *c.*1900–10, on this page. The bow was a transitional motif between the Belle Époque 'garland' style and the Art Deco geometric styles.

Chaumet was the Parisian tiara-creator of choice in
the late nineteenth and early twentieth centuries.
Chaumet made models of its tiaras and these models
form an important part of the *maison*'s archive.
The models above, all 1890–1930, are made of
maillechort (or nickel silver, an alloy of copper,
nickel and zinc, and here silver plated) with gouache-
painted replica sapphires and varnish.

THE LOUIS XIV
GRAND SAPPHIRE

This magnificent 135.74-carat 'violet' (or today considered medium-blue with violet hues) Ceylon sapphire was added to the French crown jewels during the reign of King Louis XIV around 1669. It was possibly cut in India before being sold to a European gem merchant who brought it to France. It has often been confused with another sapphire from the French crown jewels called the 'Ruspoli', but the cuts are completely different: the Grand Sapphire is a six-sided oblong stone, while the Ruspoli Sapphire was an oval faceted stone.

In 1792, at the height of the French Revolution, the royal treasury was looted and most of the jewels, including the Grand Sapphire, were stolen and hidden. One of the thieves was caught and, in exchange for his life, he revealed where the jewels were secreted. Upon its return, the Grand Sapphire was deemed an uncut stone and therefore given not to the Louvre but to the National Museum of Natural History, where it still resides today. The Ruspoli Sapphire was sold at auction in 1813 and was acquired by the French jeweller David Archard, who sold it to the notable gem collector, philanthropist and banker Henry Philip Hope (1774–1839). It is thought that Tsar Nicholas I then acquired the sapphire and later Cartier set it in a Russian head ornament, called a *kokoshnik*, in 1909.

Queen Marie of Romania is seen wearing the Cartier tiara with the believed Ruspoli Sapphire at its centre in the painting by Philip Alexius de László, but it was her daughter, Princess Ileana of Romania, who sold the tiara to a jeweller in the United States in the 1950s. Neither the *kokoshnik* nor the Ruspoli Sapphire have been heard of since. A 2015 study in *Gems and Gemology*, the quarterly scientific journal of the Gemological Institute of America (GIA), has clarified the identity of the two famed sapphires, and it is likely the confusion over their identities stemmed from their similar carat weights: the authors of the study brought together evidence to suggest that the Ruspoli Sapphire weighed 137.20 carats, while the Grand Sapphire weighs 135.74 carats.

A 1924 portrait, top, by Philip de László, of Queen Marie of Romania wearing the supposed Ruspoli Sapphire at the centre of the Cartier *kokoshnik* tiara, made in 1909 for the Grand Duchess Vladimir. Formerly of the French crown jewels, the Ruspoli Sapphire, 137.20 carats, has often been confused with another stone from the royal collection, the Grand Sapphire, centre, of 135.74 carats.

The story of the Romanov family has been written about in great detail and their sumptuous and decadent lifestyles told through their jewellery. In this book, I concentrate on the sapphire jewels that were included in their vast inventory.

The Romanov family were the last imperial dynasty to rule Russia, from 1613 to 1917. Following the Revolution of 1917, many of the Romanov family were executed by the Bolshevik troops, ending the monarchy. Eighteen Romanovs held the Russian throne during these 300 years, includ- ing, most notably, Peter the Great, Catherine the Great, Tsar Alexander I and Tsar Nicholas II. Their jewellery collections were exceptional — and today we can only imagine their true extent.

Russia was ruled by three empresses during the eighteenth century: Anne (r. 1730–40), Elizabeth (r. 1741–61) and Catherine the Great (r. 1762–96), whose reign was the longest of any Russian ruler. They each understood the value of having an important jewellery collection, and word of their love of jewels reached Parisian goldsmiths, who travelled to work in Russia, bringing with them the current Parisian rococo style. Catherine the Great was a huge supporter and patron of the arts and she adopted western European philosophies and culture. She had an extensive jewellery collection and she ensured that jewels became an integral part of the Russian court tradition — the art of intimidation was achieved through the wearing of priceless gems and jewels that no other European court could rival. Catherine converted an imperial bedchamber in the Winter Palace into the 'Brilliant Room'; it was lined with glass cabinets containing numer- ous pieces of jewellery, insignia, gold boxes, watches and chains, including treasures that she chose to give away as gifts.

When it came to jewels, the Romanovs wanted the biggest and the best, and they had the funds to make it possible. Trips to Paris were made on a regular basis, and, when Empress Maria Feodorovna first visited Cartier at Rue de la Paix in 1907, the Cartier brothers decided to launch 'Operation Russia', a strategy to cater to the Romanovs' jewellery needs. This strategy included Pierre Cartier visiting Russia to ask one of their best clients, Grand Duchess Vladimir, if they could use one of her buildings to showcase their work and to meet with clients. Their trips were successful to the point that they made the list of official suppliers to the imperial family. Back in Paris, Louis Cartier hung the royal warrant in the showroom for all to see.

The imperial Russian families continued to acquire some incredible jewels, and sapphires played an important role in the imperial jewellery collections. The main members of the royal household that had sapphires in their collection and that are mentioned in this chapter include Catherine the Great, Empress Maria Feodorovna, Grand Duchess Vladimir, Empress Alexandra Feodorovna, Queen Marie of Romania and Grand Duchess Victoria Melita.

Very few jewels have survived from these vast collections: the majority were melted down, broken up or sold by the Bolsheviks. The few that remained in the imperial treasury finally met their fate as they were sold by the Soviet government to a syndicate in order to raise money for food and armaments. The jewels were then dispersed at a Christie's auction in 1927. It is extremely rare to find any imperial Russian jewellery from that auction; all the pieces were entered into an inven- tory — drawn up by the brother of famed jeweller Carl Fabergé — which is an invaluable resource when it comes to authenticating pieces. The charming pair of mid-eighteenth-century sapphire and diamond flower bouquet sprays (on page 119), each tied with a bow and worn pinned to the hair or sewn onto the bodice, are from the imperial collection; they are rare because very few Russian jewels from this period exist today.

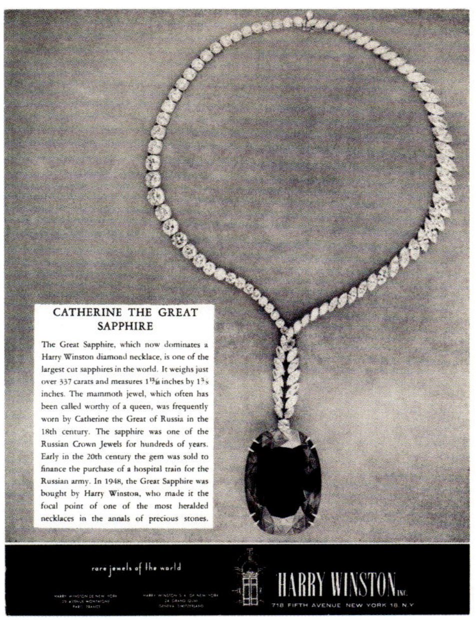

Empress Catherine the Great took pride in her gem and snuff box collection. This sapphire-set snuff box, above, made in 1776 in the workshop of Johann Gottlieb Scharff, features a later 1790–1 enamel miniature by Gavriil Kozlov of a monument to the Empress built in 1790. She frequently wore her 337.10-carat sapphire, which was sold in the early twentieth century to finance an army medical train. It was bought in 1948 by Harry Winston, set in a necklace and exhibited; this advert is from 1959.

A photograph, taken in the 1920s, of Grand Duchess Victoria Melita (1876–1936) wearing her Cartier necklace (left), a Christmas gift in 1911 from her husband, Grand Duke Kirill Vladimirovich of Russia, first cousin of Tsar Nicholas II. The diamond-set chain necklace suspends two cabochon sapphires.

A portrait by Konstantin Makovsky of young Empress Maria Feodorovna (1847–1928) wearing large sapphire-set necklaces, tiara and stomacher. Maria Feodorovna is known to have also worn the 197.80-carat Romanov Sapphire sewn onto her gown (see page 286).

In 1874 Duchess Marie of Mecklenburg-Schwerin (1854–1920) married the third son of Alexander II of Russia and became the Grand Duchess Vladimir, Grand Duchess Maria Pavlovna. She also became known for her sumptuous parties, which rivalled those of the imperial court, and very soon the salons of the Vladimir Palace were the epicentre of St Petersburg high society. Jewels had to be big enough to make an impression.

In 1909, the year of her husband's death, Cartier completed the spectacular sapphire and diamond *kokoshnik* tiara for the Grand Duchess. At the centre of the tiara was the Ruspoli Sapphire, weighing 137.20 carats, which the Duchess had supplied herself, along with six other sapphires weighing a total of 102.16 carats. Most tiaras were transformable, and this jewel was no exception: the motifs could be detached and worn as brooches. Louis Cartier personally went to St Petersburg to deliver the tiara. In 1910, the Grand Duchess decided she would like a sapphire and diamond *devant de corsage* to complement the sapphire and diamond tiara and her existing sapphire and diamond necklace. Sapphires from her own collection were used, including the central oval sapphire of 162 carats, which was even bigger than the Ruspoli Sapphire in the tiara. Later, the *devant de corsage* was reset with pearls.

Remarkably, the Duchess managed to escape St Petersburg in 1917, during the start of the Revolution, on her own train to the Caucasus, hoping to return. She had left her jewels hidden in a secret room in the palace and, when a trusted friend, Albert Stopford, visited her, she told him of their whereabouts; he promised her he would retrieve them. Back in St Petersburg, there was looting and chaos as the Bolsheviks ransacked the palaces, but Stopford disguised himself as a workman and entered the Vladimir Palace. He found the jewels, wrapped them in newspaper, put them in two bags and managed to smuggle them out of Russia. A few days later, the whole palace was ransacked and nothing was left unturned.

Eventually, the Duchess had to leave the Caucasus but succeeded in escaping to France, where she died in 1920. Stopford was true to his word and returned the jewels to her children, who eventually sold them over the years.

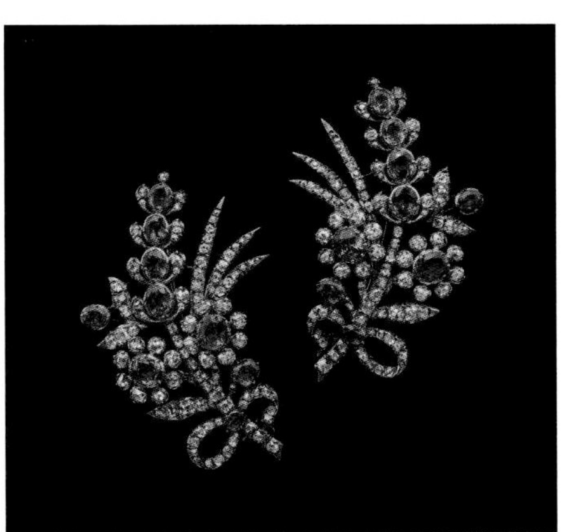

Top: Archive photographs of Grand Duchess Vladimir's sapphire- and diamond-set *kokoshnik* tiara, 1909, and *devant de corsage* brooch, 1910, both made for her by Cartier Paris.

Above: Russian diamond-set floral sprays, *c.*1740–70, with sapphire buds and flower heads, to wear in the hair or pinned to a bodice, once part of the Russian imperial jewels and sold at Christie's, London, in 1927.

In 1893, Princess Marie of Edinburgh (1875–1938), the granddaughter of Alexander II and of Queen Victoria and daughter of Grand Duchess Maria Alexandrovna of Russia, married Prince Ferdinand of Romania (1865–1927). Becoming Queen of Romania when her husband became King Ferdinand I in 1914, she adored jewellery and was not shy when it came to wearing great quantities of it all at the same time. At the outbreak of World War I, she decided to send her jewels to Russia for safe-keeping, only for them to be stolen.

To compensate for the loss of her jewels, Ferdinand I decided, at the end of the war, to help replenish her collection. This entailed many visits to Cartier. Louis Cartier had a spectacular sapphire weighing 478.68 carats that he had shown at the firm's autumn show in San Sebastián in 1919, but which had remained unsold. It was subsequently remounted in a long diamond and platinum sautoir that King Ferdinand bought for 1.38 million francs, payable in four instalments. Queen Marie wore the sautoir often, especially with the *kokoshnik* sapphire and diamond tiara that she had bought from her sister's mother-in-law, the Grand Duchess Vladimir.

Above left: Queen Marie of Romania in Belgrade, 1924, wearing her Cartier sautoir, suspending a 478.68-carat cushion-shaped sapphire (facing page) from a diamond-set chain and bought in 1921 by her husband, King Ferdinand.

Above right: Queen Frederica of Greece in Madrid, 1966, wearing Queen Marie's sapphire on a shorter diamond and pearl necklace. The Romanian royal family had sold this sapphire to Harry Winston, from whom it was supposedly bought by a Greek client, who gifted it to Queen Frederica.

Queen Marie of Romania's sapphire,
a 478.68-carat cushion-shaped sapphire
purchased and set into a pendant by Cartier,
*c.*1913; later chosen by King Ferdinand of Romania
in 1921 for his wife Marie, who wore the stone
hanging from a Cartier diamond-set
sautoir to his coronation in 1922.

A portrait by Friedrich August
von Kaulbach of Alexandra Feodorovna, born
Princess Alix of Hesse (1872–1918), wife of Nicholas II
and the last Empress of Russia. She wears the Köchli
Tiara, named after the Swiss jeweller Friedrich Köchli
who made the jewel for Emperor Alexander III and
Empress Maria Feodorovna to gift to Alexandra in 1894,
the year of her marriage to their son, Nicholas. The
tiara was set with cushion-shaped sapphires linked by
diamond-set curls and surmounted by diamond-set
rays. Its location is unknown; it is likely
that the tiara was broken apart.

HANOVERIAN BLUE

A brooch by Cartier in the shape of a leaf (left), set with curved cabochon Ceylon sapphires from a pavé-set diamond vein and interspersed with emeralds, amethysts and a ruby within a baguette-cut diamond frame. The Duke of York, later King George VI, bought this brooch for his wife, Elizabeth, in 1928. The Duchess of York wore the brooch on her hat. She then gifted the brooch as a birthday present to her daughter, Princess Elizabeth, during World War II. Princess Elizabeth was photographed wearing the brooch in 1946 (top), for a visit to Nottingham.

Hanoverian Blue

Royals looking for a gemstone to demonstrate their love and affection for another have very often tended to favour blue sapphires. Queen Victoria (1819–1901), the last and longest-reigning ruler from the House of Hanover, is renowned for the importance she placed on jewellery and the sentiments it can convey. Her sapphire jewels, whether gifts or personal commissions, are laden with meaning. Victoria and her husband, Prince Albert of Saxe-Coburg and Gotha (1819–1861), enjoyed using jewels to convey their love and devotion to each other; they understood and played with the language of flowers and the meaning of gemstones to create jewels that encapsulated their affections. Sapphire, as the gem that represents constancy and the heavens and is revered for its powers of protection, served its purpose well for Albert. Together, they became patrons of the arts and commissioned many jewels; with the wealthy British aristocracy following suit, enthusiasm for jewellery grew during Victoria's reign.

Victoria's influence established sapphire's status as the royal gemstone. Taking the lead from her commissions, later British royals also turned to sapphires to help mark important events, particularly marriage. Their patronage has supported jewellers across the world, including the jewellery houses of Garrard and Cartier.

The latter half of Victoria's reign saw a surge in the amount and variety of sapphires available in Europe. The mines of Ceylon (now Sri Lanka), Kashmir (in northern India) and Montana (in the US) helped to propel the use of dazzling sapphire colour combinations. The increased availability of sapphire also aligned with movements across design and society that were reacting to the developing systems of Victorian mechanized mass production. The Arts and Crafts jewels in Britain and the early jewels of the European and American Art Nouveau movement spoke to designers' and makers' desires to transform the role of design in everyday life; more unconventional and often less expensive materials became a key principle, and sapphire's availability and relatively unknown range of colours made it a suitable and exciting choice.

Jewellery is part of a monarch's working attire and demonstrates their duty, dignity and power, but the jewellery that Prince Albert gifted to Victoria was very much a private message of deep love and devotion. The day before their wedding in 1840, Albert gave Victoria a brooch set with a large oval sapphire surrounded by diamonds. Victoria is depicted wearing this brooch in Franz Xaver Winterhalter's painting of the Queen on her wedding day (above right). Prince Albert commissioned the brooch from the jewellers R & S Garrard, who soon afterwards – in 1843 – were appointed by Victoria as official crown jewellers.

For her wedding, the Queen wore a simple garland of white orange blossom in her hair, a white satin gown with Honiton lace and her Turkish diamond necklace and chandelier earrings. The only colour to stand out would have been the blue of the sapphire: nothing was going to overshadow Albert's brooch. She recorded her choice in her journal on the day of her wedding, 'I wore my Turkish diamond necklace and earrings and dear Albert's beautiful sapphire brooch.' Victoria was so attached to the brooch that she bequeathed it to the Crown so that it would remain with the nation's most important jewels, permitted to be worn only by future queens. As a mark of her enduring devotion to Albert, Victoria stopped wearing this brooch after his death in 1861.

In 1845, for Victoria's 26th birthday, Prince Albert gave the Queen a smaller blue sapphire and diamond brooch, which she described in her journal as 'a beautiful single sapphire brooch, set round in diamonds, much like the beauty he gave at our marriage but not quite so large'. In her will, Victoria left this brooch to her third daughter, Princess Helena. It is unclear where Prince Albert bought the second sapphire, but he was known, in the early years of their marriage, to spend considerable sums of money on jewellery from well-known jewellers of the time, such as Hunt & Roskell, Garrard and Kitching & Abud.

Top: The sapphire and diamond cluster brooch
commissioned by Prince Albert from R & S Garrard
for his bride, Queen Victoria. Victoria wore the brooch
to their wedding in 1840, as depicted in the 1847
portrait by Franz Xaver Winterhalter (above right),
commissioned by Victoria as an anniversary gift to
Albert. Queen Elizabeth II frequently chose
to wear it, as above on a visit to Saudi Arabia.

The Crimean War (1853–56) came about as a result of Russian threats to control access to religious sites in the Holy Land, including Bethlehem, then under Ottoman rule. These threats jeopardized European trade and strategic interests in the Middle East and India and gave rise to tensions between Catholic France and Orthodox Russia. Eventually, the war was brought to an end by Russia's acceptance of defeat against the alliance of Britain, France, the Ottoman Empire and Sardinia. The parties marked this acceptance by signing the Treaty of Paris in 1856.

Prince Albert gave this Alliance Flag brooch (above) to Queen Victoria at Christmas 1855 to mark the victory – although the Treaty of Paris had yet to be be signed, Sebastopol had already fallen to the French. During the conflict, the press had criticized Queen Victoria for not taking much interest in international affairs, but in September, when Sebastopol fell, victory bonfires were lit at Balmoral and the Queen cut short her trip to Scotland to return to London for discussions on the immediate situation.

This brooch was made by one of Garrard & Co's outworkers, John Linnit. The design includes the British Union Jack flag, the French flag and the star-and-crescent motif of the Ottoman Empire, set with sapphires, rubies, diamonds and emeralds with blue enamel decoration. Prince Albert paid £90 for the brooch on 30 August 1855.

Top: A ring, *c.*1837, set with an octagonal step-cut sapphire above what appears to be a lock of hair and with two diamonds. Queen Victoria gifted the ring to one of her coronation train bearers, and it is inscribed with her coronation date. The ring returned to the Royal Collection in 1925 as a gift from Queen Mary.

Above: A ring, 1689–94, belonging to Queen Mary II (1662–1694), set with a Ceylon sapphire intaglio bearing the royal coat of arms, a crown and Mary's MR monogram; the ring's shoulders chased with Tudor rose motifs. The collector and 'art referee' for the South Kensington Museum (now the Victoria and Albert Museum) Charles Drury Fortnum gifted the ring to Queen Victoria in 1897.

In 2015 a buying battle ensued over one of Queen Victoria's most treasured jewels: her sapphire and diamond coronet, designed for her by Prince Albert. A dealer had sold the coronet to an overseas buyer but, in an attempt to keep this historic jewel from leaving the United Kingdom, the British government blocked its export. It was feared that, if the coronet left the country, it would probably never be seen on British soil again. This temporary export ban imposed a deadline of 27 December 2016 to find serious offers commensurate with the coronet's £5 million value, plus £1 million in tax. It was eventually the generosity of the Bollinger family that secured the jewel for the nation. They gifted the coronet to the Victoria and Albert Museum for display in the existing William and Judith Bollinger Gallery, which houses the museum's incredible jewellery collection of more than 3,500 items.

Queen Victoria's will described the coronet in detail: 'Designed by H.R.H. Prince Albert and set by Kitching. Most of the stones belonging to a former present of King William IV & Queen Adelaide 1840.' The decoration mirrors the Saxon *Rautenkranz* (circlet of rue), the band of stylized trefoil leaves that runs diagonally across the shield of Prince Albert's coat of arms. It is beautifully made and fully articulates because of its 23 hinged sections. The hinges allow for the jewel to be opened out so that it can be worn as a tiara, or alternatively closed right up into a coronet. The coronet is mounted in silver and gold with cushion-shaped diamonds and 11 step-cut sapphires of octagonal and trapeze-shaped mixed cuts — further evidence that the stones were sourced from other jewels (matching stones would have indicated that they had been chosen specifically for this particular coronet). Prince Albert loved repurposing stones from Victoria's existing jewellery, and a journal entry by Victoria, from 22 February 1843, confirms Albert's role in rearranging her jewels: 'We were very busy looking over various pieces of old jewellery of mine, settling to have some reset, in order to add to my fine "parures".' Sapphires were clearly Albert's chosen gem, for he had already given Victoria the sapphire and diamond brooch the day before their wedding. During that same year, Victoria bought a pair of sapphire and diamond earrings and a bracelet, forming an impressive parure.

Victoria appointed Kitching and Abud 'Jewellers to the Queen' in 1837; by 1840 they were a major supplier. In Victoria's private account book she listed 'The Small Coronet' of 1842 as having a cost of £415, the equivalent to just under £50,000 today.

Queen Victoria loved to wear the coronet and did so throughout her early years as queen, as seen in the 1842 portrait by Franz Xaver Winterhalter in which she is wearing the coronet around her fashionable chignon hairstyle. After Albert died in 1861, at the age of 42, Victoria very seldom wore any coloured gemstone, except on one rare occasion in 1866 when she wore the sapphire coronet to open Parliament, the first time she had felt able to undertake this duty since Albert's death. She wore the coronet in place of the heavy state crown, which was relegated to a cushion that was carried next to the Queen instead. Queen Victoria was nervous to attend the grand occasion, but by wearing the coronet she possibly felt that Prince Albert was with her in spirit. An 1874 portrait of Victoria by Richard Graves shows the Queen wearing the coronet as a means of securing her widow's veil.

Following Victoria's death, the coronet was passed to King Edward VII and then to King George V, who gifted it to his daughter, Princess Mary, at the time of her marriage, in 1922, to Viscount Lascelles, later the 6th Earl of Harewood. In the 1920s, Princess Mary wore the jewel open as a fashionable bandeau, low over her forehead. She later passed the coronet to her son, the 7th Earl, who was a first cousin of Queen Elizabeth II. His second wife, who was an Australian violinist and whom he married in 1967, wore the coronet on many occasions until 2011, when it was eventually sold to a dealer.

After its acquisition for the nation, the coronet first went on public display — in the William and Judith Bollinger Jewellery Gallery — in 2019, the bicentenary year of the birth of both Victoria and Albert.

Top and facing page: Queen Victoria's coronet,
1842, designed by Prince Albert to mimic a motif
from his coat of arms and made by 'Jewellers to
the Queen' Kitching and Abud using sapphires and
diamonds from Victoria's other jewels. This piece
is hinged so that Victoria could wear it in different
ways, as depicted in the portraits above: left, an
1842 portrait by Franz Xaver Winterhalter; and
right, an 1874 portrait by Richard Graves.

Queen Victoria became Empress of India in 1876 after a strengthening of relationships between the two countries following the royal tour of the subcontinent made in 1875 by Edward (1841–1910), Prince of Wales, the eldest son of Victoria and later King Edward VII. (In the course of her long reign, Queen Victoria herself never visited the country.)

Before this new period of relations, Britain's interactions with the Indian subcontinent had been managed via one extremely powerful company: the East India Company, founded in 1600 to control the lucrative trade in spices and fight off competition from the Dutch. Before it was eventually dissolved in 1874 (following abuses of power that included illegal exports of opium to China), the Company grew to become the world's most powerful trading company. With an army of 260,000 soldiers, at its height the Company ruled nearly all of the Indian subcontinent and eventually expanded into the Persian Gulf, China and the rest of Asia, becoming virtually the sole trader worldwide of tea.

Meanwhile, trade with the subcontinent throughout the centuries had generated great interest in the exotic – Orientalism, as it was later termed – which influenced European artists and designers. The influence resulted in a display of fabulous gemstones and lavish enamelled jewels at the Great Exhibition in 1851, a collaboration between Prince Albert and Henry Cole, who later became the first director of the South Kensington Museum (the precursor to the Victoria and Albert Museum).

The British Raj, as direct imperial rule by the British Crown was known after power was transferred from the East India Company following the Indian Mutiny of 1857–8, extended across two-thirds of the Indian subcontinent from 1858 until the independence of India and Pakistan in 1947. During that time the cross-fertilization of ideas deriving from cultural exchanges was admired and desired in the arts of both East and West. The luxurious jewels and the opulent displays of wealth of the rulers of the 600 or more princely states were witnessed in great abundance when the Prince of Wales visited India in late 1875 and early 1876.

Ceylon, 'jewel box of the Indian Ocean', had not originally been part of the itinerary, but was hastily included after the Prince of Wales demanded to know why the island had been left off the map of the forthcoming tour that his royal advisors showed him. An explanation of the incident appears in the diary of the tour that was published in 1877 by William Howard Russell (1820–1907), a former war correspondent and then private secretary to the Prince of Wales, who had accompanied the Prince on his trip: 'When the map of India was laid on the table, immediately there came to the surface the difficulty of getting many places within the limits of the time which the Prince could devote to his visit.'

Dr Joseph Fayrer, the Prince's physician, had wanted to arrange the trip around the weather and climate so as to reduce the risk of his coming into contact with diseases such as malaria or cholera, believing it essential that the Prince 'not be exposed to long journeys at unhealthy seasons, and to rapid transitions from cold to heat'. Dr Fayrer recommended that the trip coincide with the beginning of November, the start of the cooler months, but trying to visit both Kashmir and Ceylon during these allocated months was proving difficult, which is why Ceylon was left off the map.

> Several attempts were made to divert the Prince from his purpose of visiting Ceylon, but he was inexorable as well as penetrating; and it is said that once, at one little 'Indian Council' at Marlborough House, there was a map produced in which Ceylon did not appear, when the routes were being laid down and discussed, but that a Royal demand, 'Where is Ceylon?' rendered the stratagem, if such it were of no avail … but to Ceylon the Prince adhered with invincible firmness, undeterred by 'sanitary considerations' and medical reports, which, sooth to say were damaged in their authority by the very opposite opinions of the *cognoscenti*.

Contrary to the doctor's fears, it seems – from a chronicle of the tour published by George Wheeler, special correspondent of the *Central News*, in 1876 – that Ceylon's climate and scenes very much appealed to the Prince.

> The Prince has seldom looked better or more pleased. No doubt his healthy appearance was due to the climate, which in Ceylon is most tolerable. He has never seen such rich vegetation or such luxuriant forests of flowers as are visible everywhere. Not only in the flat town itself, but also up among the neighbouring hills, whose peaks are shrouded in clouds, does one find the fullest justification for the exclamation of Prince Soltykoff that Colombo is no more than a botanic garden of the loveliest kind, and on the most gigantic scale. Between the flowers and the fruits and the odd, strangely bright attire of the people, the traveller on landing knows not where to rest his eyes first.

Interestingly, Russell's account of their trip to Ceylon and his references to their rail journey to the sapphire-mining region of Kandy make no mention of sapphires being gifted to the Prince or the display of sapphire jewellery, only describing coffee and tea plantations. Nevertheless, the Prince received many gifts, including jewellery, on his travels and, given Ceylon's rich gemstone deposits, particularly of sapphire, it is likely that sapphires were presented to the Prince. In the early nineteenth century, Ceylon sapphires had been hard to transport away from the island because the ruler of Kandy had imposed restrictions on the stones' export.

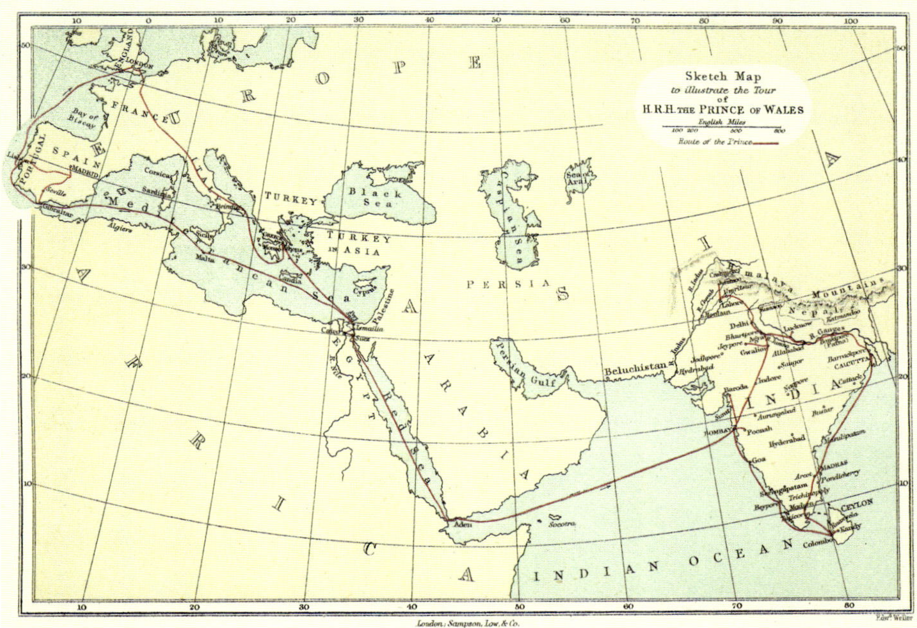

After this trip, sapphires became more abundant in European jewellery, likely promoted by diplomatic gifts to the Prince that he brought back to Britain. Paris hosted many world's fairs in the late nineteenth century at which the Portuguese, French and British displayed artefacts from their colonies – including, notably, jewellery set with gemstones sourced from afar.

Ceylon remained under British rule until 1948. Both before and after its independence, many attempts were made to control the country's mining. Eminent jewellery historian Dr Jack Ogden has explained that the detailed 1939 *Report of the Sub-Committee of the Executive Committee of Labour, Industry and Commerce on the Marketing and Cutting of Ceylon Gems* was an effort by the British government to determine how best to regulate gem mining in Sri Lanka, including mining's impact on the environment – a concern that was, in some respects, ahead of its time. Before the report could be acted upon, the outbreak of World War II took over world events and nothing came of the recommendations; in fact, mining in Sri Lanka today is conducted in very much the same manner as it has been for over 2,000 years.

Top: A map of the Prince of Wales' route on his 1875–6 tour of the Indian subcontinent, by Sydney P Hall, in William Howard Russell's record of the tour.

Above left and right: Two photographs by official tour photographers Bourne & Shepherd documenting the 'View of the Ceylon Railway Incline', from which the Prince admired the views, and a triumphal welcome arch erected in bamboo and decorated with fruit for the Prince's arrival in the sapphire-mining region Kandy.

Centre: An illustration from *The Illustrated London News*, 8 January 1876, depicting the Prince's involvement with the annual Esala Perahera festival in Kandy to honour the Buddha's Sacred Tooth Relic.

GARRARD
OF CALCUTTA

The British jewellery and silversmithing company Garrard is synonymous with British royalty. In 1735 Garrard received its first royal commission from Frederick, Prince of Wales (1707–1751), and in 1737–8 a request for a service of plate costing over £3,000; another purchase, an elaborate silver-gilt table centrepiece, is still in the Royal Collection. The business was established in 1722 by George Wickes (1698–1761), an accomplished rococo silver-smith and jeweller based in Threadneedle Street, London, and soon moved to new premises in Panton Street, Haymarket, in 1735. When Wickes and his business partner, Samuel Netherton, retired in 1760, the running of the company was taken over by two employees, John Wakelin and William Tayler. Eventually, through succession, in 1792 the firm began to be operated as a partner-ship between Wakelin and Robert Garrard. In 1802, sole control of the company passed to Robert Garrard. Throughout its changes of ownership, which until 1946 included the wider Garrard family, the company maintained its royal patronage. Garrard gained the high-est accolade, that of 'Crown Jewellers in Ordinary', from another goldsmithing firm, Rundell, Bridge & Company (formerly Rundell, Bridge & Rundell), which had been crown jewellers from 1797 until Queen Victoria appointed Garrard in 1843. Garrard held this royal appointment for 164 years, until 2007.

During its time as crown jewellers, and especially during Victoria's reign, Garrard received many royal orders. Victoria and Albert's love for the decorative arts and craftsmanship was coupled with their promotion of methods of improved manufacture. The 1851 Great Exhibition held in Hyde Park and inspired by Prince Albert was the first of many international exhibitions at which Garrard exhibited. Under the direction of Robert and Sebastian Garrard, R & S Garrard showcased mainly stock and commis-sioned work, rather than making specific pieces for the show. During this period, shops differed very little in appearance from the residential properties that surrounded them; but, once inside a shop, the visitor would have been impressed by the sheer amount of silver and jewellery crammed into glass cabinets, as it was usual to carry enormous stock which might not sell for years. Having such large reserves of stock was feasible because there were such negligible levels of inflation.

The Great Exhibition was a huge success, and the profits from it went towards financing the creation of the South Kensington Museum, later to become the Victoria and Albert Museum. Over six million people attended the exhibition, and Queen Victoria herself visited the jewellery displays many times. The Queen greatly admired the magnificent jewels from India, the 'tasteful' jewels from Russia – though the Russian exhibits arrived some

time after the opening as they had been held up in the port due to snow and ice – as well as medieval-inspired jewels by A W N Pugin. There are many entries in catalogues of the exhibition about the gems on display, but there is hardly any mention of sapphires – an apparent lack that may have been a result of the Sinhalese rulers' restrictions on the export of Sri Lankan sapphires. The jewels on display were nevertheless breathtaking: Queen Victoria herself lent the Koh-i-Noor Diamond; the East India Company lent the Daria-i-Noor ('Sea of Light') Diamond; Hunt and Roskell exhib-ited part of the famous collection of gems belonging to merchant banker Henry Hope; and Garrard exhibited jewels inspired by the sculptures from Nineveh.

Jewellery was worn for purposes other than mere adornment. The trend set by Victoria and Albert, who took great pride in designing and rearranging jewels to convey messages and symbolic mean-ings, was very much embraced by the wider Victorian audience. This belief in the power and importance of jewellery was also rein-forced by the next generation of Victoria's family; in 1863, the heir to the throne, the Prince of Wales, married the beautiful Alexandra of Denmark, who also loved jewellery and was a leader of fashion. Anything worn by the Princess of Wales created great demand for similar styles and was quickly copied throughout the jewellery and fashion industries.

Garrard's many regal creations have been worn at the coronation of every monarch since that of Victoria's son, King Edward VII, and his wife Alexandra in 1902. Garrard designed and made the Imper-ial Crown of India for King George V's inauguration as Emperor of India in Delhi in 1911. In fact, 1911 was a busy year for Garrard: in June there was the coronation of King George V, for which its jewellers had to make a new crown for Queen Mary; in July it was on hand for the newly reinstated investiture of their son David (later Edward VIII) as Prince of Wales; then, in November, the crown jewellers travelled to the Delhi Durbar.

As crowns could not be taken from the shores of Great Britain, a new crown had to be made for the Delhi Durbar. The new crown was funded by Indian revenues with the understanding that it could be kept in the Tower of London so long as India remained in the Commonwealth. The Imperial Crown of India had to be half constructed in India so that the King could try it on for size and it could be adjusted at the same time. To facilitate this crown-making, Garrard opened their Calcutta branch on Dalhousie Square in 1911 and, soon after, another shop in the foothills of the Himalayas in Simla (a destination that many British families journeyed to each year to escape the hot and humid weather of the plains), and later

another branch in Delhi. The advantage of establishing branches in India was that the Garrard directors did not have to rely on moving stock back and forth between Britain and India through travelling representatives, who would have expected commission. After the Delhi Durbar, the Imperial Crown was placed on display, to be viewed by appointment only, in the Calcutta showroom.

Garrard was a particular favourite amongst the preferred jewellers of the maharajahs; to encourage their continued support, the London shop had an exclusive space named the 'Indian, Colonial and Country Department'. Many of the commissions were inspired by traditional Indian designs but incorporated Western techniques, such as the use of platinum. The metal was not immediately accepted, however – even though the maharajahs wanted to embrace the 'new European look', they could not easily let go of the association between wealth and yellow gold that is so ingrained in Indian tradition. A Garrard senior sales assistant reported that their good client Maharajah Umaid Singh 'was afraid he will be seen as a silver prince and not a gold one'.

It was not until the 1920s that the white metal platinum was trusted as the metal of choice for Indian commissions.

Calcutta became an important centre for colonial silversmiths, who set up workshops and also employed and trained Indian craftsmen in British silver traditions. The first of these workshops was that of Hamilton & Co. With the anticipated Calcutta International Exhibition in 1883, a silversmithing style called the 'Calcutta Style' evolved, and silverware depicted rural Indian scenes or religious stories and folklore.

After the success of the 1911 Delhi Durbar, Garrard & Co made the Imperial State Crown for the coronation of King George VI on 12 May 1937. The Garrard designers kept the crown's design very similar to the one that was made for Queen Victoria's coronation in 1838 by Rundell, Bridge & Rundell. Garrard & Co continued to make jewels for the British royal family, with sapphires a prominent feature as the chosen royal gemstone.

Above, from top: Illustrations of a jewelled set, *c.*1910–25, including a choker necklace, brooch or compact, ring, earrings and an aigrette, all featuring grape and flower motifs with blue vases (the page is labelled 'Garrard & Co Ltd, Goldsmiths to the Crown, London, Calcutta'); and three early twentieth-century designs by Garrard for sapphire- and diamond-set aigrettes, or turban ornaments, in white metal.

A necklace set with oval, cushion- and
pear-shaped Ceylon sapphires with clusters
and ribbons of circular-, single- and rose-cut
diamonds. The necklace was made *c.*1880, shortly
after the Prince of Wales' trip to Sri Lanka and the
strengthening of ties between the island and Europe
that brought sapphires back into focus
in European jewellery design.

SAPPHIRES FOR A ROYAL WEDDING —
AN ENDURING TRADITION

Blue sapphires have long been favoured for their capacity to demonstrate royal love and devotion in matrimony. Many members of the British royal family have either chosen or been gifted sapphire jewellery.

{1923}   Elizabeth Bowes-Lyon, later Queen Elizabeth the Queen Mother, chose a sapphire and diamond ring set with a Kashmir sapphire in a platinum mount when she married the future King George VI.

{1934}   Prince George, Duke of Kent and Princess Marina of Greece and Denmark selected a square-cut 7-carat Kashmir sapphire in a ring with baguette-cut diamond shoulders for their wedding. Princess Michael of Kent inherited the sapphire ring from Prince Michael's mother, Princess Marina.

{1935}   Prince Henry, Duke of Gloucester and Lady Alice Montagu Douglas Scott opted for a square sapphire.

{1961}   Prince Edward, Duke of Kent and Katharine, Duchess of Kent selected a sapphire and diamond engagement ring.

{1962}   Princess Alexandra chose a large three-stone sapphire and diamond ring for her engagement to Sir Angus Ogilvy.

{1973}   Princess Anne chose a sapphire and diamond ring when she got engaged to Captain Mark Phillips.

{1981}   Prince Charles, now King Charles III, and Lady Diana Spencer, later Princess Diana, chose a sapphire and diamond cluster engagement ring.

{2010}   Prince William, now Prince of Wales, presented the sapphire ring he inherited from his mother, Princess Diana, to Catherine Middleton, later the Duchess of Cambridge and then Princess of Wales.

{2018}   Mr Jack Brooksbank chose an engagement ring set with an oval mixed-cut padparadscha sapphire for Princess Eugenie.

Top: Prince George and Princess Marina shortly after their engagement in 1934. Cartier London commented in *The Guardian* in September 1934 that, 'Prince George has displayed the most modern taste in his choice both of the ring and of the setting. His selection will undoubtedly make sapphires the most popular ring for engagements this year.'

SAPPHIRE
GIFTS

The gifting of sapphire jewels within the British royal family has not been limited to engagement rings; since Queen Victoria and Prince Albert began their marriage with sapphire gifts, sapphire crystals' size and colour, plus their meaning of 'constancy', have made the stone an obvious choice for special jewels.

The unusual brooch worn by Queen Elizabeth II below has a square-shaped cabochon sapphire alongside a cushion-shaped diamond, each in a scroll frame set with 36 smaller cushion-shaped diamonds. It was originally a wedding present for Princess Victoria Mary of Teck (1867–1953), later Queen Mary, in 1893 from Empress Maria Feodorovna of Russia. Queen Mary wore the brooch horizontally high on her collar; later, when Queen Elizabeth II, Mary's granddaughter, inherited the jewel in 1953, Elizabeth took to wearing the jewel vertically on her left shoulder — she typically wore all her brooches on her left shoulder — as pictured below in 2006.

King George VI gave his daughter, then Princess Elizabeth, a sapphire bracelet for her 18th birthday in 1944. The Cartier bracelet (above) is set with 10 graduated square-cut sapphires and baguette- and brilliant-cut diamonds in a design echoing the Art Deco style.

Above: Queen Elizabeth II on a 2006 visit to HMS *Albion* wearing the impressive sapphire and diamond brooch she inherited from her grandmother Queen Mary, who had received it on her marriage in 1893; and a sapphire and diamond bracelet, by Cartier, gifted in 1944 to Princess Elizabeth by her father, King George VI, on the occasion of her 18th birthday.

In 1947, King George VI gave his daughter, Princess Elizabeth, later Queen Elizabeth II, a sapphire and diamond necklace and earrings for her wedding. The demi-parure was originally made *c.* 1850. In 1963, a new sapphire and diamond tiara and bracelet were made to match the necklace and earrings; the sapphires are thought to have come from a necklace that belonged to Princess Louise of Belgium, which is why the tiara is sometimes referred to as the Belgian Sapphire Tiara.

Queen Elizabeth II attending a 1996 banquet
in Prague wearing her wedding sapphires and the
necklace of Princess Louise of Belgium that the
Queen bought in 1963 and converted into a tiara.

One of the most iconic sapphire and diamond rings is the one that Prince Charles, now King Charles III, gave to Lady Diana Spencer to mark their engagement in February 1981. Lady Diana, only 19 years old at the time, chose the ring from a Garrard catalogue. It is set with a 12-carat oval mixed-cut blue Ceylon sapphire, surrounded by 14 brilliant-cut diamonds. Following her tragic death in 1997, her personal jewellery collection was handed to her two sons, Princes William and Harry, to be divided equally between them. It is said that the sapphire ring was initially chosen by Harry but that, when William was going to propose to Catherine Middleton, now Princess of Wales, Harry gave him the ring so that the future Queen of England would wear it — a fitting tribute to their mother. Prince William carried the ring in his backpack for three weeks before proposing to Kate in October 2010 in Kenya. When interviewed alongside Kate shortly after the announcement of the engagement, William said that Kate wearing the ring was a way for him to have his mother be very much part of their wedding and stay in everyone's memories.

Top: Diana, Princess of Wales in 1991
wearing her sapphire and diamond cluster on
a choker necklace of pearl strands and her
sapphire and diamond engagement ring.

Above: Prince William, now Prince of Wales, gifted
his mother's ring to Catherine Middleton, now
Pincess of Wales, on their engagement in 2010.
They are shown here admiring the ring at the
official announcement on 16 November 2010
at St James's Palace.

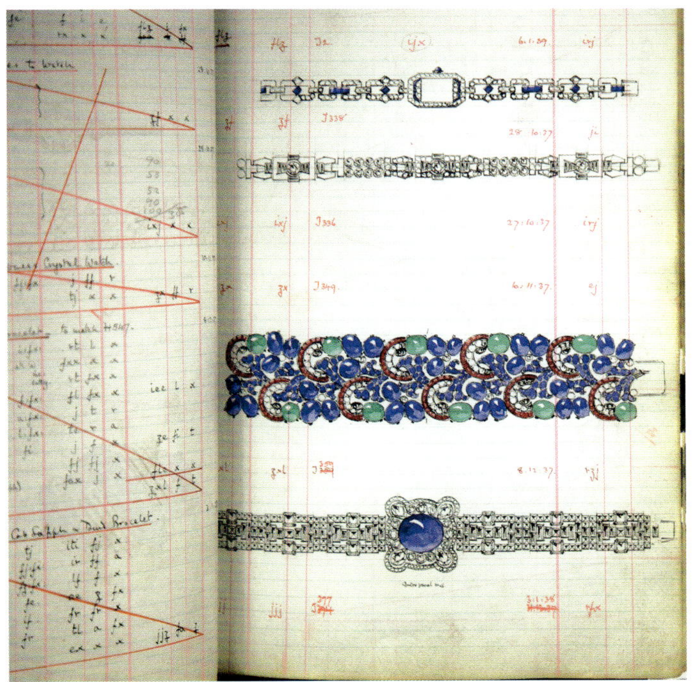

Facing page: A double-row necklace and earrings of oval and cushion-shaped sapphires within diamond-set borders. The clusters date to *c.*1850 and were once part of a longer, single row necklace that was adapted to its current design with earrings in the 1930s by Jane Ann Gordon Cory (1865–1947), a great collector of jewellery who bequeathed her collection to the Victoria and Albert Museum, where this set is on display.

Above: Two pages from the illustrated ledger books of London silversmiths and jewellers Hennell. Hennell was the jeweller to the British aristocracy; they both designed new jewels and acted as custodians for their most noble clients' tiaras and diamond parures. These pages show intricate drawings of sapphire-set jewels amongst the firm's jewelled commissions, for which it became particularly renowned in the nineteenth and twentieth centuries.

Top: Nineteenth-century brooches, from left: A cockerel with an unusually shaped sapphire body, diamond-set wings and head and enamel details; a fleur-de-lys of sapphires and diamonds, *c.*1885, the central petal set with a Kashmir sapphire of approximately 4.04 carats, and engraved '14-4-87' to the reverse; and a sapphire and diamond bumblebee with ruby eyes and a central Kashmir sapphire of approximately 2.50 carats.

Lady Llewelyn Smith's tiara, *c.*1909, by celebrated
Arts and Crafts designer and craftsperson
Henry Wilson (1864—1934), now in the Goldsmiths'
Company's Collection. A golden Orpheus and
wild beasts sit atop the band set with blue star
sapphires, rubies, moonstones, rock crystals and
pearls. White plumes should be secured in the
support at the back of the band.

During the late nineteenth century an artistic revolution was
simmering in Britain, promoting beliefs among artists that industrialization and its mechanical processes compromised and
undercut the quality of products and innovative design. Factories,
in which working conditions were known to be appalling, produced
mass-manufactured items that these artistic revolutionaries
considered to be devoid of any expression of individuality or artistic integrity. This dissatisfaction grew during and in the wake of
the Great Exhibition of 1851.

This new movement — which developed over the next few decades
into what became known as Arts and Crafts — embraced artistic
freedom, with an ethos of giving art back to the people. The driving forces behind the movement were undoubtedly John Ruskin
and William Morris. Spurred by the motto 'Hand, head, heart', they
aimed to revive 'honest' crafts and true art, which they felt had
been missing from people's experiences. They wanted everyone to
be exposed to art in the course of their everyday life.

Co-operatives and guilds were founded to propagate these ideals
and to encourage and teach amateur craftsmen the enjoyment of
making items from start to finish. With regard to jewellery, it had
been customary for goldsmiths to have their own specialist skills,
such as setting, polishing and mounting, but according to the ideals
of the Arts and Crafts movement individuals were encouraged to
design and make the whole item, which often led to amateur craftsmanship. Guilds had masters to teach the metalworking discipline,
but rarely was anything signed, and their students would create
items similar in style to their master's work, making it very hard
to tell pieces apart. The ethos of giving art back to the people also
meant that the use of precious materials was actively discouraged
— diamonds were never used, and faceted gemstones very rarely;

instead, colour was introduced through enamel work. The role
of the skilled jeweller inevitably became diluted and, while the
movement produced jewellery that looked affordable, sometimes the pieces would cost more to make than machine-made
jewellery, which was not the objective of the movement.

After the death of Queen Victoria in 1901, a period of great
change swept through society and fashion, including in the
creative arts and jewellery design. While the Arts and Crafts
movement, active in Britain from around 1890 to 1930, made
an impact in art and design, Europe and America saw the blossoming of Art Nouveau. Art Nouveau had a distinctive style
that, even though short-lived, produced jewellery with a similar intention to the Arts and Crafts jewellery of Great Britain,
as jewels were made with materials that were not necessarily
expensive. However, that is where the similarity ends: the big
difference was in the manufacture. In Art Nouveau jewellery,
unlike in Arts and Crafts, the skill of the goldsmith was held in
high regard — indeed, these jewels are still considered to be
some of the finest jewels ever made.

Aesthetic and feminist movements combined with Art Nouveau
to challenge fashion. The constrictive and very uncomfort-
able corsets that caused waists to be unrealistically tiny
and restricted movement were being replaced with more
free-flowing clothes that reflected the aesthetic movement,
a style that was keenly followed by thespians and free-thinking
individuals. The sinuous curves and the intense detail and
interpretation of nature of Art Nouveau brought a whole new
emphasis to jewellery design.

The genius of the Art Nouveau period was undoubtedly René
Lalique (1860—1945). Even though he only produced jewellery between 1885 and the early 1900s, his creations had an
enormous impact in the world of jewellery and fashion. He
used materials that were often inexpensive and sometimes
unconventional for jewels, such as horn, enamel, tortoiseshell,
pearls and even finely moulded glass, with the occasional
gemstone that was carefully picked to convey a message
or to symbolize an emotion. Lalique wanted the public to
understand that the value of his pieces lay in the artistry,
craftsmanship and design of a jewel, not just in the materials
alone. Nature was Lalique's main inspiration; he interpreted its
cycles and seasons into jewels the like of which had never been
seen before. While many of his contemporaries were making
platinum jewellery set predominantly with diamonds from the
recently discovered diamond mines of South Africa, to be worn
on stiff restricting corsets, Lalique's jewels were liberating,
like the clothes they adorned.

Jewels by famed Art Nouveau designer
René Lalique: The striking Woman with Sapphire
headband tiara, *c.*1897–1900 (top), with a cushion-
shaped sapphire, in the collection of the Lalique
Museum, Hakone, Japan; a winter woodland pendant,
*c.*1898–1900 (above left), with enamelled snow-laden
fir trees in a gold foliate frame that suspends a
sapphire drop, in the collection of Hamburg's Museum
for Arts and Crafts; and the delicately enamelled
Anemones brooch, *c.*1901–3 (above right), with
an oval sapphire and an opal pendant, also in the
collection of the Lalique Museum, Hakone.

A 2018 exhibition in the Fitzwilliam Museum, Cambridge, displayed a ring with a rich story.

During the early 1900s, the artistic climate encouraged and inspired many artisans to experiment in the art of jewellery making. One young man, Charles de Sousy Ricketts (1866–1931), applied his design knowledge to jewellery, sculpture and illustration. Ricketts was born in Switzerland and came to England with his father and sister. He was sensitive to the arts and loved visiting museums. It was at the South London Technical School of Art in Kennington, where he enrolled in 1882, that he met his life partner, Charles Haslewood Shannon (1863–1937).

In 1894, Ricketts and Shannon set up a publishing company together. Ricketts illustrated over 80 volumes of poetic works while also producing other works of art, including theatre designs, notably those for Oscar Wilde's *Salomé* in 1906. Ricketts was drawn to Renaissance jewels, which, along with his interest in sculpture, inspired him to briefly try his hand at designing jewellery. His efforts amounted to a sketchbook with 58 jewellery designs, all with a sculptural theme incorporating Renaissance-inspired enamel decoration. He designed these pieces with the intention of giving them as gifts, but he did not try to make them himself; instead, he outsourced the goldwork to the talented goldsmithing workshop of Carlo & Arthur Giuliano.

He and Shannon had two close friends, lovers Katherine Bradley and her niece Edith Cooper; collectively, they pushed the boundaries of conventional nineteenth-century society. In an age when male homosexuality was totally forbidden and punishable, their mutual friendships allowed the four to live their lives without constraint and criticism. Katherine and Edith wrote together under a single pseudonym: Michael Field. Often, they referred to themselves as 'the Poets' and to Ricketts and Shannon as 'the Painters'. In December 1903, Ricketts promised to design a ring for Bradley inspired by an unpublished 1902 play by Michael Field called 'A Messiah'. The play was based on the story of Sabbatai Sebi (1626–1676), a Jewish mystic who claimed to be the long-awaited Messiah, but who on arrival in Constantinople was imprisoned and had to either face the death penalty or convert to Islam.

This ring is in keeping with Ricketts's preferred style of sculptural design. It is inspired by the ring given to Sabbatai by the Sultan of Constantinople and is modelled on the shrine of the Mosque of Omar, its blue dome translated by Ricketts into a cabochon star sapphire. Probably made by jewellers Carlo & Arthur Giuliano, though it is not signed, the ring's hollow, mosque-shaped bezel is pierced with four doorways with a sloping roof and two windows above and long tapering 'windows' between each doorway. Ricketts intended the ring to represent the five senses: touch, smell, taste, hearing and sight. The inside holds a loose emerald intended to give flashes of green like stained-glass windows and the outside was initially smeared with ambergris, which was a strong-smelling waxy substance from the digestive system of a sperm whale that was often used as the foundation for making perfume.

In 1929, Shannon fell from a ladder and suffered such considerable brain damage that he had to receive extensive nursing. Devasted by his partner's continued suffering, and overwhelmed with the pressure of mounting nursing costs, Ricketts died of a heart attack in 1931. Shannon died in 1937.

William Burges (1827–1881) was an important Victorian architect and designer. A romantic who liked to recreate past eras, particularly the medieval period, he brought the idealized imagery of these eras into the present. Shunning both nineteenth-century industrialization and neoclassical architecture, he was drawn instead to the Gothic revival. In 1865, he received the commission of a lifetime to restore and transform the exterior and rooms of Cardiff Castle into an awe-inspiring eclectic mix of East-meets-West-meets-Gothic design.

This ring, with its strong Gothic revival undertones, was designed by William Burges around 1870. The cabochon sapphire is set within a mount that has applied trefoil-motif decoration.

Henry George Murphy (1884–1939) was a rare craftsman as he transitioned between Arts and Crafts and the later Art Deco styles. An accomplished silversmith as well as jeweller, he began work as an apprentice to Henry Wilson, who was a member of the Art Workers' Guild, and carried on working for Wilson in his workshop in Kensington and later in Wilson's studio in Kent. Murphy became a skilled goldsmith and was influenced by Wilson's predilection for Gothic and Renaissance designs. Murphy's work with bright enamels and intricate settings was of the highest calibre. Jewellery of the early twentieth-century Arts and Crafts movement was not always of the first quality as the craftsperson was expected to make a jewel from start to finish, without necessarily executing each stage with the same level of skill. Murphy, in contrast, learnt every stage of goldsmithing to a high standard, from enamelling, engraving, repoussé and niello to gem setting.

Murphy started his own business in 1912, although it was not until 1928 that he was able to open retail premises – known as the Falcon Studio – in Marylebone, London. He also shared his skills by teaching at the Royal College of Art and became principal of the Central School of Art and Design in London. Paul Atterbury and John Benjamin's 2005 book *The Jewellery and Silver of H G Murphy* is the first publication dedicated to this exceptional British craftsman, whose lifetime achievements have finally been celebrated.

This stunning gold circular pendant displays Murphy's expertise in the art of enamelling; here, he has used bright colours, reminiscent of Renaissance jewellery, in a highly intricate, voluminous frame, set with emeralds and rubies, with a cabochon blue sapphire in the centre. As a measure of great craftsmanship, the attention to detail on the reverse is just as good as it is on the front.

An early twentieth-century pendant by
H G Murphy, with a cabochon sapphire set at
the centre of concentric borders of intricate
enamelwork, the scrolled patterns and gold
strapwork spaced between enamel
plaques set with emeralds.

While Queen Victoria and her family were reviving the language of jewels and the burgeoning Arts and Crafts movement was rippling through Britain, across the Atlantic the establishment of one of the world's biggest jewellery houses was underway. This house was ready to grapple with some of the same movements affecting European design, but in its own way.

Two young friends, Charles Lewis Tiffany (1812–1902) and John Burnett Young (1815–1859), were eager to set up in business together. On 14 September 1837 they opened a small stationery and fancy goods shop at 259 Broadway in lower Manhattan. New York was certainly not at the global forefront of fashion or of entrepreneurial business at the time; on the contrary, there were frequent financial upheavals and uncertainties, banks were failing, and there was widespread anxiety leading to the flour riots in February 1837. However, austerity did not deter the two entrepreneurs, and they looked to Tiffany's wealthy cousin, Jabez Lewes Ellis III, to join them. In 1841 the business became Tiffany, Young & Ellis.

Owning sumptuous jewels was not an option for many New York women during this period, and so the business trio started to sell paste jewellery, calling it 'Palais Royal'. Riding on the back of this successful initiative, a few years later they started to sell real jewellery imported from London and Paris. Their move to gemstones and precious metals attracted the attention of keen gemmologist George McClure, who joined the firm in 1845 and

later married John Young's sister, Maria, in 1849. His expertise in diamonds and gems led him to become one of America's most respected gemmologists and helped Tiffany start to make its own gem-set jewellery.

In late 1847, John Young made the long journey to Europe with the intention of buying European 'novelties', but when he arrived in Paris in February 1848 he found chaos and civil unrest. His arrival had coincided with the abdication of King Louis-Philippe d'Orléans in favour of his eldest grandson, the Count of Paris. Louis-Philippe's wife, Queen Marie-Amélie de Bourbon, surrendered many of her jewels during the tumult, but she managed to keep two important sapphire and diamond parures, which she had enjoyed wearing during her 18 years as Queen of France. As the fall of the French monarchy paved the way for the Second Republic, Paris became a troubling place for Young, but this commotion did not deter him from grasping the opportunity to buy jewels from the fleeing aristocrats loyal to the Orléans and Bourbon families.

The Tiffany partners' astute business sense and their efforts to take advantage of the politically triggered decline in gemstone prices in Europe meant they built healthy profit margins that allowed the company to prosper. Tiffany bought out Young and Ellis, while making McClure and fellow employee Gideon Reed partners, and the firm became Tiffany & Co. Throughout the following decades Tiffany & Co came to be recognized as an innovative jewellery business. The creative talents of many highly acclaimed jewellery designers forged Tiffany's success. Edward C Moore (1827–1891), son of John C Moore, the founder of Tiffany's silver business, was chief of the design department. George Paulding Farnham (1859–1927) designed for Tiffany from 1885 to 1908 and trained under Edward C Moore; Farnham subsequently became chief designer and manager of their Diamond Jewellery Manufacturing Department. Louis Comfort Tiffany, son of Charles Tiffany, became the art director of Tiffany & Co in 1902.

Tiffany & Co exhibited at many international fairs to much acclaim, including the Expositions Universelles of 1867 and 1878 in Paris, where its reputation and dominance in the jewellery arena led to many awards. By now, Tiffany was known for its fine jewellery and gemstones – indeed, the company was referred to as one of the world's greatest jewellers – but its much trusted and knowledgeable gemmologist McClure's eyesight was deteriorating; the enthusiastic 22-year-old gemmologist George Frederick Kunz (1856–1932) took over the gemmology work in 1879.

Tiffany had a winning team of experts in their respective fields, all of whom were working tirelessly to prepare for the prestigious 1889 Exposition Universelle in Paris with its most memorable attraction – the Eiffel Tower. But then came a sudden surprise announcement: instead of showcasing the French crown jewels in the upcoming fair, the Republican French government decided they would sell them at auction. The sale was to be held in the Pavillon de Flore of the Louvre from 12 to 23 May 1887. The jewels were a nostalgic reminder of the glamour of the Bourbon and Bonaparte courts, but history was to be destroyed as the items were broken up into separate lots before the sale. Tiffany & Co had already established a presence in Paris in 1850, putting the firm in the right place to acquire some of the finest gemstones on the market and ensuring that they were on the list to receive news of the sale of the crown jewels – news revealed through a catalogue (in which the jewels were poorly photographed) that was sent to Tiffany in New York. The sale achieved a total of 6,868,050 gold francs, a sum that, even in 1887, was nowhere near the jewels' monetary worth. For Tiffany, this was a fortunate outcome, as they purchased nearly a third of the sale, including a brooch that had been split from a sapphire parure made by Bapst for the Duchess of Angoulême, the only surviving child of Louis XVI and Queen Marie Antoinette.

The man entrusted to showcase these fabulous acquisitions was Farnham, who had begun his first year's contract with Tiffany & Co in 1885. His uncle, Charles Thomas Cook, was Tiffany's right-hand man and later president of the company. Farnham's designs were based on in-depth studies he made of design motifs from earlier periods, from Louis XVI revival, Renaissance revival and Orientalism to American First Nations designs and heritage, in combination with Tiffany's ongoing fascination and love for nature. Inspirations from exotic, far-flung countries, such as Russia, Turkey, Spain, Egypt, Portugal, Java and Japan, intrigued American audiences.

Farnham worked under the guidance of chief designer Edward Moore and together they prepared for the 1889 Paris Exposition. The orchid mania that was then prevalent inspired Farnham and Moore to create themed, realistic enamel- and gem-encrusted

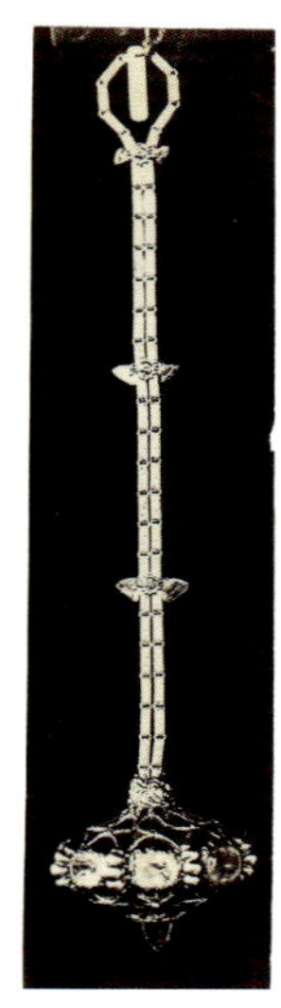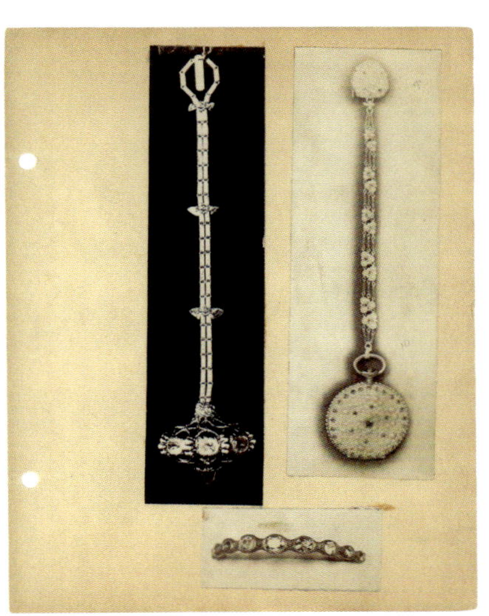

A *bonbonnière*, shown on facing page, designed by George Paulding Farnham for Tiffany & Co in 1889 and exhibited in the same year at the Exposition Universelle in Paris; the photo above from Tiffany's Exposition scrapbook. From a platinum chain set with sapphires hangs the circular *bonbonnière* with six sapphire-set segments to hold small sweets (*bonbons*) that were popular at the time.

orchid brooches. Their orchid creations were showered with praise and they were awarded a gold medal. Farnham honed his skills and swiftly rose through the firm as a designer; his role became crucial in developing Tiffany's international acclaim over the next two decades.

The world of design during this period was dominated by the interpretation of nature, especially through the work of the aesthetic movement Art Nouveau. The Tiffany archives show the company leading the way in the production of numerous nature-inspired jewels.

The dragonfly, along with the butterfly, was an iconic symbol of the Art Nouveau period. Nineteenth-century Japan, whose culture inspired many Art Nouveau designers, was called the 'Land of the Dragonfly' because the arrangements of its islands resembled the insect. The dragonfly continued to be the preferred insect motif of the twentieth and even now the twenty-first century at Tiffany.

This sapphire and diamond dragonfly brooch made *c.*1890–1900 can also be converted into an *en tremblant* hair ornament, allowing the dragonfly to quiver realistically when worn. The makers' attention to detail is evident in the careful arrangement of the diamonds to resemble the veins of the insect's wings.

This Renaissance revival brooch was more than likely inspired by the famous Italian jewellers Carlo & Arthur Giuliano, who were working at the same time as Tiffany's George Paulding Farnham in the late nineteenth century; but the brooch has the Tiffany twist, for, where there would have been enamel decoration, Farnham instead used different coloured sapphires. This brooch is a real celebration of American gemstones, as these sapphires came from the deposits in Montana. This was a very exciting moment in the history of gemmology, as the Montana deposits had only been discovered around 1865; Tiffany gemmologist George Kunz later said, 'At no locality has there ever been found so great a variety of rich colours in corundum gems as Rock Creek.' Kunz was on the committee of the US Geological Survey from 1883 to 1909 and, during this time, he produced annual reports on precious stone production. Having George Kunz on the team, alongside Farnham, meant that Tiffany was able to be at the forefront of offering exciting gemstones. This brooch was exhibited at the Exposition Universelle in Paris, in 1900. Tiffany was very proud to be able to exhibit the beautiful multi-coloured sapphires from stones that were found in America.

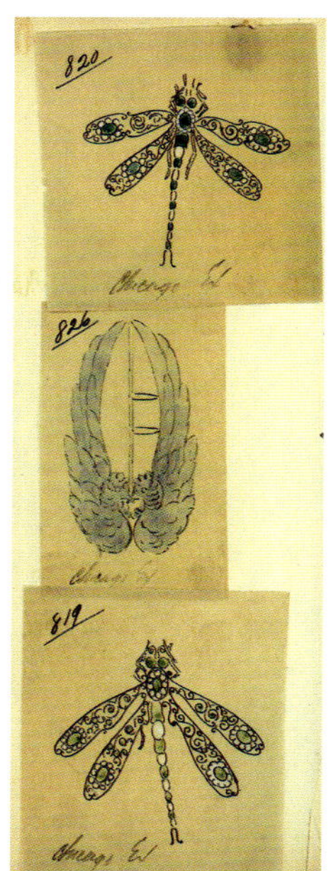

Above: Sketches by Tiffany & Co for dragonfly brooches to be set with diamond and sapphire wings, some of which were made and exhibited at the 1893 Columbian Exposition in Chicago; and, right, a sketch of Tiffany's chief designer George Paulding Farnham's Renaissance revival brooch, 1900, to be set with sapphires of every colour.

To celebrate the 400th anniversary of Christopher Columbus's 1492 landing in the New World, the World's Columbian Exposition was held in Chicago in 1893. Many companies exhibited at this international fair, but Tiffany triumphed, receiving 56 medals for excellence — a medal for every collection shown.

Tiffany exhibited a 'Collection of Gems and Semi-precious Stones' ('semi' is a term we no longer use to describe gemstones, as all gemstones are deemed precious if they have beauty, rarity and durability), which, in recognition of the collection's quality, was later purchased for the Field Museum. There was a dazzling display of eight different varieties of coloured sapphires, from Siam (Thailand), Ceylon (Sri Lanka), Montana (United States), 'Cashmere' (India), Chitanka (Czech Republic) and Rhenish Prussia (Germany). This educational assemblage of gemstones had been brought together thanks to the passionate and expert eye of Tiffany's gemmologist George Kunz. His unrivalled knowledge of gemstones, the majority of which had never been considered for jewellery before, deepened Tiffany's fascination with mineralogy and highlighted the wide design potential these stones offered designer George Paulding Farnham.

Tiffany's display of gemstones delighted and captured the imagination of many visitors, and a caption to a photograph of the collection echoed this sentiment: '… attractive to you for the magnificence of the gem collection here displayed, for these gems — strange baubles of the human race — fairly dazzle the eyes of the beholder with their magnificence and beauty'.

In celebration of this gem collection, Tiffany created 60 gem-set rings, including this ring described as 'Fancy Sapphire Ring: Spray of fancy coloured sapphires and diamonds'. The ring is made of silver and gold; set in the centre is an oval yellow sapphire with blue, purple and striated blue and purple sapphires below, surmounted with green, pink and purple sapphires. This ring is

a fabulous example of the many natural colours of sapphire, an occurrence that would not have been widely understood at the time given the familiarity and dominance of blue sapphires. The design of the ring echoes the *giardinetto* rings of the mid-eighteenth century in Europe, which derive from the Italian translation of a flower ornament, in the form of a vase or basket of flowers, and which were given as love tokens to either a friend or lover — a design Farnham referenced from the history books but, of course, executed with a Tiffany twist. The original *giardinetto* rings were set with emeralds, diamonds, rubies and blue sapphires, but Farnham, with the gemmological eye of Kunz, set his rings with largely unknown coloured gemstones.

George Kunz was able to supply Farnham with coloured gemstones that were not yet widely appreciated, and which fuelled Farnham's lavish designs. Under Farnham's guidance, the design studio achieved unparalleled confidence in using striking colour combinations and unusual gemstones to create jewels that set them apart from any of their rivals. Using coloured gems that were not expensive allowed Tiffany to make pieces that were still colourful and unique, yet affordable — a clever business tactic that helped Tiffany through the financial crisis of the mid-1890s.

Two jewels designed by George Paulding Farnham for Tiffany & Co. Above left: A gold cigarette case with sketch, 1900, its lid set with table-cut sapphires of geometric shapes arranged between diamonds and blue-enamelled banners; and, above right, a ring set with purple, green, pink, yellow and blue sapphires and with diamond-set scrolling shoulders, its design inscribed 'Chicago Ex', marking the ring for exhibition at the World's Columbian Exposition, Chicago, 1893.

Documentary photographs of sapphire mining
operations in Montana, US, in the early twentieth
century: a photograph by D B Sterrett of Yogo
Creek, above the American Sapphire Mine, 1910,
where miners also worked underground (middle);
open-pit operations by the New Mine Syndicate,
Utica, Montana (right), *c.*1925.

In the mid-1800s, America was a magnet for settlers trying to find their fortune through gold prospecting. From North Carolina in the early 1800s, to the Georgian gold rush of 1829 and Californian gold rushes of 1848–55, the gold hunters' work ignited a fever that continued to inspire prospectors to keep searching for the elusive nuggets of gold.

When prospectors started to mine in Montana, they repeatedly came across blue pebbles that kept getting in the way of mining for gold; these pebbles clogged up the sluices and would be discarded. It was only when Ed 'Sapphire' Collins reportedly sent samples of these blue pebbles to Tiffany & Co in 1865 that the true identity of these Montana stones became apparent: they were sapphires. Tiffany & Co had a few stones cut to assess the potential of the sapphires from this new deposit. London-based jeweller Edwin W Streeter and gem merchant Horatio Stewart both went to Montana to quietly inspect the Missouri River.

Dr J Lawrence Smith wrote the first scientific paper about the discovery of American sapphires: in the September 1873 issue of the *American Journal of Science*, he reported that the stones were discovered within the gravel beds of the Missouri River that runs through Montana. Twenty years later, in 1895, sapphires were then discovered in Yogo Gulch (gulch is a geological term for a deep narrow ravine that is a result of a river or stream), near the Judith River, in Montana, by Jake Hoover, who had been prospecting for gold. Two years later, Hoover set up a sapphire syndicate.

At Tiffany, George Kunz, whose reputation as a knowledgeable gemmologist was widely respected, was sent an old cigar box containing some specimens from Hoover's newly found deposit. Kunz considered these sapphires to be of very good quality and later declared, in an article for the *Saturday Evening Post* of 26 November 1927, that '[Hoover and his associates] had acquired the most valuable sapphire mine in America, yielding more wealth than all the sapphire mines in America put together, a finer quality of gem.' In Kunz's opinion, the sapphires from the 'Yogo Gulch-Judith River' deposit were more promising for jewellery than the Missouri River sapphires because of their varying tones of blue – from light to dark. Kunz was the first to recognize the full potential of the Yogo sapphires and, indeed, was the first to appreciate their high clarity and range of blues. The best of the Yogo blues Kunz described as being a 'cornflower blue'; the colour that has since been adopted to describe the most desirable blue sapphire colour. Though Americans were slow to fully appreciate these gems, it was Tiffany and Kunz who helped label these sapphires as world class.

Tiffany sent Hoover a cheque for $3,750 for the parcel of stones, before proceeding to cut them and starting to design jewels around them. One of these first designs became one of the most iconic Tiffany jewels created by Paulding Farnham: the Iris corsage brooch of 1900 (overleaf). The Iris won the gold medal when it was exhibited at the Paris Exposition Universelle that year.

Across the Atlantic, word was out about Yogo sapphires, and a London-based jeweller – Johnson, Walker and Tolhurst Ltd – took control of Hoover's New Mine Sapphire Syndicate. Johnson, Walker and Tolhurst expanded the mining rights, which then became known unofficially as the 'English Mine', and the increased area produced one million carats of both gem and industrial qualities. The flat shape of the rough blue crystals and the pale crystals from other locations in Montana meant that they were not suitable for jewellery but were perfect for using in watches. The English Mine then purchased the Yogo American Sapphire Company and dominated the sales of sapphires from Montana up until the outbreak of World War I.

A sketch from Tiffany & Co's archive of
a sautoir and another jewel, both set with blue
stones. Tiffany created two sautoirs of
Montana sapphires to exhibit at the 1889
Exposition Universelle in Paris.

Flower brooch and two watercolour illustrations designed by George Paulding Farnham for Tiffany & Co, set with American sapphires from Montana selected by Tiffany gemmologist George Kunz. The Iris corsage ornament, top, was exhibited at the 1900 Exposition Universelle in Paris and purchased by Henry Walters of Baltimore, who bequeathed it to the Walters Art Museum. The smaller flower brooch design (right), 1890–1900, is also set with Montana sapphires to its heart-shaped petals.

These American sapphires were constantly being challenged; the successful production and use of synthetic sapphires in jewellery manufacture in the 1920s, coupled with the Great Depression, the outbreak of World War I and the disastrous flash flood in 1923 that destroyed much of the mining equipment, all meant that the interest in the mine greatly waned. The Yogo Gulch mine struggled to recover and was again up for sale.

Following the Great Depression and until the start of World War II, nearly all the rough was shipped to Europe to be cut and used in jewellery manufacture, while the industrial rough was kept in America to be used for industrial applications, such as navigational instruments and compasses.

Selling the Montana sapphires on the international market was, however, always an uphill struggle: competition was fierce between the highly prized and sought-after blue sapphires from Ceylon (Sri Lanka), Kashmir and Burma (Myanmar) during this time. American sapphires were deemed less precious, with the coloured sapphires often being considered too pale and the blues too glassy in appearance. These aesthetic concerns, combined with the small size of the crystals and the high cost of mining, made it difficult for the American mines to be commercially viable. There was sometimes an attempt to market the Montana sapphires in Europe as 'Oriental' sapphires under the pretence that they were pale blue Ceylon sapphires.

Falsely marketing sapphires as Sri Lankan still happens today: sapphires from recently discovered deposits such as Madagascar, for example, are sometimes sold as Sri Lankan for financial gain, fuelling the wrongful idea that sapphires from an 'older' deposit are somehow more valuable.

Ownership of the Yogo mine has changed many times since the deposit was discovered in 1895. It has produced much gem-quality sapphire and is still mined today, but only small quantities are found. Again, the high cost of mining and the low yield makes the mine commercially unviable, no matter how beautiful the stones are.

New Horizons

A necklace designed by Louis Comfort Tiffany
and signed Tiffany & Co, *c.*1910. The pendant
is set with an oval violet Ceylon sapphire,
surrounded by gold vine leaves and berries
and suspended from a three-row neckchain
of gold beads and sapphire clusters.

New Horizons

'The women of today — Are always coming up with the most astonishing tricks — They play the banjo — They drink cocktails, drive cars — Not content with cutting their hair — They show us their calves — Right to the top, to the top, if you please.' — ARMAND LANOUX, *Paris 1925*, 1957

Looking back to the early twentieth century, it seems like there was no other period when change was quite so abrupt or so invigorating: the concept of the modern woman was emerging, and many jewellers and couturiers took full advantage of these changes in social expectations.

After the death of Queen Victoria in 1901, the Western world saw a seismic shift in fashion and jewellery designs. The Art Nouveau style in Europe was short-lived, but nevertheless resulted in jewels of undeniable beauty and elegance, displaying exemplary craftsmanship. In the United States, under the artistic direction of Louis Comfort Tiffany, Tiffany & Co initiated an American version of Art Nouveau, which became known as the 'Tiffany Style', using sapphires from deposits in Montana. Tiffany produced jewels that were freer in style than the platinum jewels being made at the same time by houses such as Cartier, Chaumet and Mellerio.

Jacques Cartier joined the family business in 1906 and spent many months working in each of the company's stone departments so that he could learn about rubies, emeralds, sapphires, pearls and diamonds by sorting them into categories. His travels to India and the Middle East — the first Cartier to attend the Delhi Durbar, which he did in 1911 — were incredible opportunities to get to know the maharajahs, who shared his passion for gemstones. Jacques introduced his brothers in Paris and New York to the fabulous stones he brought back from these trips, and he replicated in gems the stunning colour combinations he had seen in Indian enamels. These jewels were Cartier's 'Hindu Jewels', later referred to as 'Tutti Frutti'.

The Great War changed fashion, lifestyles and women's roles in society. The female form altered dramatically during the war years, when women had to engage in activities that helped the war effort. They worked in factories and hospitals, and fashion had to accommodate practical needs. The floral Art Nouveau and its representational imagery used by jewellers such as Lalique, Henri Vever and Georges Fouquet gave way to stylization, simplicity and geometric forms — the beginnings of Art Deco.

After the war ended in 1918, there was a thirst to start living again. In the Roaring Twenties, enjoyment was found in music, dance, yachts and travel. Women, who, at the start of the 1900s, were dressed in crinoline, bustles or dresses with padded hips and voluminous chignons, were now, just 20 years later, wearing 'flapper' dresses, opting for short hairstyles, applying their make-up and even smoking in public, exposing their arms and ankles. Jewellery responded to these changes, and necklaces — with gemstones wrapped all the way round the neck — became more lavish than the traditional pendant. Shorter hairstyles also meant that shoulders were on display.

LOUIS
COMFORT TIFFANY
(1848—1933)

With the continued growth of his formidable design team, headed by G Paulding Farnham and world-leading gemmologist George F Kunz, Charles Lewis Tiffany had since 1837 built a phenomenal jewellery business. In the early twentieth century, Tiffany & Co was well on its path to becoming the household name it is today; leading much of that development was the next generation of the Tiffany family. Tiffany and his wife Harriet had six children, including, in 1848, Louis Comfort Tiffany, born in New York, who went on to become one of the most celebrated Art Nouveau designers of the twentieth century.

Louis took up the arts while still in his teens. At age 18, he trained under leading American landscape painter George Inness, studying in New York at the National Academy of Design. During his studies Louis also became interested in glassmaking. In 1879, possessing the entrepreneurial genes of his father, he formed Louis C Tiffany & Associated Artists with Candace Wheeler, Samuel Colman and Lockwood de Forest, all interior designers, writers and painters. Even though this collaboration did not last more than four years, it gave Tiffany the necessary experience to strike out on his own.

With his father's influential connections and financial security, Tiffany was able to make a name for himself as an interior designer and glassmaker for New York's high society. He worked with the most prestigious names in America, including Cornelius Vanderbilt, Mark Twain and Henry O Havemeyer, a sugar industrialist, and secured high-profile commissions, such as redecorating the White House for President Chester Alan Arthur in 1882. Louis's stellar reputation in the US served as the foundation of his forthcoming international success.

In 1885, Louis formed the Tiffany Glass & Decorating Company and proceeded to win many awards at international fairs and exhibitions. These accolades and achievements, plus recognition as a leading light in the Art Nouveau movement, meant that, when his father died in 1902, Louis was given authority over the family business. In this position, Louis asserted his design superiority, which, in turn, quickly impacted Paulding Farnham's position as head designer. Farnham knew he would now have to take a step back from leading the design team, although he continued working in the company in a lesser capacity since his uncle, Charles Thomas Cook, was Tiffany's president. Cook died in 1907 and in 1908 Farnham left Tiffany to devote the last 20 years of his life to painting and sculpture.

During this time, George Kunz held on to his post as chief gemmologist at Tiffany & Co and the company continued to benefit from his unrivalled expertise and knowledge. Louis took great inspiration from each stone Kunz presented to him; both men considered a gemstone's most important feature to be its singular beauty rather than its rarity or its market value. The Arts and Crafts and Art Nouveau movements appreciated stones for their contribution to the artistry and craftsmanship of a jewel. For Louis, a jewel's design should focus on a gemstone's play of light, the palette of gem colours and the synergy between materials. His experience as a painter and his work with glass and other media inspired him to create innovative jewels. He worked with enamel designer Julia Munson (1875—1971), who had moved with him from glass to jewellery design. Louis steered Tiffany's jewellery in new directions that meant incorporating enamels and gems, like moonstones and opals, that interact with light to display optical effects. Traditionally revered diamonds, emeralds and rubies were used instead to complement other stones, such as topazes, beryls, tourmalines, garnets, aquamarines and peridots, which took centre stage in Louis's jewels. Montana sapphires played a significant role in Tiffany's selection of gemstones because of their unique array of colours. These jewels were often referred to in Europe as the 'Tiffany Style' and, indeed, the company established a separate workshop for making 'Tiffany Art Jewellery'.

With the outbreak of World War I, Art Nouveau and the 'Tiffany Style' were destined not to survive; the world had moved on and Louis, now 66 years old, was too set in his ways to embrace the new Art Deco style. He therefore soon ceased designing jewellery for Tiffany & Co and in 1918 stepped down as design director, also dissolving his glass business in 1924. Louis's work for Tiffany is still widely regarded by collectors as the North American partner, or rival, of the European Art Nouveau and Arts and Crafts movements. He remains part of a small and revered list of designers to have their names attributed to Tiffany jewels, while his designs and management style pushed Tiffany & Co to become an internationally recognized brand.

A portrait of Louis Comfort Tiffany,
likely photographed in the 1880s.

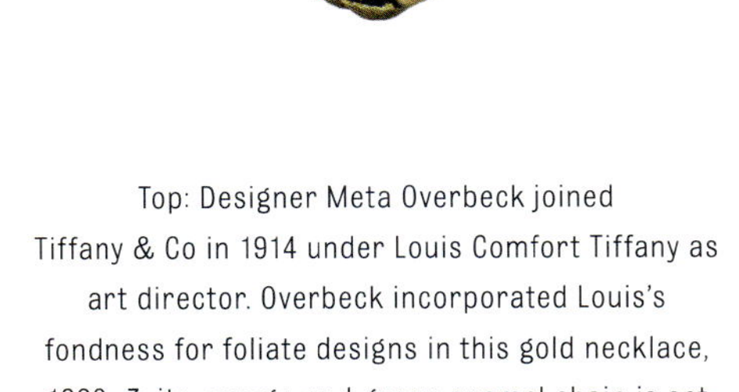

Top: Designer Meta Overbeck joined
Tiffany & Co in 1914 under Louis Comfort Tiffany as
art director. Overbeck incorporated Louis's
fondness for foliate designs in this gold necklace,
1922–3; its orange-and-green enamel chain is set
with sapphires, emeralds, coloured diamonds
and zircons and suspends a pendant set with
a Montana sapphire of nearly 40 carats –
one of the largest known Montana sapphires.

Above: A brooch, 1913–14, designed by Louis
Comfort Tiffany for Tiffany & Co, with a gold vine
and berry motif surrounding 13 Montana sapphires.

Top: A Latin cross pendant, *c.*1915, by Louis
Comfort Tiffany for Tiffany & Co, set with
circular-cut Montana sapphires that graduate
in size to meet the baguette-cut emeralds
at the points of the cross.

Above: A brooch, 1907–14, designed by
Louis Comfort Tiffany for Tiffany & Co, with
a border of intertwined gold leaves around
nine oval and round Montana sapphires.

Louis Comfort Tiffany embraced moonstone's
shimmering optical effect by combining the stone
with blue Montana sapphires for a striking colour
contrast. This necklace (top), designed by Louis
with Meta Overbeck between 1914 and 1933,
surrounds a cabochon moonstone with a border
of sapphires in a platinum filigree frame, flanked
by shield-shaped motifs set with sapphires and
two oval moonstone cabochons on a double chain.
Louis used the same stone combination in this
brooch (above), designed *c.*1915, also set with
moonstones interspersed with sapphires.

Above left: A Cartier London brooch, c.1933,
from Jacques Cartier's own design for his wife,
Nelly. The amethyst represents Nelly; the four
diamond-set squares their four children;
and the sapphire borders Jacques,
as sapphire was his birthstone.

Above: A 1922 photograph of the three Cartier
brothers, Pierre, Louis and Jacques, with their
father, Alfred, second from right.

Below: A typical page from one of
Jacques Cartier's travel diaries from the 1920s,
in which he details the shape, weight and colour
of the sapphires he had bought or considered
purchasing in Colombo, Ceylon (Sri Lanka)
from gem dealer Macan Markar.

Before 1900, the Cartier family partnered with select Parisian workshops and chose only the best of their designs to retail in the Cartier showroom. Then, around 1900, Cartier established its own design and manufacturing business and the family ensured that they had certain, trusted workshops that produced exclusively for them. Until the late 1920s, Cartier Paris continued to work with many of its favoured external masters including Charpentier, Harnichard, Lavabre, Picq, Andrey and Droguet. When Cartier's new international outposts opened in London and New York in 1902 and 1909, respectively, Cartier went further and began to set up its own in-house workshops (if these became overloaded with work, the firm would still subcontract elsewhere).

With its own design workshops in place, the three Cartier brothers then conquered the jewellery world with their Paris, London and New York showrooms. Each brother had his specialism. Jacques Cartier's field was gemstones, and his particular fondness for sapphire led to designs for wonderful jewels realized with exquisite sapphires brought back from his travels. His great-granddaughter, Francesca Cartier Brickell, explains Jacques's fascination:

> Of the three Cartier brothers, Jacques was the most knowledgeable about gemstones. He travelled through India and Ceylon tracking down some of the best gems on the planet that would later appear in Cartier's elegant creations in their Paris, London and New York showrooms. My grandfather (Jean-Jacques Cartier) told me how his father Jacques had been awed by the beauty of so many precious gems but that it had been sapphires for which he had felt a special affinity because they were his birthstone. He loved visiting the sapphire mines in Ceylon and filled pages of his diary with sketches and descriptions of the incredible sapphires he had found there. One of the few items of jewellery he wore was a sapphire ring and, when he made his adored wife Nelly a special amethyst brooch (her birthstone), he surrounded it with four diamond rectangles (representing their four children) each framed with a sapphire border, representing himself.

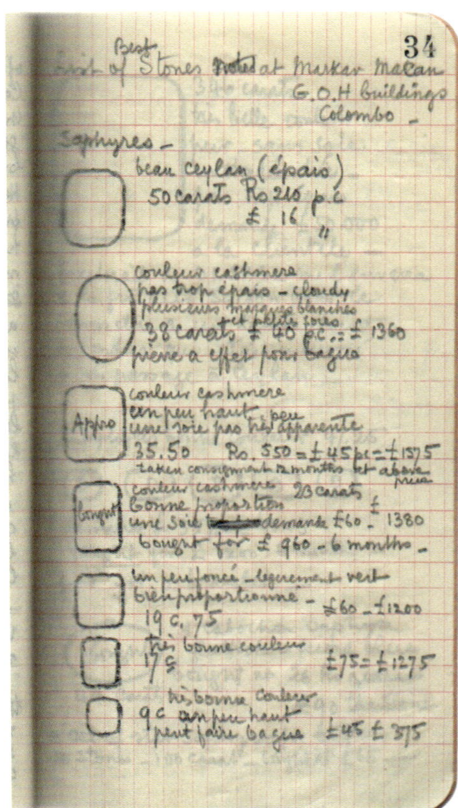

This 1912 brooch-pendant (below, described by Cartier as a *broche 'draperie'* in reference to the 'drape' effect of the garland-style jewel) was designed by Cartier head designer Charles Jacqueau (1885–1968) and made in the Picq workshops in Paris to be part of Cartier's stock collection. Such brooches – a form that Jacqueau used many times from around 1910 – were suspended from two separate elements joined by a pin; the pin is invisible when worn. Centrally set with a lozenge-shaped Ceylon sapphire weighing 28.70 carats, this design depicts the 'evil eye', which was a talismanic protection motif used in ancient civilizations. The sapphire sits within a diamond rectangular frame set with calibré-cut sapphires; the frame is suspended from a swing chain, millegrain-set with circular-cut diamonds, which terminates on each side with a cabochon sapphire. The pin runs through two semi-circular loops set with calibré-cut sapphires. The design is very bold for the period, as many other jewellery houses were producing jewels with the sinuous shapes of the Art Nouveau movement; this jewel highlights how Cartier's designers always pushed boundaries.

Below: A *broche 'draperie'* brooch-pendant by
the Cartier Paris head designer Charles Jacqueau
and made by the Picq workshops in 1912.
It is set with a buff-top 28.70-carat
lozenge-shaped Ceylon sapphire.

This sapphire *broche 'draperie'*, above, was made in the Droguet workshops for Cartier Paris as a special commission for the Countess Pálffy-Daun of Hungary. In 1913, Count Joseph-Guillaume Pálffy-Daun married Marie-Amélie, Countess Esterházy, and this jewel may have been ordered as a wedding gift for the new Countess Pálffy-Daun, or to be worn by the existing Countess Pálffy-Daun, Joseph-Guillaume's mother. The sapphires are recorded in the Cartier order as being repurposed from a necklace and pendant of the client. Resetting jewels was a relatively common undertaking for Cartier's biggest clients: as fashions changed, so did the jewels. The jewel is made of platinum and the diamonds and sapphires are millegrain-set, which was a popular setting style of the early 1900s. The total weight of the five sapphires is 39.66 carats.

Above: A *broche 'draperie'* commissioned
from Cartier Paris for the Countess Pálffy-Daun
of Hungary in 1913 and set with five
sapphires totalling 39.66 carats.

A Cartier Paris pendant from 1912 with
carved rock-crystal surround and two
cabochon sapphires of 13.45 and 1.98 carats,
respectively. The central section can be worn
as a brooch. Its plaster cast, left, was kept
for the Cartier archives.

The Cartier designers often looked to the art and architecture of previous centuries for inspiration. The Renaissance, for example, was a rich source of ideas, especially objets d'art like the sixteenth- and seventeenth-century rock-crystal vessels produced in Milan and Prague for the courts of Europe. Many of these carved stone vessels had been given as gifts to the French royal courts and so had found a home in the Louvre, where they could be viewed and studied by Cartier personnel who frequently visited the museum.

From these vessels came the idea of using carved rock crystal in the jewellery designs of diadems and pendants. All the rock-crystal pieces were beautifully carved and engraved by the Berquin-Varangoz workshop before being handed to the gold-smiths at Cartier to make into jewels. This brooch-pendant (facing page) was made in 1912 in the Picq workshops to be part of the Cartier Paris stock. Its intricately carved rock-crystal surround depicts the heads of two fauns, and the pendant was one of Cartier's first rock-crystal jewels. Charles Jacqueau designed the central oval plaque to be removable so that it could be worn separately as a brooch. The plaque is set with a cabochon sapphire, weighing 13.45 carats, within a platinum pierced foliate border, set with circular-cut diamonds and two natural pearls. The surmount is also set with a cabochon sapphire, weighing 1.98 carats.

The engineering involved in the jewel is visible when the sections are separated and shows the high level of craftsmanship needed to make the tiny screws that set the central motif in place. The jewel would have come complete with a mini screwdriver, secreted in a section underneath the bottom lining of the fitted box.

Though archives are now most often digitized, this is only a recent way to document past works. Cartier documented all its production between 1910 and 1920 by making plaster casts of each jewel to store in its archives. Notice that the plaster cast of this jewel shows the reverse imprint of the jewel, as well as the jewel's relief, so you can see how high the stones sit. The plaster cast was taken after the stones had been set, otherwise there would be too many vacant holes in the cast. Hundreds of plaster casts of Cartier's complete jewels are stored in Paris, along with the corresponding hand-drawn and painted designs.

A Cartier Paris brooch with frosted rock-crystal surround and a central sapphire cabochon of approximately 57.60 carats, originally set in 1924 with a carved emerald.

Cartier's use of carved rock crystal developed at the same time as its discovery of Indian carved gemstones. Splendours from the Indian subcontinent have always fascinated Europeans, ever since seventeenth-century travellers, such as gem merchant Jean-Baptiste Tavernier and English ambassador to the Mughal courts Sir Thomas Roe, brought back stories of the incredible wealth and riches they witnessed in these far-away, exotic lands. Until the early twentieth century, however, European jewellers had yet to familiarize themselves with the heritage of Indian jewellery or ingratiate themselves with its owners.

The 1911 Delhi Durbar attracted huge attention as, for the first time, members of the British royal family would be participating, notably the recently crowned George V and Queen Mary. Well aware that the great and the good of India would be there in all their finery, many jewellers saw it as a great networking opportunity. Indian princes were held in great esteem by both European high society and the jewellers that catered to them: a high-society dinner party would scarcely be deemed worthy of attending if there was not an Indian prince at the table.

The three Cartier brothers understood the significance of getting to know these splendid rulers. Securing jewellery commissions from such patrons was important, not only for the revenue they brought but also for associating the Cartier name with the 'exotic' subcontinent. As the networking event of the century, the Delhi Durbar of 1911 presented an opportunity the Cartier brothers knew they could not miss. Cartier had sent sales representatives to India in the past, but it was quickly decided that one of the Cartier brothers must attend this grand royal affair in person.

At the time, Jacques Cartier (1884–1941), the youngest of the three brothers, ran the London showroom. He had become acquainted with many of the Indian rulers, whose displays of impressive wealth were becoming legendary, when they had visited Cartier on their trips to London. Of the Cartier brothers, Jacques seemed the obvious choice to represent the *maison* at the Delhi Durbar.

This trip also came at a fortuitous moment for Jacques personally, as his future father-in-law had asked him to prove his love for his future bride, Nelly, by waiting a year without seeing her; if, after 12 months, they were still in love, then the marriage could go ahead. With both business and personal pressure, the 27-year-old

Jacques Cartier set sail for India, via the Middle East, along with Cartier's pearl expert, Maurice Richard, and a suitcase of jewels to tempt the maharajahs.

The Delhi Durbar was a spectacular event. In her book *The Cartiers: The Untold Story of the Family Behind the Jewellery Empire*, Francesca Cartier Brickell includes a vivid description of the celebration, as seen by her great-grandfather: 'Each ruling prince had his own vast tent area consisting of multiple tents and exquisite gardens. There were "beautiful avenues lit by electricity in the evening" and even a specially built railway to get around.' The nine-day event was attended by 12,000 people, and Jacques Cartier busied himself getting to know as many Indian rulers as he could. Ultimately, the trip was a huge success and brought Cartier several new prestigious clients. It was also the beginning of many trips to India for Jacques with his wife, Nelly.

The richness of colour combinations in India's applied arts and jewellery did not go unnoticed by Jacques: every time he came back from India he would always bring 'souvenirs'. Jacques Cartier's eye for design was inherent: he instantly knew the designs and colours that would go well together. He had a real appreciation for beautiful objects and a keen eye for detail, so much so that his mother pretty much left him to decorate his parents' house and he also oversaw the refurbishment of the Cartier showrooms. In 1912, Jacques held an exhibition in London entitled *Oriental Jewels and Objets d'Art Recently Collected in India*; a similar exhibition was held in New York the following year.

When Jacques went to India in 1911, he was impressed with enamels from Lucknow that used green and blue colours together. This combination reminded him of French *cloisonné* enamels of the 1880s that paired the same colours. Setting boldly contrasting colours together was Cartier's signature; indeed, they had already been experimenting with opposing colours on small items as early as 1903–6.

A Cartier Paris pendant, made on special order
in 1923, set with carved emeralds and two
Sri Lankan sapphire cabochons, including one
of 121.03 carats supplied by the client.

A 1927 Cartier New York bracelet set with
an engraved hexagonal sapphire of 59.39 carats,
surrounded by sapphires, emeralds and diamonds,
and sold to Mrs William Randolph Hearst.

Jacques's eldest brother, Louis Cartier (1875–1942), called this blue and green combination the 'peacock pattern'. Cartier used this pairing in jewels such as this bracelet made in 1927, which was sold to Millicent Hearst, wife of the legendary American newspaper publisher William Randolph Hearst and a vaudeville performer in the years prior to her marriage.

Carved gems were prized in Indian jewellery, and Jacques chose beautiful carved sapphires, emeralds and rubies of all sizes for some of Cartier London's exciting new jewels. Cartier sourced gemstones from India that were carved into flowers, leaves and berries, which were arranged in platinum like growing vines. Diamonds were used only to highlight the coloured gemstones and enhance the design. Charles Jacqueau was an avid follower of the Ballets Russes and was greatly inspired by their stage designs. His work incorporated baskets, garlands of flowers and bunches of grapes, which he interpreted using the carved gemstones.

The large engraved sapphire, weighing 59.39 carats, in this 1927 Cartier New York bracelet was originally set by the jeweller in 1924 in a pendant suspended from a necklace of emerald beads and sapphire rondelles. Cartier then reset the sapphire as just a pendant in 1925. In its current home in this bracelet, the hexagonal sapphire of engraved floral design is flanked either side by carved sapphires graduated in size and completed with cabochon, buff-top and calibré-cut sapphires and emeralds, with just a sprinkling of brilliant-cut diamonds. Since many of Cartier's princely Indian clients would have been circumspect about wearing a carved sapphire – on account of the Hindu belief that the stone should only be worn with great caution because of its association with the planet Saturn – carved blue sapphires were predominantly only set in jewels for the Western market.

These stones, which were never of 'gem' quality, were carved in order to add surface decoration. What they lacked in clarity, they made up for in colour: their bright colours gave the jewels great depth and impact. The 'Indian' style was very popular with Cartier's international clients during the interwar years; it was also a great period for commissions from Indian princes not only for Cartier, but also for Boucheron, Chaumet, Mauboussin and Van Cleef & Arpels. Ironically, these jewels, which were made mainly during the Great Depression in America, were quite affordable at the time due to the lower quality of the carved stones. Now, Tutti Frutti jewels are so sought after that a fine example can achieve a price at auction well over $1 million.

Jeanne Toussaint (1887–1978) was in her mid-twenties when Louis Cartier, the eldest of the three brothers, first met her in Paris before World War I. She was a stylish lady, always immaculately dressed, who mixed with 'colourful' Parisian ladies and was never short of wealthy admirers. Her search for style may have been a reaction to her harsh upbringing as the youngest of five children from a poor family. Her father died when she was only seven years old. As soon as Jeanne was able, she left the family home to make a better life for herself in Paris.

Louis Cartier was mesmerized by Toussaint. He adored her from their first meeting and over their lifetimes they became soulmates. Marriage to her was not an option for Louis, due to family pressures that likely related to Toussaint's background and Louis's position as eldest in the family. Louis employed Toussaint to work in the handbag department of the showroom. Her critical eye and her understanding of fashion meant that she saw jewellery from a different perspective; she scrutinized the Cartier jewellery designs and saw the ever-changing haute couture fashions as perfect backdrops for jewellery. Toussaint never designed items but, in 1933, she was promoted to overseer of fine jewellery production. For the next four decades she became integral to Cartier's success, even after Louis died in 1942. She contributed to jewellery ideas and worked with head designer Charles Jacqueau, creating jewellery that became highly sought after the world over and that is now emblematic of Cartier's brilliance.

Toussaint had been ahead of fashion by wearing Indian jewellery since the 1910s, but, importantly, she knew how to wear it. India was a popular destination for wealthy American and European 'tourists' in the 1920s and '30s, and many of Cartier's clients would return with expensive jewellery 'souvenirs' from their trips, which were then transformed and remounted by Cartier under the guidance of Toussaint. The Cartier Tutti Frutti jewels became increasingly popular in the 1920s and '30s, mainly due to one woman: Daisy Fellowes (1890–1962). Fellowes, heiress to the Singer sewing machine empire, could single-handedly make a fashion house rise in popularity. The French poet, filmmaker and designer Jean Cocteau remarked that 'she launched more fashions than any other woman in the world'. She was, for a short while between 1933 and 1935, the Paris correspondent for the influential *Harper's Bazaar* magazine, a role that gave her a platform of authority.

As Jacques Cartier recalled, Paris in 1935 was 'overrun with Maharajahs, casually wearing fabulous jewels'. It was against such a backdrop that Daisy Fellowes organized an Orient-themed fancy-dress party. All attendees made sure they were bedecked in sumptuous jewels. *Vogue* magazine did a feature on 'Eastern splendor' and referred to Cartier's Indian-inspired jewels as 'barbaric' because of their use of carved gemstones that looked rough in comparison to the exact cuts of faceted stones. Perhaps fuelled by this article, and certainly motivated by her own love of being different and ever the centre of attention, Fellowes commissioned Cartier in 1936 to make a 'Hindu necklace' that became a showstopper: the Collier Hindou. Cartier used gemstones from three earlier jewels it had made for Fellowes: a necklace of 1928, a bracelet of 1929 and another unidentified bracelet. The necklace was constructed by the Lavabre workshop for Cartier Paris. Using the carved sapphires, emerald beads, carved ruby leaves and marquise-cut diamonds, they created a necklace that is the pinnacle of Indian-inspired jewels of this period.

I remember handling the Collier Hindou and was struck both by the impact of colour created by the gemstones and by the quality of craftsmanship as evidenced by the way the necklace articulated. The necklace has volume as the emerald beads are set at a slight angle, but this arrangement makes the necklace so tactile. The prominent stones are blue sapphires, which, with their reputation in India as gems to be avoided or worn with caution, mark the necklace as a jewel for the Western market. Originally, the front of the necklace had a removable clip/brooch in the centre, set with the two large, carved sapphires. These two sapphires now form the clasp, but the necklace was secured with a traditional black silk cord before this alteration. It was altered to its present form at the request of Fellowes's eldest daughter, the Comtesse de Castéja, in 1963. Today, the necklace sits beautifully around the neck, though, if it had been left in its original design with its silk cord fastening, it would have been more versatile as it could be easily adjusted to fit any neck size.

I have a fond memory from years ago, when I met Francesca Cartier Brickell and we travelled to Geneva with her mother, Veronica, to view the Cartier Collection. It was a special moment to see a family legacy come to life as they held jewels that Francesca's great-grandfather and her mother's grandfather, Jacques Cartier, had created with his brothers. Francesca tried on the Collier Hindou and looked stunning, proving that jewels, however important, should be worn and enjoyed.

The Collier Hindou and its graphite and gouache design, left, made by Cartier Paris in 1936 for Daisy Fellowes, heiress of the Singer sewing machine fortune, and remounted in 1963 upon her daughter's request to remove the original black cord fastening. With 13 briolette-cut sapphires, collectively weighing 146.9 carats, and two leaf-shaped carved sapphires, weighing 50.80 and 42.45 carats, respectively, with sapphire, emerald and ruby beads interspersed with diamonds. The sapphire briolettes and emerald beads were reused from an earlier Cartier Paris necklace, design right, made in 1928 and bought by Fellowes.

Mona Bismarck (1897–1983) came from humble beginnings in Louisville, Kentucky, but her beauty and intelligence enabled her to gain the wealth and high social standing she sought. These desires came at a price. By 1926, Mona had married for the third time. Her new husband, the incredibly rich businessman Harrison Williams, ensconced her in the wealthy high-society circles of New York. She became a close friend of the photographer Cecil Beaton, who supposedly likened her to a goddess made of rock crystal, with eyes that reminded him of aquamarines.

Along with her passion for art, Mona acquired some magnificent jewels from her three husbands. When styles changed, she often had her jewellery altered to suit the new fashions. On many occasions, she turned to renowned jewellery house Cartier.

It is possible that Harrison bought this particular sapphire (facing page) for her on one of their trips to Asia, as the large gem, weighing 98.57 carats, is from Myanmar (Burma). Alternatively, it may have been a stone selected and bought by Cartier, as sapphires were Jacques Cartier's favourite stone. A Burmese sapphire of this size, exceptional colour and transparency is incredibly rare: factors that in combination make this one of the most famous sapphires in the world. The Bismarck Sapphire, as it is now known, was originally set horizontally into a necklace designed and made by Cartier around 1927. The sensational centre stone was accented by 312 baguette- and brilliant-cut diamonds and eight square-cut sapphires.

After a long illness, Williams died in 1953, at the age of 80. In a little over a year, Mona married again, this time to her secretary, Count Edward von Bismarck, grandson of Otto von Bismarck, Chancellor of Germany. Edward was the same secretary that had helped Mona to restore Il Fortino, her villa on the island of Capri. The Bismarcks moved permanently to Europe, buying a mansion in Paris. Meanwhile, there was still thought of changing the sapphire necklace. In 1959, Mona had the design altered by Cartier so that the sapphire was set vertically in the necklace. She donated the necklace to the Smithsonian Institution in 1967.

After another marriage, her fifth, Mona retired to Paris, where she died at home in 1983 at the age of 86. A foundation was established in her name to support Franco-American friendships in the arts and culture.

Top: A design by Cartier New York, *c.*1927,
for the necklace commissioned by Mona Bismarck;
this design depicts the large sapphire's original
horizontal setting and the necklace's sapphire-
and diamond-set links.

Above: A 1930s photograph by Horst P Horst
of Mona Bismarck wearing sapphire jewels.

The Bismarck Sapphire Necklace, originally made
for Mona Bismarck by Cartier in 1927 with the
98.57-carat Burmese sapphire set horizontally
in a necklace of sapphire- and diamond-set links,
and altered by Cartier upon her request in 1959
to set the stone vertically in a new diamond
necklace. The sapphire and its diamond and
sapphire surround can be detached to be
worn as a clip brooch.

Documentation of gemstones that have passed through a jewellery house from 1930 to the 1980s makes for fascinating reading, especially when the jewellery house in question is one of the most famous in the Place Vendôme. The gemstones detailed in Boucheron's Book of Stones were all recorded while being set or reset in the *maison*. Boucheron's stone team documented each significant stone with a gouache sketch and information about the stone's size and weight, plus details of the buyers and sellers and the stone's price, noted in code.

Entries include some well-known stones with long histories. One such example is the 291-carat oval sapphire that had been stolen from Count Branicki (1816–1879), a Polish nobleman and political exile who took up French nationality, the entry about which mentions the famous gem dealer Raphael Esmerian (1903–1976), who had acquired it at a later date. Also listed on the same page is a very large Ceylon sapphire of 478 carats that had belonged to Queen Marie of Romania, introduced in the chapter At Court, and was passed down to her grandson King Michael, who abdicated in 1947. The stone apparently passed through Boucheron in 1957 with the possible idea to sell it, though it was eventually bought by Harry Winston who, in turn, sold it to a client to gift to Queen Frederica of Greece.

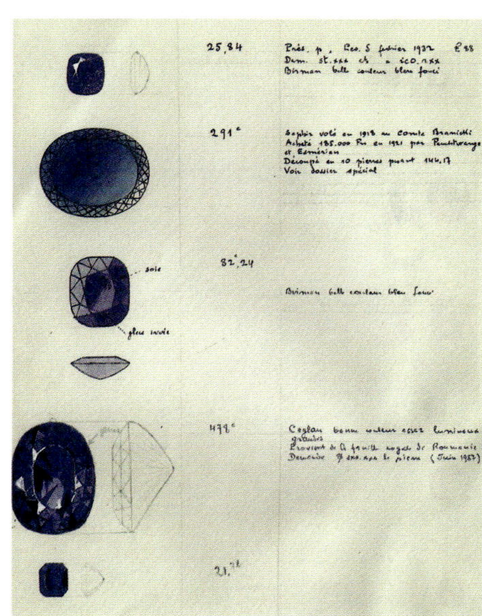

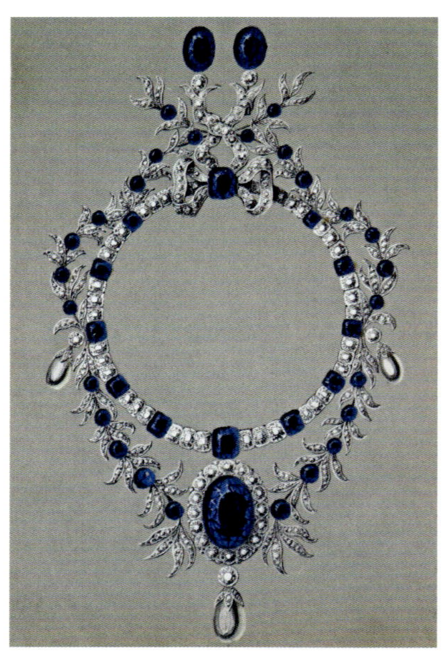

From left: A page of gouache sketches of sapphires, including the 478-carat sapphire of Queen Marie of Romania, from Boucheron's Book of Stones; the 159-carat oval sapphire, another large sapphire to pass through the *maison*, set as a pendant to a double necklace of sapphires and diamonds for Mrs Marie-Louise Mackay in 1878, shown in a photograph from 1890 and its original design.

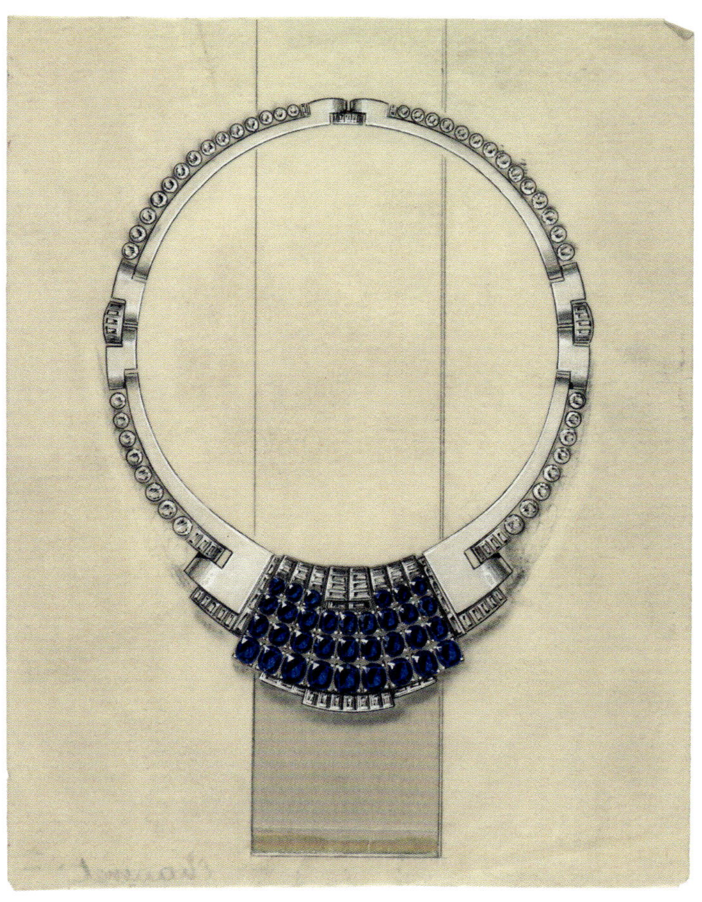 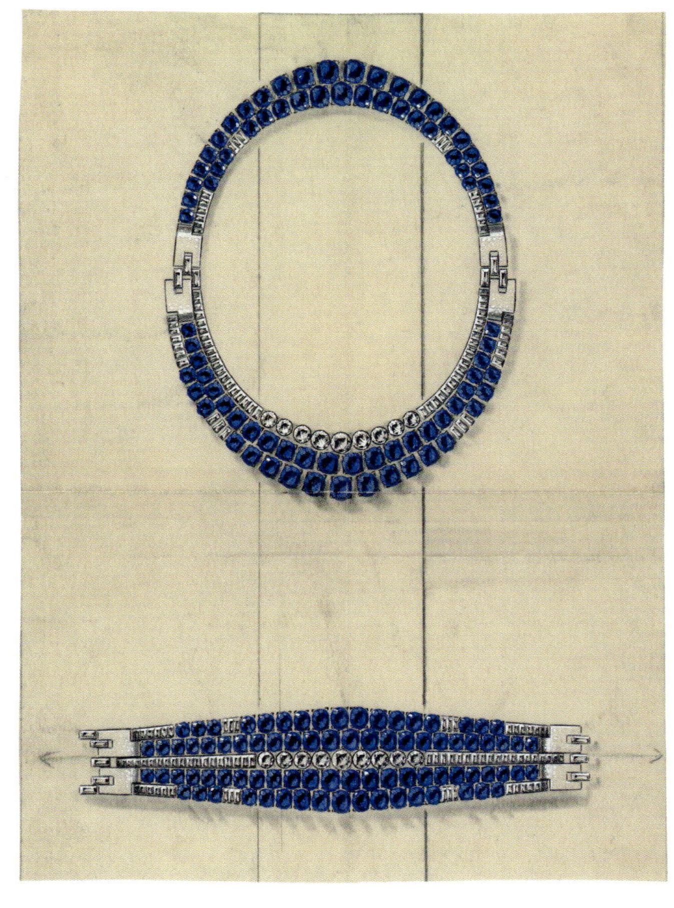

Parisian jewellery houses capitalized on the 1920s and '30s fashion for bold jewels by sourcing the best and the brightest gemstones and setting them together in large designs. For their monochromatic colour schemes of rock crystal and platinum against a single colour of gemstone, sapphire was a popular choice. Sapphires come in a variety of hues and tones, so some jewellery houses used this variety to their advantage by arranging and matching different tones of blue together, which added to the jewels' impact.

Simplicity was key. 'Odeonesque' designs echoed the style of the architecture of the Odeon cinemas, which opened around Britain from 1928. Technology in diamond cutting had also advanced so that the new baguette-, tapered- and square-cut diamonds were now complementing the faceted coloured gemstone. By mixing different stone shapes, these designs still made an impact on the eye, without relying on lots of colours, so that attention stayed on the sapphires. Women's hairstyles were now shorter as part of the boyish *garçonne* look and so their necklace fittings became more visible. The jewellery houses responded to this change in style and set gemstones all the way round the necklace, emphasizing that jewels should be as impactful from the back as from the front.

Two designs from the Parisian *maison* Chaumet for necklaces of the 1930s, set with oval-shaped sapphires and round and baguette diamonds all round the neck, so as to be visible when worn with the fashionable short hairstyles. The design on the right appears to be for an adaptable necklace-bracelet.

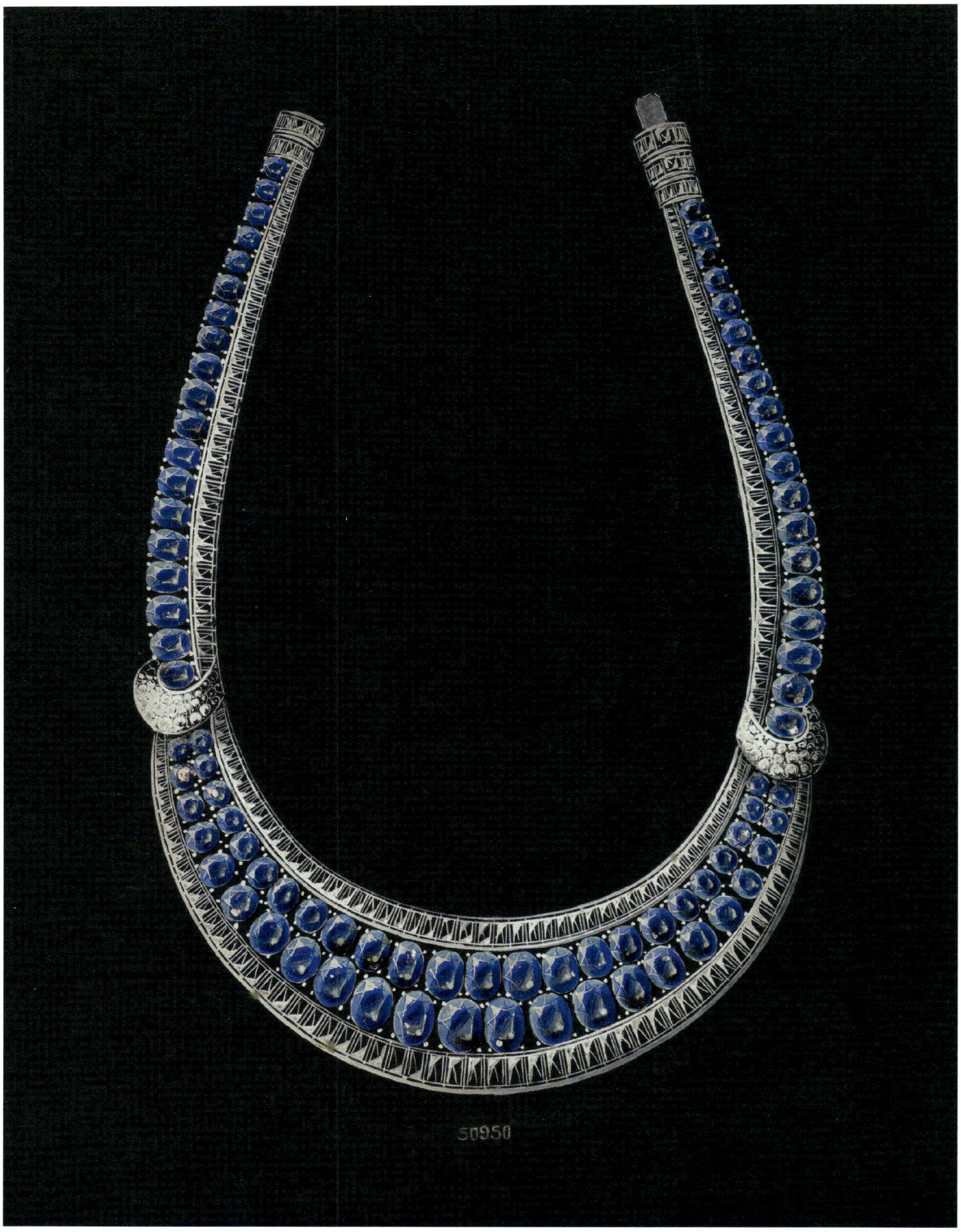

Three 1930s designs for necklaces by Van Cleef & Arpels. The combination of rich blue sapphires and white diamonds lent itself well to the late 1920s and '30s trend for monochromatic colour pairings, but it is no mean feat to find enough matching sapphires to create these jewels. The design, left, for the necklace of light and royal blue sapphire flowers demonstrates Van Cleef & Arpels' appreciation of sapphire's natural colour range and the striking effects it can be used to create.

A sapphire and diamond bracelet and
collarette necklace by Van Cleef & Arpels, both
dated 1937. The precise graduation in size of
the sapphires in the collarette and the colour
matching of the sapphires in the bracelet are
examples of the attention to detail for which
Van Cleef & Arpels is known.

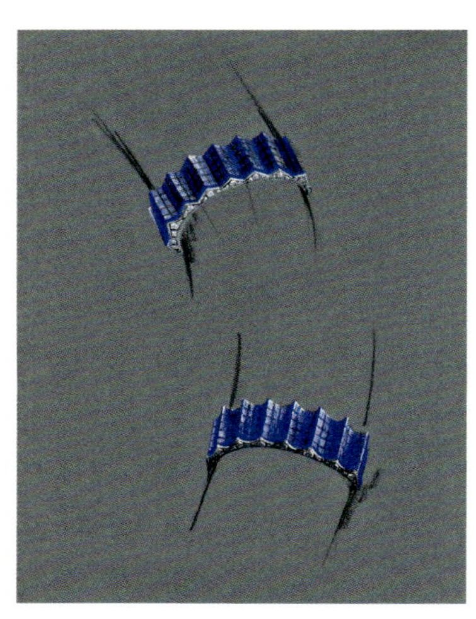

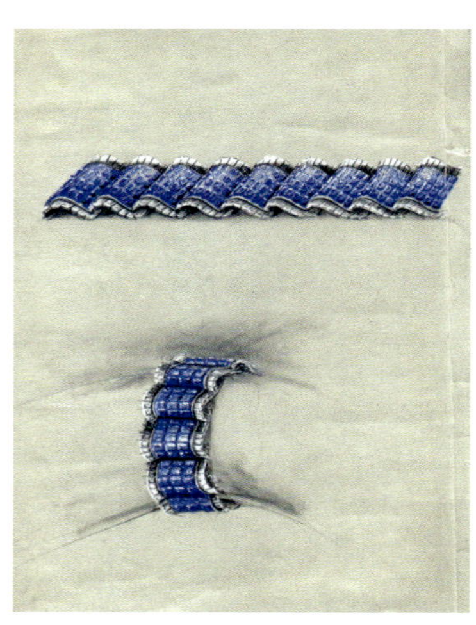

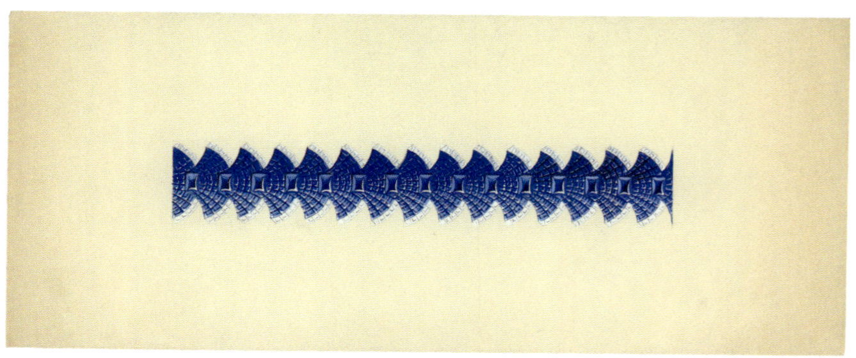

Advancements in lapidary gave way to new and exciting developments in setting gemstones. Van Cleef & Arpels patented the Mystery Set technique in 1933 and became the master of this craft. The Mystery Set jewel gives an illusion of uninterrupted blocks of coloured gemstones without any visible metal settings. Each small, square-cut sapphire, ruby or emerald has grooves cut underneath that allow the stone to be slotted onto a lattice of precious metal rails in the body of the jewel. When the stones are set and the jewel is viewed from the front, all you see is a blanket of one colour without any metal claws.

This technique requires exceptionally skilled and time-consuming work, not only in cutting the gems, but also in sourcing stones that match in hue and tone. Sapphires are notorious for varying in colour, so each Mystery Set sapphire jewel takes a long time to produce. As the Van Cleef & Arpels craftspeople perfected the technique, they also mastered setting the stones on undulating curves that allowed the designs, like those above, to come alive.

Left: A design for a bracelet of rolling peaks of
Mystery Set sapphires, by Van Cleef & Arpels, *c.*1930.

Centre: Two 1930s designs by Van Cleef & Arpels for
a necklace and a bracelet, both with the same Mystery
Set sapphire fan, or clam shell, repeated motif.

Right: A design for a bracelet of Mystery Set
sapphires, set to flow like ribbon or waves.
Van Cleef & Arpels, *c.*1938.

A diamond and sapphire bracelet with
a front buckle-style curved band of Mystery
Set sapphires by Van Cleef & Arpels, 1937. The
seamless bracelet of interlocking hexagons,
known as Ludo, was a pioneering technology
of the *maison*.

Two archival illustrations for Van Cleef & Arpels
brooches that use the flow of Mystery Set
sapphires to depict the movement of a butterfly's
wings and the softness of a flower's petals; the
butterfly design is from the mid-1930s and the
product card of the flower brooch is dated 1937.

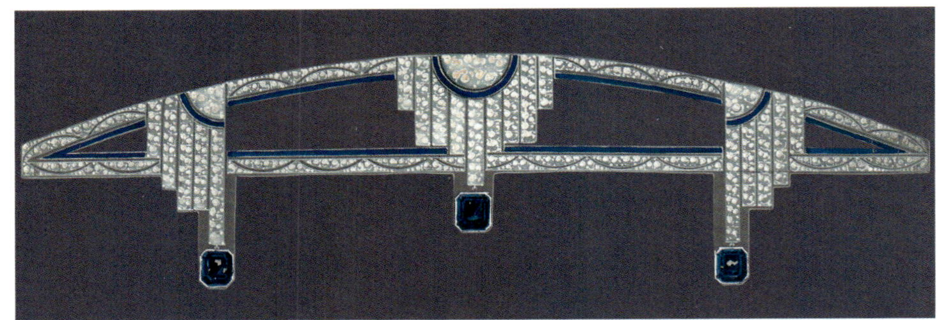

By the 1900s, the *maison* of Mellerio had already celebrated its 300th anniversary, surviving many wars and revolutions. One of the main markers of a jewellery house's success is its ability to keep pace with changes in contemporary fashions and life-style and to do it well. During this period, following World War I, many of the successful nineteenth-century jewellers had failed. As tiara-wearing occasions were evolving, Mellerio ensured that its designs reflected the modern style. The Art Deco tiara designs here encompass the trend for bandeau-style head ornaments, with one featuring sapphires on diamond cascades, a design that almost flips the traditional tiara shape upside down. This tiara was realized in diamonds and sapphires in 1929 and was presented at the Exposición Internacional in Barcelona that year.

From top: A design for a tiara by Mellerio, made in 1929 with sapphires and diamonds and exhibited at the Exposición Internacional in Barcelona in the same year, as seen in the top right corner of the photograph of Mellerio's display at the fair, and a 1930s design by Mellerio for a tiara set with diamonds and sapphires, including a large octagonal or cushion-shaped sapphire.

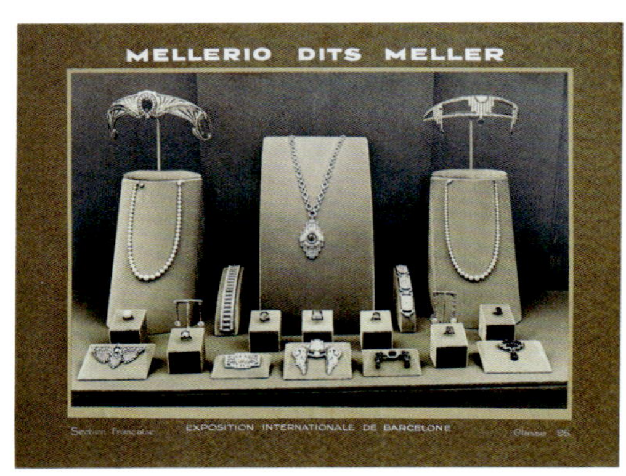

The Exposition des Arts Décoratifs et Industriels Modernes was held in Paris in 1925. At this historic event, exhibitors showcased their new works of art and jewellery, many of which were influenced and inspired by geometric and architectural forms. The emergence of this style became so linked with the fair that it was later called Art Deco. Art Deco designs, together with carved Indian gemstones, produced some of the most iconic jewels of the twentieth century. In 1925, jewellery displayed bold, monochromatic, geometric designs, using unusual materials such as carved rock crystal, a perfect stone to emphasize the brightness of blue sapphires.

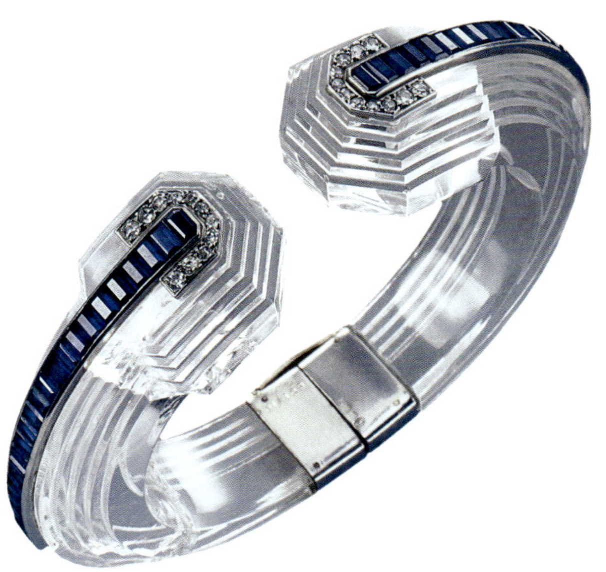

Three jewels of carved rock crystal from the collection of Pforzheim Jewellery Museum, set with calibré- and step-cut sapphires, two with pavé-set diamonds. Right: A bracelet, *c.*1925, by René Boivin, Paris. Top and left: A clip and brooch, respectively, both *c.*1925, Paris.

Though we today associate the desirable jewels of Boivin with its founder, Monsieur Jules René Boivin (1864–1917), most of these jewels were, in fact, made by a team of women – unusual in the early twentieth century – led by Boivin's wife, Jeanne, who pushed the company to make and sell exciting jewels by nurturing rising design talent. I profile these women in more detail in the chapter Jewels of Impact.

René Boivin was born in Paris in 1864. At the age of 17, he started as an apprentice in his brother's goldsmithing workshop, where he learnt the skills of the trade and became an accomplished engraver. He went to art school to learn to draw, and it soon became apparent that he had a talent for design. He left his brother's workshop to set up on his own in 1890 and, in 1893, not only did he register his maker's marks, he also married Jeanne Poiret, sister of the famous couturier Paul Poiret. Marrying Jeanne introduced him to the world of fashion, which he embraced. In his designs, he rejected the popular Art Nouveau jewels in favour of chunky jewels inspired by the Middle East and Asia. While producing jewels for wholesale to retailers such as Mellerio and Boucheron, he also started to gain his own private clients.

His untimely death in 1917, when he was 53 years old, was closely followed by that of his only son, which left Jeanne to carry on the business alone. Against all odds, Madame René Boivin, as she wanted to be known after her husband's death, expanded the firm and made Boivin the success it is still regarded as today. Madame Boivin at first carried on designing in her late husband's taste and style, but as time went on she started to develop her own ideas and employed mainly female designers to help her create the jewels that have become synonymous with the Boivin style. One of these designers, Suzanne Belperron, née Vuillerme, began her career with Madame Boivin as a young sales girl, but progressed through the company as a designer, until she left to pursue her own famed design career.

Madame René Boivin's circle of friends were from the worlds of fashion, dance and art; since each championed individuality, it is no surprise that Boivin's jewels were fearless, bold and almost sculptural in their relief. The Boivin style also added curves to the geometric lines of Art Deco. Madame Boivin used many different colourful gemstones – some of which were not well known – in both faceted and cabochon cuts, resulting in jewels that were so distinctive they did not often need a signature to be recognized. Her jewellery was intended for women who wanted to stand out from the crowd with style and grace.

Above and top: Boivin Pont brooch, *c.*1930, of *bombé* shape with a central row of cushion-shaped coloured sapphires between two lines of circular-cut diamonds and carved rock-crystal edges, signed 'R. Boivin' under the leadership of Madame Boivin.

STAR SAPPHIRES

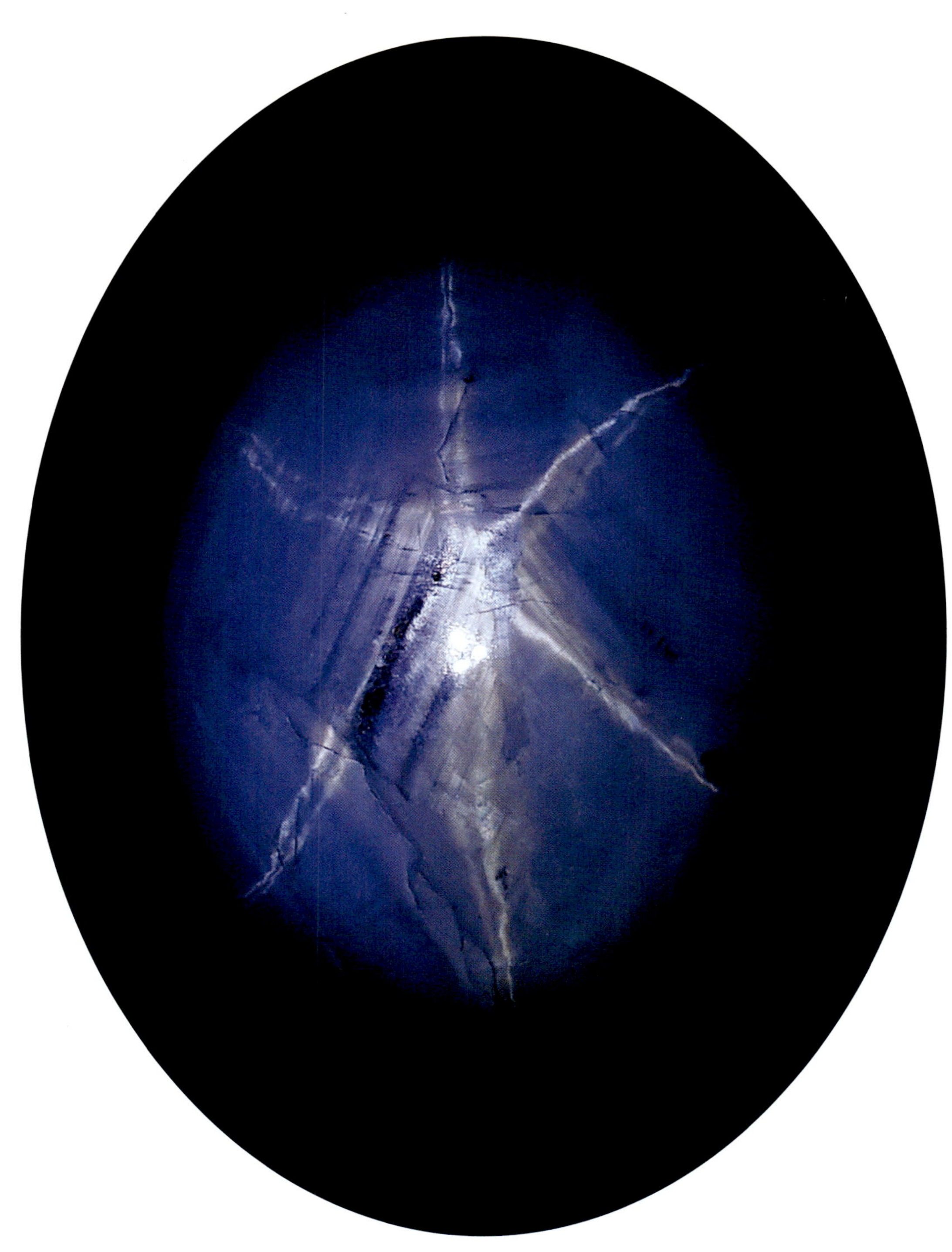

The Star of Adam, a 1,404.49-carat pale
blue star sapphire from Sri Lanka. Since its
publicized discovery in January 2016, this stone
has been celebrated as the world's largest known
star sapphire. Its name honours the mountain
Adam's Peak at the centre of Sri Lanka.

Star Sapphires

Lights, action, sparkle – the women of 1930s Hollywood loved their big and bold jewels. Many of the leading actresses wore their own jewellery, but, on camera or on stage, jewellery needed to be seen, so size was important. The larger natural crystals of the sapphire made the stone a strong contender for on-screen jewellery, and the star sapphire – a variety little known amongst the stars of stage and screen before this time – became the stone of choice. Spotlights on set revealed the stones' optical effect of asterism (rays of light that reflect in a star shape over the stone's surface), and camera flashes picked up the stars shining from large sapphire cabochons.

The jewellers to the stars of the 1930s and '40s, such as Paul Flato, Raymond C Yard, Trabert & Hoeffer-Mauboussin, Oscar Heyman & Brothers and Fulco di Verdura, were all aware of the opportunities presented by the star sapphire. It was a relatively affordable stone at the time, even during the austerity years, and so it was ideal to include in jewellery designs. Designs emphasized each stone's asterism display, which gave jewels a unique look in comparison to faceted, transparent gemstones. The large star sapphire cabochons used in these jewels presented an increase in scale that was replicated in the booming costume jewellery market. Audiences wanted to feel that they, too, could live a life of luxury; costume jewellery was an affordable means to recreate a fantasy.

In star stones, asterism is seen because of the combination of small, needle-like 'silk' inclusions and the lapidary's careful cutting in a specific orientation. Star sapphires displaying six-rayed asterism are more common than those displaying 12-rayed asterism. In a six-rayed star, fine fibres of the mineral rutile lie in three orientations at 60/120-degree angles to each other. A layering of rutile and hematite-ilmenite needles can produce a 12-rayed star. Importantly, a star can only be seen if the stone is cut as a cabochon (one of the oldest ways of cutting and polishing a stone), and in a specific orientation – the cabochon may be the simplest of cuts, but it requires skill to find the correct orientation. It is crucial that the cutter orientates the crystal to look down its c-axis – a direction relating to the stone's crystallographic planes – so that the inclusions lie in the right direction to produce a star when cut. With a strong, direct light, the 'star' will travel around the top of the cabochon dome as you move the stone. A good star sapphire needs the star to sit in the centre of the cabochon, with each ray of equal length running from the top to the base of the cabochon, and to be clearly defined. If the star is too perfect and seems to stay in one position when the stone is moved, it could be a man-made star sapphire.

Not all cabochon-cut sapphires will display asterism, as it is the presence of the silk inclusions that provides the opportunity for a star. Transparent star sapphires are incredibly rare as the needles generally form in large amounts, making most stones opaque. If these translucent or opaque stones were faceted, they would look very cloudy and dull, with no asterism. Star sapphires are usually unheated because heat treatment generally improves a stone's colour and transparency by burning out the rutile inclusions needed for the star effect. With prices of faceted stones increasing, more crystals have been subjected to heat treatment to burn out their needles, which improves their colour and clarity ready for faceting. As a result, fewer star sapphires are coming onto the market and so dealers are keen to acquire those stones that are available. Star sapphires can occur in nearly all colours, including black sapphires from Cambodia and India due to the dark colour of the mineral hematite. Natural yellow, orange and green star sapphires are very uncommon. The majority of blue to blue-grey star sapphires are found in Sri Lanka, but fine blue star sapphires have come from Myanmar (Burma). Star sapphires displaying good asterism, a rich blue colour and of good translucency are extremely rare and very sought after if ever found.

Cabochon star sapphires have always been deemed particularly special – used as a talisman to protect against the 'evil eye', with the three crossing bands of light representing faith, hope and destiny. Their popularity was certainly enhanced by their association with Hollywood celebrities, as author Beth Bernstein has shown in her book *If These Jewels Could Talk: The Legends Behind Celebrity Gems* on the jewellery collections of these Hollywood stars. The star sapphires in these personal collections were very large, and the owners were not afraid to wear them, despite their relative youth; on the contrary, bigger was obviously better. Bernstein highlighted another common link between these stars: William Powell (1892–1984), an actor who worked alongside most of them – in 14 films with Myrna Loy alone – and who also seemed to favour blue star sapphires. He purchased star sapphires for both Carole Lombard and Jean Harlow. To own fabulous jewels places these stars of the silver screen in stark contrast to today's actresses, whose big 'red carpet' jewels are normally borrowed and very rarely owned.

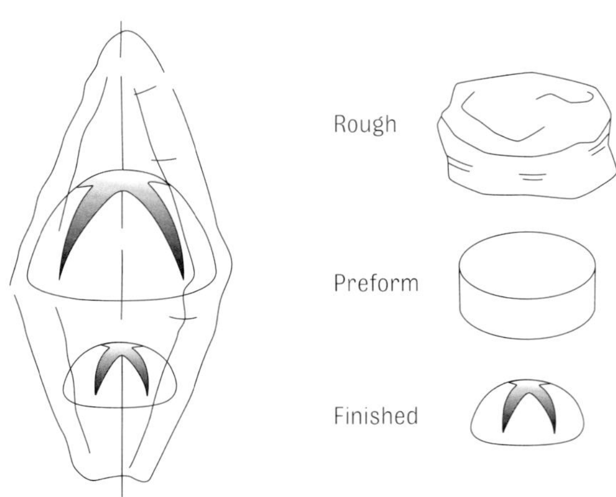

An illustration of a bipyramid crystal
and tablet-shaped crystal with the orientation
and possible proportions a lapidary considers for
cutting circular cabochons that will achieve a star
in sapphires that display asterism.

The Star of India, a 563.35-carat
star sapphire from Ceylon, in the collection
of the American Museum of Natural History.
In 1964, the *New York Times* photographed the
smashed glass cases in the museum after the
Star of India and other gems were stolen
(and eventually recovered).

One of the most famous star sapphires is the Star of India, a 563.35-carat blue cabochon. The Star of India is not, in fact, from India but from Ceylon (Sri Lanka). Its misleading name likely reflects the stone's journey after its discovery over 300 years ago, as it would have been traded through India without any note of its origin. That changed when Tiffany's gemstone expert, George Kunz, bought the stone in the late 1880s. Kunz was commissioned by the wealthy banker and philanthropist J P Morgan to amass a world-class collection of gemstones for the Paris Exposition Universelle of 1900. Morgan later donated all the gems in Kunz's collection, including the Star of India, to the American Museum of Natural History. In 1964, the Star of India was stolen, along with more than 20 other precious gemstones and diamonds, from the museum. Luckily, the thieves were caught and one of them later led the police to a bus locker in Miami, where the Star of India and some of the other gems were recovered. Many of the stolen stones, including a famous diamond, were never found. The Star of India is a fine example of a star stone, not only because of its size, but also because it has been cut to display the six-rayed asterism on both its top and bottom domed sides.

CAROLE
LOMBARD

Carole Lombard (1908–1942) began acting when she was only 12 years old and went on to become one of the highest-paid American actresses of the 1930s. At a time when men often had greater control over a woman's career than she did herself, Lombard's reputation grew considerably through her link to William Powell, whom she met at the age of 22, with him 16 years her senior. Powell was an American actor, renowned for being suave and cosmopolitan, who had leading roles alongside some of the most iconic film actresses of the 1930s and 1940s, including Jean Harlow, Joan Crawford, Bette Davis, Myrna Loy and Rosalind Russell. He appeared in 97 films and was nominated three times for an Oscar, though was never lucky enough to win. Lombard and Powell married in 1931, but their marriage lasted little over two years. Despite their personal split, their acting careers remained connected; only a couple of years later, they starred together in *My Man Godfrey* (Universal, 1936), which earned Lombard an Academy Award nomination for Best Actress.

Powell had given Lombard a large star sapphire engagement ring weighing approximately 80 carats. She continued to wear this after their divorce, pairing it with another star sapphire of nearly double the size (150 carats), which she bought for herself. She commissioned a transformable jewel for this latter sapphire so that she could wear it as a brooch, ring or pendant.

Lombard married another leading actor, Clark Gable, in 1939, but the marriage came to an untimely end. In the early hours of 16 January 1942, Lombard, along with her mother and Clark Gable's press agent, Otto Winkler, tragically died as their plane crashed on their return to California from Las Vegas.

Carole Lombard photographed in 1934,
wearing her star sapphire engagement ring
of approximately 80 carats, gifted from
William Powell, which she continued
to wear after their divorce in 1933.

Jean Harlow (1911–1937) lived a very short life but nevertheless made quite an impact in Hollywood thanks to her comedic talent. She also managed to fit in three marriages. Harlow understood the power of seduction and wore striking jewels in an unconventional way to lure eyes in her direction. She would pin brooches in her hair, around her waist, on a daringly low-fronted neckline, or to the V of a backless gown – wearing jewels in this way had never been seen before.

Harlow was known for her love of sapphires, their size and lustre making them perfect for catching the attention of admirers. One of those admirers was fellow actor William Powell, recently divorced from Carole Lombard, who initially proposed in 1935 with a diamond ring. Rumour has it that Harlow was happy to accept Powell's proposal, but was less enamoured with the diamond ring, which she asked to be exchanged for a large star sapphire. In the photograph below – taken on the set of her 1937 film *Personal Property* – Harlow is seen wearing her star sapphire ring. She never took the ring off, but they never married; Harlow died unexpectedly at age 26 from health complications. In 1937, while on set working on what was to be her final film, *Saratoga*, Harlow fell ill and called for Powell to come from his own filmset and take her home. In a matter of days Harlow died, with Powell at her side.

Jean Harlow wearing her impressive
star sapphire ring on the set of the film
Personal Property (MGM, 1937).

Myrna Loy, with Keenan Wynn and long-time
co-star William Powell, right, in the film *Song of
the Thin Man* (MGM, 1947) wearing a cabochon
ring that could be her star sapphire ring, top,
with diamond-set claws and baguette-
cut diamond shoulders.

Myrna Loy (1905–1993) was one of Hollywood's highest-paid actresses and starred alongside some of its biggest names: Clark Gable, James Stewart and Cary Grant. She made 14 films with William Powell alone. Loy married and divorced four husbands in her lifetime, but that seemed to be considered the norm amongst A-list celebrities during Hollywood's Golden Era. In 1991, Loy was awarded an honorary Academy Award for her acting and charitable work, including her role in the Red Cross during the World War II.

Loy loved star sapphires and owned a star sapphire ring, above, that was reputedly made by Paul Flato, who was revered as the jewellery designer to the stars.

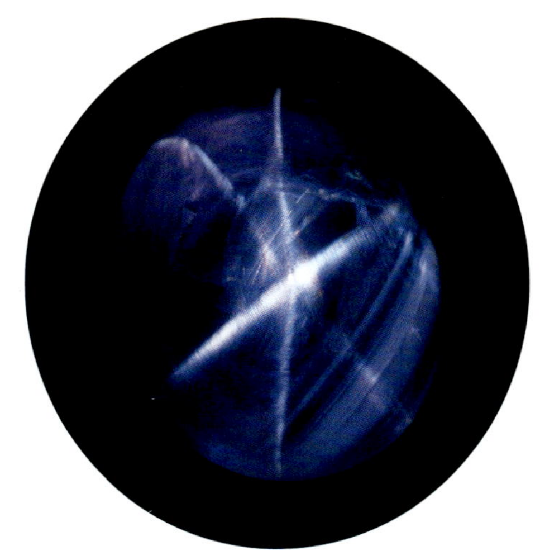

Actress Mary Pickford with her new husband
Buddy Rogers at their wedding in 1937; Pickford's
clip is set with a large cabochon that could be
her Star of Bombay, a 181.82-carat violet-blue star
sapphire, illustrated above. The sapphire is now
on display at the Smithsonian's National
Museum of Natural History.

Mary Pickford (1892–1979) was the 'Queen of the Movies'. A star of many successful silent features, by 1920 Pickford was one of the most famous women in cinema, with a real life that matched her glamorous on-screen one: she met the famous silent film actor Douglas Fairbanks Sr and they started an affair, much like the storyline of many of their films, before divorcing their spouses to marry each other on 28 March 1920. Soon after their wedding, Fairbanks gave Pickford a platinum ring set with an enormous 181.82-carat violet-blue star sapphire called – despite its Sri Lankan origins – the Star of Bombay. Fairbanks had bought the ring from the Hollywood jewellers Trabert & Hoeffer-Mauboussin, who had originally acquired this rare six-rayed star gem. For the next eight years, Fairbanks and Pickford seemed inseparable, never wanting to be far apart. They were known as the King and Queen of Hollywood, and theirs was deemed the marriage of the century.

Pickford did not wear big jewels on screen, instead favouring a string of pearls, but her success and business acumen enabled her to indulge her passion for jewels off screen. In the 1920s it was fashionable to wear a number of gem-set bracelets together, and Pickford was often seen wearing a few bracelets from her collection at the same time; each bracelet represented an important event in her life. Her favoured jeweller was Trabert & Hoeffer-Mauboussin, and she owned one of their wide diamond bracelets set with a large cabochon sapphire.

Fairbanks and Pickford made their first talkie film, *The Taming of the Shrew*, together as a joint venture in 1929, during the Great Depression, but the film was not well received. Around the same time Fairbanks started a relationship with Sylvia, Lady Ashley, an English model, and he and Pickford separated in 1933. In 1936, Fairbanks and Lady Ashley married, but in December 1939, at the age of 56, he suffered a heart attack and died.

Pickford kept the 'Star of Bombay' and enjoyed wearing it for over 60 years until she died in 1979, aged 87. She bequeathed the sapphire to the Smithsonian Institution in Washington, DC, and today it is on display in the Janet Annenberg Hooker Hall of Geology, Gems and Minerals of the Smithsonian's National Museum of Natural History.

JOAN
CRAWFORD

Blue sapphire was Joan Crawford's (1904–1977) favourite stone, so much so that Crawford and her jewellery collection supposedly earned the nickname 'Joan Blue' from the media. Crawford was always seen wearing large, fine jewels – 'big and bold' was a motto Crawford lived by when it came to her jewels. She acquired a large 72-carat emerald-cut sapphire ring during her first marriage, in 1929, to Douglas Fairbanks Jr, whose father was married to fellow Hollywood actress Mary Pickford. The marriage ended in 1933 and, soon afterwards, on the set of *Today We Live*, she met Franchot Tone, a producer, director and actor who became her co-star for a number of films. Two years later, Tone presented Crawford with an engagement ring, a 70-carat cabochon star sapphire in a platinum Raymond C Yard mount. Crawford would often be seen wearing both sapphire rings together, along with brooches clipped to her hat or 'turban' head-scarf, or on necklaces set with large gemstones.

Crawford had an impressive jewellery collection, but one particularly well-known jewel was her Raymond Yard wide Art Deco platinum bracelet, encrusted with baguette-, half moon-, marquise-shaped and brilliant-cut diamonds set with three large cabochon star sapphires weighing 73.15 carats, 63.61 carats and 57.65 carats.

Crawford was the top box-office star, a position and work expectation that put a strain on her marriage to Tone. Though their marriage ended in 1939, they maintained a friendship, and Crawford looked after him up to his death from lung cancer in 1968. Crawford had four marriages in total, three of which ended in divorce; the fourth, to Alfred Steele, ended with his death in 1959. These events did not hinder Crawford's rise to fame, which first rivalled and then surpassed that of other MGM actresses, such as Greta Garbo and Norma Shearer. Crawford clearly did not want to go unnoticed and she wore her jewellery to make an impact. She made her jewels work just as hard as she did to achieve her goals and ambitions, to make sure she was never forgotten by Hollywood producers and directors.

Joan Crawford, renowned for her love of sapphires, photographed for her title role in Otto Preminger's film *Daisy Kenyon* (20th Century Fox, 1947), wearing what appears to be one of the multiple star sapphire-set jewels in her collection.

Only a handful of gem merchants today can trace their family businesses back to the Silk Road trade of 500 years ago. In Sri Lanka, the Maricar family still runs one of these few age-old organizations, led now by gemstone dealer Salih Maricar, who recounted to me the history of his family enterprise and their gemstone, spices and tea-trading heritage.

Originally from Chennai in South India, Salih's ancestors had travelled from the Arabian Peninsula to trade horses for the spices of South India, Sri Lanka and the Indonesian/Malaysian archipelago – the traders Marco Polo encountered on his travels. In fact, the name Maricar is an anglicized version of the Tamil 'Marakkalarayar', meaning seafaring merchants, and it particularly refers to traders that ploughed the seas in wooden ships known as Arabian dhows, or *dhoni* in Tamil. Their time in Sri Lanka introduced the Maricars to the wonders of the island's gemstones, and their well-established trading connections and knowledge of the trade routes helped them to become the gemstone traders they are revered as today.

Salih's great-grandfather, Syed Ahmed Maricar, was a well-known diamond and pearl dealer, but he was particularly famed for his emeralds – indeed, he was considered the king of emeralds. He travelled to Yemen, Oman, Sri Lanka, Burma and the Far East along the sea-trading routes to source and sell stones, particularly to European merchants. Though he lost everything in the 1920s financial crash, his son, Syed Mohamed Maricar, managed to continue trading. Syed initially dealt in spices and tea, but he soon set up the Kohinoor Trading Company to focus on gemstones. He had the respected family name and was fluent in Tamil, Sinhala and English, which helped international sales. By the 1940s, Syed was a trusted dealer.

Family friend Naina Muhammad Sahib, from a Sri Lankan family of merchants and miners, asked Syed if he would handle a very valuable transaction for them: two very special star sapphires. The two star sapphires were cut from the same Sri Lankan crystal in the late 1940s; once cut, the two stones weighed 330 carats and 200 carats, respectively. The larger sapphire was reputed to be the largest sapphire in the world at the time – Syed described it as 'larger than a golf ball but smaller than a billiard ball, with a beautiful velvet deep blue and without a flaw', and said it was

probably worth more than $100,000. Syed took both stones to the United States for experts to evaluate and endorse. He also wanted to show the stones at the 1950 US International Trade Fair in Chicago to highlight the occurrence of important gems in Ceylon, as it was so often believed that India yielded better stones. The stones attracted much media attention whilst exhibited at the fair, as well as some less desired attention – luckily, plans to blow up Syed in the Congress Hotel, where he was staying, and steal the stones were foiled by the FBI and Syed left for New York with the stones.

Frederick H. Pough, curator of physical geology and mineralogy at the American Museum of Natural History in New York, inspected both stones and wrote in a letter to Syed dated 1 June 1952: 'I appreciate the opportunity to examine the two fine star sapphires that you have generously afforded us. The 329-ct round stone and the 200 ct oval stone seem to me to be unequalled among star sapphires in their combination of rich colour, star perfection, and freedom from flaws in stones of their size. You are very fortunate indeed to have two stones of such wonderful quality.'

The larger star sapphire was eventually sold to a private buyer, who later exchanged it (in lieu of tax) with the Smithsonian National Museum of Natural History, where it is currently displayed as the Star of Asia. The 200-carat sapphire was also sold, but there is no record of its buyer.

While in Sri Lanka, Syed met two brothers, Edward and William, who were gemstone traders from Wales. They had founded their company, E & W Hopkins Ltd, in 1938. Syed, Edward and William became friends and colleagues, and Syed acted as translator in Sri Lanka during their deals to export gemstones to Europe. When the brothers retired in the late 1960s, they handed their entire company over to Syed. Syed's son, grandson Salih and Salih's nephew continue to trade in some of the finest gemstones to this day. Salih remembers his childhood spent largely in his father's office, playing with parcels of sapphires of all different colours like they were pebbles: 'There were so many sapphires being found during the 1970s, and then those stones were only worth $40–50 per carat' – a far cry from their value today.

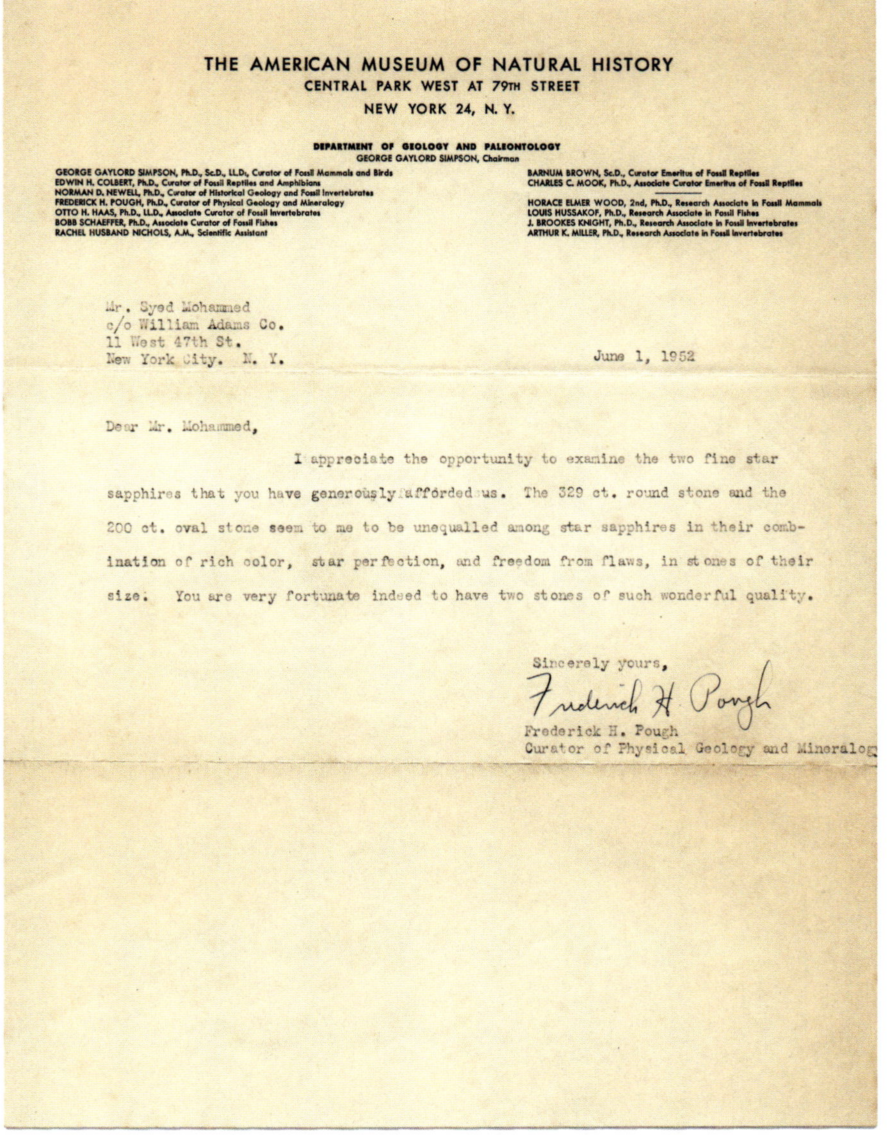

THE AMERICAN MUSEUM OF NATURAL HISTORY
CENTRAL PARK WEST AT 79TH STREET
NEW YORK 24, N. Y.

DEPARTMENT OF GEOLOGY AND PALEONTOLOGY
GEORGE GAYLORD SIMPSON, Chairman

GEORGE GAYLORD SIMPSON, Ph.D., Sc.D., LL.D., Curator of Fossil Mammals and Birds
EDWIN H. COLBERT, Ph.D., Curator of Fossil Reptiles and Amphibians
NORMAN D. NEWELL, Ph.D., Curator of Historical Geology and Fossil Invertebrates
FREDERICK H. POUGH, Ph.D., Curator of Physical Geology and Mineralogy
OTTO H. HAAS, Ph.D., LL.D., Associate Curator of Fossil Invertebrates
BOBB SCHAEFFER, Ph.D., Associate Curator of Fossil Fishes
RACHEL HUSBAND NICHOLS, A.M., Scientific Assistant

BARNUM BROWN, Sc.D., Curator Emeritus of Fossil Reptiles
CHARLES C. MOOK, Ph.D., Associate Curator Emeritus of Fossil Reptiles
HORACE ELMER WOOD, 2nd, Ph.D., Research Associate in Fossil Mammals
LOUIS HUSSAKOF, Ph.D., Research Associate in Fossil Fishes
J. BROOKES KNIGHT, Ph.D., Research Associate in Fossil Invertebrates
ARTHUR K. MILLER, Ph.D., Research Associate in Fossil Invertebrates

Mr. Syed Mohammed
c/o William Adams Co.
11 West 47th St.
New York City. N. Y.

June 1, 1952

Dear Mr. Mohammed,

I appreciate the opportunity to examine the two fine star sapphires that you have generously afforded us. The 329 ct. round stone and the 200 ct. oval stone seem to me to be unequalled among star sapphires in their combination of rich color, star perfection, and freedom from flaws, in stones of their size. You are very fortunate indeed to have two stones of such wonderful quality.

Sincerely yours,

Frederick H. Pough

Frederick H. Pough
Curator of Physical Geology and Mineralogy

World's Largest

A SILVER DOLLAR is held in her right hand by Julie Regal of Chicago to illustrate the size of a huge gem, said to be the world's largest star sapphire, held in her left hand. The gem was brought to this country by Syed Mohammed of Colombo, Ceylon, and will dazzle visitors at the U. S. International Trade fair in Chicago. *(International)*

The Star of Asia, facing page, a 330-carat milky blue Ceylon star sapphire brought to the world's attention at the 1950 US International Trade Fair, Chicago, by Syed Mohamed Maricar of the Maricar dynasty of gem dealers, Syed shown in a photograph, left, with the renowned dealers Edward and William Hopkins. Promotional photographs from the fair show the stone's size (far left), including, in a newspaper clipping, its size in relation to a silver-dollar coin (above right). The letter from the American Museum of Natural History (above left) thanks Syed Maricar for bringing the stone to the museum's attention.

As its name suggests, the Black Star of Queensland is an Australian star sapphire. Weighing in at 733 carats, this stone was the largest known star sapphire until the more recent discovery of the Star of Adam (page 178).

The large size of the Black Star makes plausible its surprising history. As a child, sapphire prospector Roy Spencer followed his father, Harry, around the Anakie gem fields of Queensland hunting for sapphires; one day, a large, dark rock caught Roy's eye and they returned home with it to use as a doorstop. It was not until 10 years later that Harry paused to inspect the stone properly, at which point he recognized it to be a huge single sapphire crystal. The Spencers sold the stone in the 1940s to American dealer Harry Kazanjian, who had recognized the tell-tale inclusions that suggested the stone's star potential. Kazanjian had the sapphire carefully cut into a cabochon to reveal its star. It was hearing Roy's story of the Black Star of Queensland that confirmed for Mike Nuell that he had made the right decision to move his young family to Australia's sapphire fields from Britain in 1969, as his wife Shirley and son Mark recounted to me (on page 303 in 21st Century and page 317 in Sapphire Discoveries).

In the United States, the Black Star of Queensland quickly gained prominence after its 'launch' into the public eye in 1948. It was celebrated on museum tours around the country and worn by celebrities; in 1971 the Kazanjian family even agreed for singer Cher (b. 1946) to wear it for her new television programme *The Sonny & Cher Comedy Hour*. Its ownership has been the subject of court battles, but now the Black Star rests in a private collection, likely no longer used as a doorstop.

The 733-carat Black Star of Queensland,
a black star sapphire set in a diamond surround
that has been worn by celebrities including Cher,
pictured above with the stone in 1971.

JEWELS OF IMPACT

Until the horrors of World War II brought the party to an abrupt end, Paris (along with Berlin) was the epicentre of hedonism, with its fancy-dress balls, café society and literary debates, not to mention the great expressions of visual and performance art, including the Ballets Russes. The Nazi occupation of France destroyed families and businesses and limited freedom of expression. Austerity, as well as restrictions on mobility, severely reduced jewellery production in Europe. Yet, with no battlefront in continental North America, trade in the United States was much less constrained. New York became a gathering place for American and European designers, who recognized its opportunities for both safety and business.

The war also had a profound effect on the availability of material. In Europe, precious metals were in short supply, so jewels were lighter in weight to use less gold, but no smaller in size – women embraced bold jewellery as a symbol of their hope and optimism. Lesser-known gemstones, including pink, yellow and pale blue sapphires, were easier to source and were used to achieve magnificent cocktail jewels at a relatively lower cost. The cocktail party had become popular in the 1920s and continued to be the social event of choice throughout the 1940s, as more people could be entertained in style without it costing a fortune. These parties were about being seen and heard, which meant fashion had to respond accordingly and jewels needed to be daring and bold.

1940s jewellery is regarded as the second stage of the Art Deco style, sometimes known as Retro Modern in recognition of its own distinctive style. The war years had generated a fascination with machinery, which became the dominant influence on jewels such as Van Cleef & Arpels' 'gas pipe' link necklaces and Mauboussin's *polonaise* link neck chain. Clasps and ingenious fittings were brilliantly engineered and this use of precious metals is beautifully illustrated in Van Cleef & Arpels' Passe-Partout jewels. Lapidary also benefited from developments in artistry, as seen in Boivin's and Belperron's carved hardstones, which allowed the stones to take on softer, curved forms that were contrasted in jewels with other textures and colours.

While the jewellery industry was still very much dominated by men, some exceptional women were gaining influence. Jeanne Boivin took over her husband's business in 1917, after his sudden death, and employed a young Suzanne Belperron, née Vuillerme, to be a sales assistant and designer. Together, they created a jewellery brand that was stylish and highly sought after. Belperron went on to make her own name in the industry and became a jeweller to the stars. Jeanne Toussaint was another fiercely independent woman who, as Cartier's artistic director, worked closely with the *maison*'s designer Peter Lemarchand to produce striking jewels that reflected its clients' desire for conversation-starting jewellery. At Van Cleef & Arpels, artistic director Renée Puissant set new, higher expectations for the creations of her craftspeople.

Facing page: A Mélange bracelet and 'bib' necklace,
1940s, of cabochon sapphires interspersed with
brilliant-cut diamonds by Suzanne Belperron, made
by Groëné et Darde for Herz-Belperron. To achieve the
effortless appearance of the colour blends, Belperron
is rumoured to have tipped the sapphires onto a table
and then carefully plotted them on the design.

The focus of this chapter is France and America, as Great Britain had barely been able to keep its jewellery trade afloat. So many of the jewellery orders in the 1940s and '50s came from America that the great Parisian jewellery houses, such as Van Cleef & Arpels, Cartier, Chaumet and Mauboussin, all maintained branches and boutiques in America. These long-established European jewellers had been used to resetting old jewels that had gone out of fashion, but this practice of remodelling jewels and seeing stones as portable wealth was unheard of in America, where buying new jewels was an important statement. For European jewellers, designing new jewels that were bold, colourful and different resulted in a far easier sell and spurred them to move with the times. New York, with its clubs and bars, and the indulgent lifestyle of its residents and visitors, combined with the influence of the Hollywood glitterati, gave designers such as Fulco di Verdura and Paul Flato a licence to design jewels that were noticeable and daring.

Dove of Peace brooch, 1939, designed by
Juliette Moutard, after an idea by Germaine
Boivin, signed René Boivin. The dove is set with
circular-cut and cabochon sapphires to
articulated wings and tail.

Above, left to right: René Boivin archive drawings
for three sparrow brooches, the first two in white
gold with faceted sapphires and diamonds and the
last with yellow gold, star sapphire and diamonds.

Though named after its male founder, the jewellery house of René Boivin was steered by a formidable woman: Jeanne Poiret was the true mastermind behind the firm's continued success.

Jeanne Poiret (1871–1959) came from a family of talented creatives. Both her sisters, Nicole Groult and Germaine Bongard, became successful designers, while Nicole's husband, André Groult, was a prominent furniture and interior designer in the Art Deco style. Their brother was the famous fashion designer Paul Poiret, who was known as a trailblazer for his role in freeing women from the constraints of the corset and voluminous petticoats. Indeed, he made his name with designs for loose-fitting clothes for a slim figure. He was inspired by the 1910s trend for Orientalism, including the fantasy world of sultan's harems, and translated these influences into his designs for harem pants, sultana skirts and the controversial kimono coat. His revolutionary shapes for clothing, combined with his use of vibrant colours, meant that Poiret was regarded as a sartorial genius.

In 1893, Jeanne Poiret married Jules René Boivin (1864–1917), an accomplished draughtsman and engraver and a talented jewellery designer. Jeanne not only helped her husband run his burgeoning jewellery business, she opened doors to avant-garde society through her family's involvement with the visual arts. Her brother's success allowed him to open his own couture house in 1903 and, instead of paying for advertising, he marketed his creations by holding extravagant and theatrical themed parties. Poiret's parties became a great opportunity for Boivin to showcase his jewels to those who appreciated creativity.

Boivin formed a close relationship with Poiret and was inspired by his fashion work to create jewels that were bold, stylish and impactful. Above all, these jewels were different. Many of the London and Parisian jewellery houses were making jewels with platinum and diamonds, which gave a white, colourless look, whereas Boivin was a great advocate of colour. He did not follow the Art Nouveau style, but was, nevertheless, inspired by nature

and his earlier work reflected naturalistic botanical themes. He ensured that many of his jewels were articulated to give movement to the design and more closely reflect nature. Boivin and Poiret gained a following of mutual clients, many of whom became friends and loyal supporters.

After Boivin's death and the death of their only son the following year, Jeanne Boivin would have been expected to fold the business, especially as a woman running any business at this time was virtually unheard of. However, at the age of 46, she decided to continue her husband's work by fulfilling the outstanding orders and creating jewels in her husband's style. Jeanne — or Madame René Boivin, as she wanted to be known — had a great eye for style. Even though she was not a designer, she had worked closely with her husband and, for 14 years, they had lived across the road from their workshop in Paris; her close proximity to the firm meant that she knew the nature of the business and understood the ethos of Boivin.

Jeanne was inspired by her late husband's and brother's use of colour, and she incorporated coloured gemstones of different cuts to play with texture and colour in her predominantly yellow-gold jewels. Her family were used to bucking the trend, and Jeanne was no different: she used materials and gemstones, such as rock crystal, wood, peridot, lapis lazuli, amethyst and tourmaline, that her clients were not used to seeing in jewellery. For Jeanne, all these elements were equally important components of the finished jewel. According to Françoise Cailles, in her history of the house of Boivin, Jeanne 'was not impressed by the ostentatious displays of gems that characterized the 1930s', recognizing instead that 'a jewel was an object with its own identity, reflecting the personality of the wearer. She liked jewels to bring out the best in women, but she also liked women to bring out the best in jewels.' With their unique combinations of materials, Jeanne felt there was no need to sign her jewellery; their individual style was enough

A Feuille brooch, 1934, by René Boivin,
unsigned, the articulated leaf motif in gold
set with circular-cut sapphires to cascading
rows of bezel-set oval sapphires.

to prompt recognition. Though her reasoning is understandable, it has made identifying Boivin pieces today that much harder. Jeanne consented to some pieces being signed, but only upon a client's specific request.

Her siblings' connections in the creative world provided initial support for Jeanne's work, but she still needed a lead designer. Importantly, this designer needed to be someone that did not yet have their own unique or established style, but that could instead echo the *maison*'s aesthetics. Jeanne hired the young student Suzanne Vuillerme in 1919, initially as a salesperson or model maker, but Suzanne quickly showed her talents as a draftswoman. Jeanne gave Suzanne the space and support to develop her designs and, for the next 13 years, their partnership enabled Boivin to grow into a jewellery business known for its distinctive and unconventional style and its eclectic, discerning clientele. Jeanne taught Suzanne how to manage the business and how to command respect from male counterparts, who still very much dominated the industry. Bringing a renewed energy to the *maison*, Suzanne worked closely with the 18-strong team of craftspeople to execute her designs. She became co-director of the firm around 1923. Even though Jeanne had developed her own style, it was during Suzanne's years with Boivin, before she departed in 1932, that the characteristic Boivin signature was cemented.

Juliette Moutard, a trained jewellery designer, replaced Suzanne in 1933 and worked with Boivin until 1970. Juliette was joined in 1938 by Jeanne and René's daughter, Germaine. This member of the next Boivin generation had worked with her uncle in the fashion world before joining the family business as a designer. Germaine continued to run the company after her mother's death in 1959. In 1976, Germaine sold the company to another Boivin designer, Jacques Bernard, who continued to produce jewels in the established Boivin style. The Asprey Group acquired the business in 1991 and, very soon after, Boivin ceased trading until the recent purchase of the company.

Ring owned by jewellery designer Elizabeth Gage, with her grandmother's 8.11-carat sapphire mounted in a domed sapphire-set ring commissioned by Elizabeth's mother from Boivin.

It is very rare today for a jewellery designer to create for more than 50 years and for their work to retain a strong following of collectors and connoisseurs. Elizabeth Gage is a jewellery designer that is part of this rare crowd. When I was with Elizabeth recently, I noticed that she was wearing amongst her jewels a ring that was not one of her own designs; when I asked who made the ring, I was delighted to hear that it was by another great female designer: Madame René Boivin.

Elizabeth told me that when her mother was young, she and her mother, Elizabeth's grandmother, travelled around India, which was then under British rule. Whilst on these travels, Elizabeth's grandmother bought this sapphire, which had apparently come from the eye of a monkey statue. Elizabeth's grandmother later passed the sapphire down to her daughter, and Elizabeth recalls accompanying her mother to Paris, when she was about 12 years old, to have the stone set by Boivin.

Sitting with Elizabeth in London for lunch between 2020 pandemic lockdowns, the sapphire ring was the main talking point. It is a beautiful stone; such an electric, intense, vibrant and rich blue that glistens when you move the ring in your hand. Weighing 8.11 carats, the sapphire is proudly set in an 18-carat gold domed mount with raised, diamond-shaped scalloped decorations, each set with a circular mixed-cut sapphire. These smaller sapphires have all been carefully matched in colour to each other and to the central stone. The ring has a wonderful weight to it. It is not signed, as is the custom for many Boivin jewels. Seeing Elizabeth arrayed in her own jewels, I could not think of a more appropriate designer than Boivin to be allowed to sit alongside Elizabeth's own creations – a compliment indeed.

A brooch, 1936–40, by René Boivin, unsigned, set with oval and cushion-shaped sapphires between a curve and radiating lines of calibré-cut rubies.

SUZANNE
BELPERRON

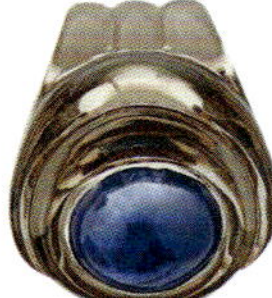

Top: Suzanne Belperron photographed
in the 1940s.

Centre: A demi-parure of a ring and
two clips of carved smoky quartz set with
cabochon sapphires, designed by Belperron,
with the maker's mark of workshop Groëné
et Darde for Herz. Belperron called the curved
design 'Bibendum' after the name of the
Michelin tyre mascot, illustrated here in
a poster from 1898 by O'Galop.

Suzanne Vuillerme (1900–1983), the celebrated creative lead at Boivin in the 1920s, was one of the fortunate young women to be supported in their ambition to work in the arts. Suzanne's mother encouraged her daughter to attend the School of Fine Arts in Besançon and agreed to her proposed move to Paris, where she began working for Boivin in 1919. She married Jean Belperron, an engineer, in 1924 and from then on worked under the name Suzanne Belperron.

As the main designer at Boivin, Belperron had the opportunity to experiment with her designs. She concentrated on abstract values of composition, exploring colour and light with structure and texture. Belperron was inspired by her surroundings and the changing environment of Paris. Advertisements and marketing campaigns would fuel her creativity. A poster by the French cartoonist O'Galop of 'Bibendum', the mascot for the Michelin tyre company, inspired her curvaceous and sculptural designs for rings and clips. She named these her Bibendum designs after the cartoon.

Blocks of lesser-known stones and materials that were not considered gem quality, including amethyst, rock crystal, rose quartz and blue chalcedony, were carved into bracelets, clips, earrings and rings. Belperron often created her stone combinations by laying gems out on the table in a palette of colour, blending pastel hues like a painter mixing oils. She was not concerned about the quality of the gems; instead, the stones would need to 'wink' to catch her attention as she turned over the irregular gems.

Through Jeanne Boivin's connections with the avant-garde visual arts crowd, the young Belperron found herself surrounded by people that promoted individuality. At this time, however, it was not customary to acknowledge a jewel's designer, only the jewellery house from which the jewel came, and it must have been frustrating for Belperron not to receive the public recognition she deserved. Belperron's designs were causing a stir within jewellery circles and she was often asked by other firms if she would leave Boivin and join them. She refused these offers until an unlikely offer was made to her from Bernard Herz, a diamond and pearl dealer, which she accepted in 1932. There is much speculation about why she left Boivin, but the most likely reason is her frustration at not receiving public recognition for her work because, as soon as she joined Herz, her name started to appear with his in *Vogue* and *Harper's Bazaar* – recognition at last.

As a well-known stone dealer, Herz would have been familiar to Belperron. Even though, rumour has it, she had been offered double to work for Van Cleef & Arpels, her loyalty to Herz was already solidified. Herz had to sacrifice his trading account with Boivin when he took on Belperron, but it was a risk he was

prepared to take as he converted from dealer to jeweller. With her departure from Boivin fixed, Belperron took the opportunity to ask several of the *maison*'s suppliers and jewellers to work for her, and many swapped their allegiance. Three polishers, two setters and her indispensable jewellers Émile Darde and Maurice Groëné, along with their workshop of 27 craftspeople, followed her move to Herz, such was their confidence in Belperron and her work.

Even though she continued in a similar vein to Boivin by not signing her jewels – she famously said, 'My style is my signature' – her jewels included the hallmark of Groëné & Darde, who made nearly all her creations and were renowned for their superb craftsmanship. Belperron would sit with the craftspeople daily, meticulously going through each piece to make sure every stage was executed to her satisfaction. Belperron was a stylish lady, known for her perfectly manicured nails, painted in red-yellow Oja polish, which she used to scoop up gems or to hold them up to the light.

Belperron had successfully put art into jewellery and had revolutionized people's perceptions of good jewellery. For the next six years, she dominated the international fashion publications, which were quick to reference the array of celebrities that were seen wearing a Belperron jewel. Fashion shoots for haute couture from names such as Augusta Bernard, Louise Boulanger, Nina Ricci and Worth were accessorized with Belperron jewels, except shoots for Jacques Fath, who was adamant that his dresses not be photographed with Belperron creations out of concern that people would look at them instead of at his dresses.

One of her most influential clients was Edward VIII who, to mark his fateful Mediterranean cruise during the summer of 1936 that brought to light his relationship with the then-married Wallis Simpson, designed and had made a half moon-shaped compact, the lid encrusted with an array of colourful gems and the bottom with a map detailing the places they had visited. Wallis Simpson later made regular independent visits to Belperron. Their patronage was early acknowledgement that Belperron was a style setter.

The outbreak of World War II marked the beginning of a new era for Belperron. When France became the focus of Nazi actions, the future of the Herz business was threatened as Bernard Herz was Jewish. Their close friendship, which had developed into a deeper involvement, was also threatened because Belperron, a Roman Catholic, came under scrutiny for her association with a Jew. To protect assets and the family, Herz entrusted Belperron with the business, and, as the Nazis started confiscating companies in 1941, Belperron incorporated a new company, Suzanne Belperron SARL, to comply with the newly instituted racial laws.

As the war continued, gold and silver were considered strategic metals, so clients had to supply the equivalent amount of precious metal to the weight of their new jewel as a form of exchange. Platinum incurred an additional tax. Belperron, who had studied during World War I, knew how to adapt to austerity and kept designing jewels to be lighter in weight.

Top: An illustration by Lady Erickson for the cover of the January 1934 edition of American *Vogue* depicting a brooch designed by Belperron. The brooch's cabochon sapphire is centrally set above tiers of carved chalcedony in the form of a fan and flanked either side by a line of calibré-cut sapphires.

Above: Sakura brooch, 1945, designed and worn by Suzanne Belperron, of one faceted and two cabochon sapphires arranged with diamonds in the form of a cherry blossom branch.

Challenges were always round the corner. Jean, Herz's son, was captured at Saint-Dié-des-Vosges and became a prisoner of war; miraculously, he managed to get someone to stand in for him at the medical examination so that his captors would not see that he had been circumcized. The Nazi occupation was getting tougher for French nationals, and, on 12 December 1941, Bernard Herz was one of the hundreds of Jews that were rounded up and taken to the internment camp near Compiègne. Belperron eventually managed to bribe her way to securing his release – just in time, as that same day the first 'special convoy' left for Auschwitz. It was in this year that Herz's wife passed away and Belperron took on more of a marital role. A few years later, in 1943, a couple and a friend visited Belperron and Herz wanting to purchase a jewel, but because they did not have the required weight in gold in exchange Belperron had to turn them away. The next day both Belperron and Herz were arrested. Belperron was released but Herz was taken to the Drancy internment camp, where he had his papers stamped 'not to be released'. Later that year, Herz was placed on a convoy to Auschwitz. Belperron did not learn of his death until after the war and, in the meantime, had joined the French Resistance.

Herz's son, Jean, was eventually free to return in 1945, but he came back to nothing: a wife who wanted a divorce, a son who did not recognize him, a mistress who had another lover, confiscated property and a family business belonging to someone else. However, Belperron had never thought of the business as her own and she offered it back to Jean. To show his gratitude he made Belperron a partner and the new company was founded: Jean Herz-Suzanne Belperron SARL. Rebuilding the company in post-war Paris was a struggle, but the difficulties turned into opportunities; indeed, Jean and Suzanne built a successful working relationship that became a 30-year partnership.

The business could have boasted of their famous clientele, including Hollywood stars Frank Sinatra and Lauren Bacall, European royalty and French artists of stage and screen François Périer and Jean-Claude Brialy, but discretion was always paramount. In 1974, Jean and Suzanne decided to close the business. Belperron was 74 years old, Jean ten years her junior; together, the families had enjoyed a partnership of 42 years. Belperron's husband had died in 1970 and she passed away in May 1983. Jean lived until 2007.

Belperron was one of the truly great designers of the twentieth century. Her unique jewels are the epitome of a strength of style that was never compromised, always pushing boundaries and setting fashions.

Three sets of jewels by Suzanne Belperron, the top two made by Groëné et Darde, the bottom by Darde et Fils, for Herz-Belperron. From top: Two branched *sakura* brooches with cabochon and faceted sapphires, respectively; a pair of double feather brooches with pear-shaped sapphires of differing hues; and a brooch and pair of earclips with cabochon sapphires interspersed with smaller cabochon emeralds.

Arlequin Zip necklace, 2013, shown
'zipped' together to wear as a bracelet,
and Colombine Zip necklace, 2013, shown here
'opened' as a necklace, both by Van Cleef & Arpels
based on the original ideas of Renée Puissant.
Both necklaces have tassels of sapphire
beads surmounted by step- and circular-cut
sapphires, respectively, to a zip edge of
sapphires and diamonds.

RENÉE
PUISSANT

Van Cleef & Arpels is a world-renowned jewellery house, synonymous with craftsmanship, innovation, style and elegance. In the early twentieth century, much of the *maison*'s innovative work stemmed from the inspiration of a member of the Van Cleef & Arpels family: Renée Puissant (1896–1942).

When Alfred Van Cleef (1873–1938) married his cousin, Estelle Arpels (1877–1960), they brought together two families of the jewellery industry. Alfred's father, Charles Van Cleef, was a lapidary and a diamond broker from Amsterdam and Estelle's father, Léon Salomon Arpels, was a gemstone dealer from Ghent. With a shared passion for gemstones, the two families joined in business in 1906, after Alfred had served six years as a jeweller's apprentice. Alfred was initially joined by Estelle's brother, Charles, to form Van Cleef & Arpels at 22 Place Vendôme, Paris, where it is still located today. Estelle's brothers Julien and Louis joined in 1908 and 1912, respectively. Louis became known as the 'marketing genius' and the firm enjoyed early success.

World War I loomed in the summer of 1914 and Alfred Van Cleef sent his wife and their only child, Renée, to the Côte d'Azur to get away from the military activities in Paris. Estelle helped in the nearby hospital and it was here that Renée met a young officer, 26-year-old Émile Puissant, who was being treated for his war wounds. They fell in love and married in 1918. When the war was over, he joined Van Cleef & Arpels as the administrative director and devised a new way of promoting sales through seasonal discounts, which had not been seen before in jewellery houses. In 1926, Émile Puissant died in a car accident. Renée never married again and devoted her short life to being the artistic director of the *maison*, which she formally joined in the same year.

Puissant was an elegant lady who knew how to wear jewels, and she brought a great sensitivity to the *maison*'s designs through combinations of gemstones. She was self-taught in design but worked closely with the designer René Sim Lacaze, who interpreted her ideas into jewels that pushed the skills of their goldsmiths to the limits. Over the next 13 years, they produced fabulous, innovative jewels. Attention to detail was paramount and clients recognized the care involved. Their jewellery was highly sought after.

The world of jewellery had never seen such creations. In 1933 they patented the Serti Mystérieux (Mystery Set), the technique of setting small square-cut stones close together so that no metal claws are visible from the front. Large wide bracelets were the fashion in the 1920s and into the mid-1930s, and Van Cleef & Arpels launched the successful *Ludo-hexagone* style, which entailed joining very small hexagonal gold elements together to make a flexible bracelet band. Each gold hexagon was star-set with a diamond or coloured gemstone. The iconic Ludo model was named after Louis because his nickname was Ludovic or Ludo. Puissant was the inspiration behind many of the Duchess of Windsor's jewels, including the Zip necklace, designed when zips were becoming an elegant feature of late 1930s fashion. Though conceived in the late 1930s, the making of the Zip necklace was delayed until the 1950s because of World War II and has since enjoyed continued success.

Renée Puissant was dedicated to her work, and very little is known about her private life. As World War II loomed, it felt prudent for the family, who were of Jewish ancestry, to relocate to the United States, where they concentrated on developing the American branches of the business. They opened boutiques in Palm Beach in 1940 and on Fifth Avenue in New York in 1942, where they used the workshops of jewellers Oscar Heyman and John Rubel. Puissant, however, decided to remain in France. In December 1942, a month after the Germans invaded the Free Zone, she took her own life at the age of 46. Knowing what was happening to others in the industry, including Bernard Herz and Suzanne Belperron, the stress and fear may have proved too much. Puissant had drafted a will, four years earlier, in which she made sure that each member of the family was named as a beneficiary.

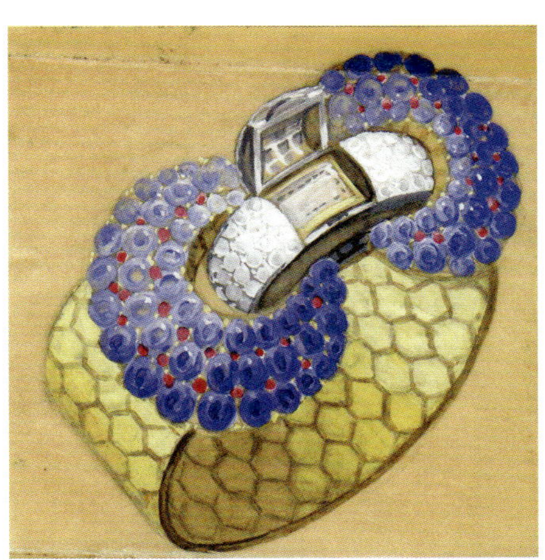

A design, 1941, by Van Cleef & Arpels for a 'secret' watch set with bands of diamonds, sapphires and rubies on the *maison*'s Ludo bracelet of interlocking gold hexagons.

The precious metals restrictions introduced during World War II prompted jewellers to swap white metals in geometric shapes for more curved and stylized floral designs in yellow gold that were lighter in design and so used less metal. Gemstones were also difficult to source, so bold, colourful, lesser-known gemstones were used to create fashionable statement jewels. These stones were not of the highest colour grade and were therefore more affordable. Women were conscious of maximizing the wearability of their jewels by having them serve several functions. Necklaces could come apart to be worn as two bracelets; brooches could separate to become two earclips or two brooch clips; and clasps could be detached to form a brooch or clips. Highly skilled technicians were needed to achieve the engineering behind these fastenings – goldsmiths were almost becoming engineers.

Van Cleef & Arpels' transformable jewels in gold and bold pastel-coloured gemstones became emblematic of the period and a trademark of the *maison*. The ingenious design for the Passe-Partout ('Go Anywhere') necklace was patented in France in 1938 and exhibited at New York's 1939 World's Fair. The jewel usually consisted of two big flower clips set with large, pale blue and yellow sapphires with ruby stamens, which attached to a flexible necklace of tubular 'gas pipe' linking. The linking could contract and expand so that the jewel could be worn as an adjustable long-length necklace, a choker, even wound around the wrist a few times as a bracelet, or indeed as a belt. The flower clips act both as clasps and as brooches when removed from the linking. Creating these jewels was a technical challenge, and Van Cleef & Arpels developed many ingenious mechanisms, including 'secret hinges' – invisible articulations that allowed a jewel to be transformed effortlessly – in order to accomplish the desired look. The flat or rounded 'gas pipe' tubular linking was very popular in the 1940s and was also used by other jewellery houses.

Fashion model Hélène Ostrowska married the youngest Arpels brother, Louis, in 1933. In 1949, she was voted one of the world's 10 best-dressed women. Many compared her with the Duchess of Windsor and she became an influential style muse for fashion houses, principally Worth and Schiaparelli. She also established her own successful luxury shoe business, based in New York. Wherever Hélène Arpels was seen, she was always wearing a Van Cleef & Arpels creation, which endorsed the *maison*'s jewels. In this photograph taken at the races in Chantilly, France, in June 1939, she wears a complete Passe-Partout parure of earclips, necklace and bracelet.

Van Cleef & Arpels' Passe-Partout necklace, with its flexible 'gas pipe' linking, is transformable to wear as a bracelet, as shown in the design, c.1940, centre, or as a belt. The pale blue-and-yellow sapphire-set flowers can be removed to wear as clips. The two necklaces above date to 1940 and 1939, respectively, the secret watch and bracelet, left, both 1939.

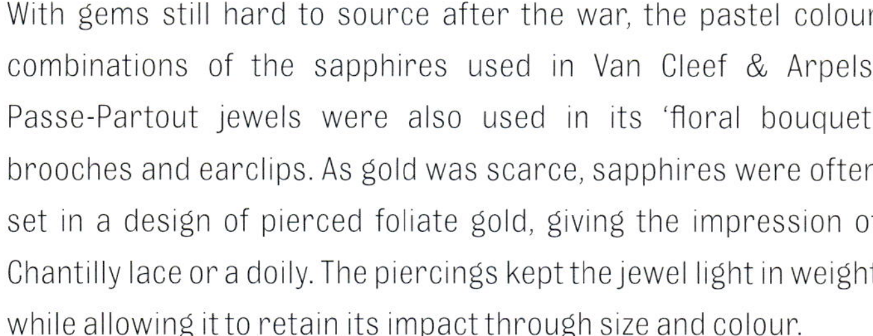

With gems still hard to source after the war, the pastel colour combinations of the sapphires used in Van Cleef & Arpels' Passe-Partout jewels were also used in its 'floral bouquet' brooches and earclips. As gold was scarce, sapphires were often set in a design of pierced foliate gold, giving the impression of Chantilly lace or a doily. The piercings kept the jewel light in weight while allowing it to retain its impact through size and colour.

Another Van Cleef & Arpels series that sometimes used the 'gas pipe' linking was the Hawaii collection. These jewels were set with small sapphires, rubies and diamonds – the colours of the French flag for wartime patriotism – clustered together as florets. Necklaces, brooches, earrings and rings had florets scattered across their surface. Fine examples of a Hawaii bracelet and brooch are featured in the chapter The Art of Collecting.

In the 1940s and '50s, animal designs joined floral designs in high popularity. These motifs were made small in size so that the wearer could play with arrangements of multiple brooches, known as 'scatter pins'. By the end of the 1940s, and continuing into the 1950s, jewels inspired by birds were very popular. Often, these jewels were set with carved gemstones from India. These jewels also had a political message: Boivin used caged bird motifs, as did Cartier, to represent the occupation of France by the Germans. To celebrate France's release from occupation, brooches featuring birds with their wings spread outside the cages appeared from the mid-1940s. Van Cleef & Arpels became known for its scatter pins and its whimsical animal brooches, which included many birds.

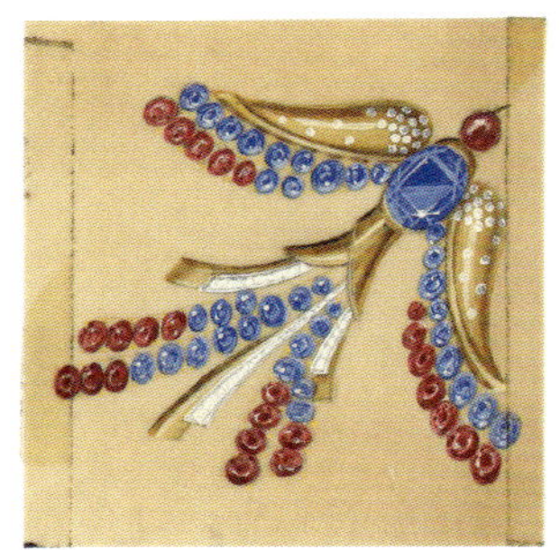

Van Cleef & Arpels' sapphire-set jewels in the
form of floral bouquets and birds. From top left:
Bouquet clip, 1940; Two Flowers clip, 1940;
Bouquet clip, *c.*1947, with sapphire petals and gold
'lace'. From top right: Inséparables clip, 1946, with
a mother and her chicks; Oiseau de Paradis clip,
1942, with calibré-cut sapphire and ruby plumes;
a flower ring, 1947. Above: Two designs for bird
clips, the left, with Mystery Set sapphires, from
the 1950s, the right from the 1940s.

BALLERINA
BROOCHES

Van Cleef & Arpels has always embraced cultural exchanges between the visual arts as these provide its designers with fresh sources of inspiration. Louis Arpels, a great ballet enthusiast, would often take his nephew, Claude, to the Paris opera house to watch the Ballet Russes in the 1920s and images of dancers appeared on Van Cleef & Arpels boxes from this time. When Claude later worked in the New York office of Van Cleef & Arpels in the 1940s, he remembered his uncle's passion for ballet, and from these memories emerged the designs of the first Ballerina clips. Van Cleef & Arpels made its first Ballerina clips in 1941 and went on to produce the charismatic jewels in both its New York and Paris workshops, as the *maison* still does today, such was their success.

The Ballerinas became an iconic signature of Van Cleef & Arpels and were highly sought after by fashionable women of the time such as Marjorie Merriweather Post, Jessie Woolworth Donahue and Barbara Hutton. Van Cleef & Arpels has made Ballerinas in both yellow gold and platinum, their skirts either sprinkled with gems or set completely with gemstones to reflect the movement of fabric. Most Ballerinas have a rose-cut diamond face.

Van Cleef & Arpels' connection to ballet strengthened in the 1960s. George Balanchine (1904–1983), the co-founder, artistic director and chief choreographer of the New York City Ballet, was introduced to Claude Arpels after World War II by a mutual friend, the celebrated violinist Nathan Milstein (1904–1992). Balanchine must have seen Van Cleef & Arpels' Ballerina jewels in the window of its Fifth Avenue store on his walks through the city and so, inspired by their mutual love of ballet, Balachine and Claude decided to collaborate. This exchange resulted in the ballet *Jewels*, in three acts: Emeralds, Rubies and Diamonds, which was created in 1967.

A Ballerina clip, 1946, by Van Cleef & Arpels,
with a skirt of sapphire-set gold discs, and
another, by Van Cleef & Arpels New York, *c.*1951,
with a large mixed-cut sapphire bodice
and a swirling skirt of diamonds.

CHAUMET

Long associated with beautifully made, naturalistic jewels of flowers, insects and birds, Chaumet was the first of the luxury jewellery houses to establish a showroom and workshops in Place Vendôme and is still there today. When Joseph Chaumet (1852–1928) died, his son Marcel (1886–1964) took over the *maison* and continued the business with the same vision and care as his father. Both these gouache drawings of necklaces date to the 1950s, when Marcel was in charge. The necklaces use blue sapphires in flower clusters in keeping with the artisans' floral talents, but the designs are a very clear response to the demand for statement jewels. Chaumet successfully moved with the times — indeed, the ability to adapt to changes in fashion over the centuries is a particular strength of this great jewellery house.

Two 1950s designs by Chaumet for necklace
with motifs of twisted gold leaves interspersed by
flowers with sapphire petals and accents of
red and green gems and diamonds.

As its name suggests, Oscar Heyman & Brothers was formed as a family business; in fact, all eight Heyman siblings joined the firm. The eldest brothers, Nathan and Oscar (1888–1970), undertook five-year apprenticeships at its great-uncle's workshop in Kharkov, Latvia (now Ukraine), where exceptional craftsmanship came as standard as clients included the great imperial jeweller Carl Fabergé.

In 1906, Nathan and Oscar emigrated to America, followed a year later by brother Harry. Nathan became a jewellery toolmaker, whilst Oscar was employed by the Cartier New York workshop, a great honour as Oscar was its first non-French goldsmith. In 1912, the three brothers joined forces to establish Oscar Heyman & Bros Inc, based in the centre of New York's jewellery district,

and soon drew their siblings Louis, William, George, Lena and Frances to the firm too. With their combined skills, they made fine jewellery for many prestigious retailers, including Cartier, Tiffany & Co and Black, Starr & Frost. Oscar Heyman & Brothers embraced all coloured gemstones and particularly favoured pastel-coloured sapphires and unconventional stone cuts, as seen in this stunning necklace and bracelet for Tiffany & Co.

In 1939, Van Cleef & Arpels entrusted the firm with making all the American Serti Mystérieux (Mystery Set) jewels, a role it maintained to the end of the century. Amidst this success, the Heyman family stayed behind the scenes and allowed the retailers to be the 'shop window' for their work, the strength of their skills and attitude earning them the moniker 'the jeweller's jeweller'.

A matching necklace and bracelet by
Oscar Heyman & Brothers, c.1960, of asymmetric
clusters of flowers with large tapered step-cut
sapphire petals, totalling approximately 400
carats, and emerald and diamond centres.

VERDURA

From top: Verdura Pleiades brooch, 1945, of seven diamond-set stars to a cabochon sapphire centre, an example of which is illustrated in a 1945 advert also featuring a Ceylon sapphire ring and a sapphire-set bracelet by Van Cleef & Arpels, complemented by Revlon's Fatal Apple lipstick and nail polish, 'blue-tinged red, wonderful with sapphires'; a 1937 photograph of Fulco di Verdura and Coco Chanel considering one of Verdura's Maltese Cross cuffs designed for Chanel.

Renowned for inexcusably bold and voluptuous jewels, Verdura is a name associated with Hollywood glamour.

Fulco Santostefano della Cerda, Duke of Verdura, was born in 1899 in Palermo, Sicily. His privileged aristocratic background introduced him to the world of art, glamorous European high society and American glitterati. His childhood was idyllic, but circumstances quickly changed: the world was at war and, like many other 17-year-olds, he was called up to serve in a depleted Italian army. Promoted to lieutenant, he was sent to the front line, where Italy suffered a massive defeat. Verdura's shoulder was badly hurt, he was discharged and sent home.

When the war was over, Verdura travelled regularly, visiting his eclectic society friends, including Cole and Linda Porter. The Porters spent their summers in Venice, and it was on one of Verdura's visits to the lagoon city that they introduced him to Coco Chanel, who was then 44 years old. According to the 29-year-old Verdura, she was 'the most chic woman I ever met'. The pair instantly connected, and he landed himself a job with Chanel as a textile designer. This career path was very unusual, as it had not been customary for Sicilian aristocrats to work for a living: but Verdura wanted to widen his horizons and leave Sicilian society behind. Chanel hired people with noble backgrounds so that they could act as her eyes and ears at all the important parties, dinners and fancy-dress balls – always being well informed was crucial to staying one step ahead of your competitors.

Chanel was impressed with Verdura's sense of style, and when she needed someone to design either her costume or her fine jewels, she never had any hesitation in charging Verdura with the responsibility. Chanel believed in wearing her fake jewels alongside her real. As cited by Patricia Corbett in her volume on Verdura, it was Chanel's belief that 'a woman should mix fake and real. …I love fakes because I find such jewelry provocative, and I find it disgraceful to walk around with millions around your neck just because you're rich. The point of jewelry isn't to make a woman look rich but to adorn her; not the same thing.' Chanel encouraged Verdura to play with gemstones of all types and colours, laying them together until the right balance of form and colour was achieved. Costume jewellery was a vehicle for introducing colour and bold design, and such jewels stood out from the 'great white silence' of 1920s diamond and platinum jewels, with their geometric lines. Inspired by the opportunity of using colour and form, Verdura transferred these ideas to the design of real jewels.

Over time, it became clear that Verdura needed to move out from under the shadow of Chanel. In 1934 he left Paris for New York, where his seemingly endless connections enabled him to meet with celebrated fashion journalist, later editor of *Vogue*, Diana Vreeland, who introduced him to the jeweller Paul Flato. Verdura started designing jewels for Flato, jeweller to Hollywood's glitterati and New York's high society, on a freelance basis. Both Verdura and Flato embraced bold, curvaceous forms with blocks of strong colour. Their early collaboration in this style was marketed as 'Verdura for Flato'. Flato never designed himself, but he knew how to convey his ideas and thoughts to his team of designers. What he was exceptionally good at was self-promotion.

It was only a matter of time before Verdura wanted to branch out on his own. Now, with an influential client list and a loan of $20,000 from his good friend Cole Porter and Vincent Astor, he opened his own company at 712 Fifth Avenue in 1939. Even though Flato and Verdura were now competitors, their relationship remained amicable.

In the 1940s, diamonds made way for colour and Verdura experimented with many different stones, his favourites being sapphire, topaz, opal and moonstone. The jeweller had always had a fascination with astrology and religion, and during the 1940s his designs were based on constellations. Elaborating on his previous designs of starburst brooches for Chanel, he created the magnificent Pleiades brooch featuring seven cabochon sapphires, weighing 54 carats in total, each set in diamond-encrusted starbursts. Even though Verdura had designed the brooch in 1941, it was not made until 1945, when he felt his American clients were ready to embrace extravagance again after the war.

The 1950s was the era for big and bold. Verdura jewellery came into its own as the mood for 'expressions of confidence' translated to the wearing of large jewels, which were no longer seen as vulgar but as a sign of celebration and prosperity. Fulco di Verdura's jewels became irresistible to stars of stage and screen since they were very photogenic and could be seen from a distance. He designed for the Duchess of Windsor, Marlene Dietrich, Joan Crawford, Greta Garbo and the socialite Babe Paley, to name but a few.

Verdura divided his time between New York and his apartment in Paris, where he now had his much-awaited salon to receive clients. He yearned to be near the Parisian craftsmen, as Paris was deemed to be home to some of the best jewellery workshops in the world. Not having a shop front on the street suited his well-known clients, as they appreciated anonymity and discretion. Verdura retired to London in the 1970s and, after writing his autobiography, passed away in 1978.

Mid-century pieces by Verdura: Raja ring, 1958, set with a Ceylon sapphire in a surround of carved turquoise and pavé-set diamond claws; Hen brooch, 1965, in the form of a diamond-set bird cradling cabochon sapphires; Aloe brooch, with sapphire spikes and emerald-set leaves 'growing' from a star sapphire; and X necklace, *c.*1949, with cabochon sapphires intertwined with crossed ribbons of baguette-cut sapphires and diamonds.

After Louis Comfort Tiffany stepped down as art director of Tiffany & Co in 1918, the company faced a tough period. It battled through the Great Depression of the 1930s and, even though sales were few and far between, its designs were successful in using inexpensive, but still colourful decorative stones. However, there was naturally little interest in the decorative arts and jewellery in the immediate post-war period, and the company's net income fell. After more than 100 years of family ownership, Tiffany & Co was sold to businessman Walter Hoving in 1955. He recognized that Tiffany had lost its trademark style, and so he immediately employed the former president of Parsons School of Design, Van Day Truex, as design director. Soon after, Hoving saw an opportunity to engage the well-known and talented jewellery designer Jean Schlumberger, inviting him to join Tiffany & Co in 1956.

Jean Schlumberger (1907–1987) was from a wealthy Alsatian family whose business was in textile manufacturing, and there wan an expectation that he would develop a career in the financial sector. Instead, he was drawn to the artistic and glamorous life of 1930s Parisian café society. Once Schlumberger started to design jewellery, he attracted the attention of fashion designer Elsa Schiaparelli. His flair for integrating unusual materials into his asymmetric designs excited Schiaparelli, and he started to make trend-setting fashion jewellery for Schiaparelli. World War II loomed and Schlumberger joined the armed forces but, when the war was over, he went to New York and set up his own jewellery business. His connections among the beau monde of Paris gave him an entrée into the upper echelons of New York society, where he soon became familiar with people such as Diana Vreeland, Elsa Maxwell, Rachel Lambert 'Bunny' Mellon, Gloria Guinness and later Jacqueline Kennedy and, pictured here wearing a famous brooch, Elizabeth Taylor. No fashioniable woman's outfit was complete without a Schlumberger jewel.

With access to Tiffany's amazing gem collecton, Schlumberger's imagination was free of any restraints. His whimiscal interpretation of nature entwined with his love of colour, his wit and elegance, and his use of different materials, resonated in his jewellery designs. Through Schlumberger's work, Tiffany & Co found itself once again among the foremost jewellery houses.

Schlumberger's Thorns clip, made in 1947 for oil fortune heiress Millicent Rogers, with its manufacturing card, top, detailing that the 46.60-carat sapphire and the 14 yellow diamonds may have been provided by Rogers for the jewel.

Hollywood icon Elizabeth Taylor (1932–2011) added multiple Schlumberger jewels to her enviable jewellery collection. The Fleur de Mer brooch was just one of Taylor's collection of sapphire jewels (as also featured in the chapter High Society). Taylor's first Schlumberger jewel was a diamond-set Dolphin brooch, gifted to her by her soon-to-be husband, Richard Burton. Burton intended this gift to mark the 1964 opening of his film *The Night of the Iguana*, based on the play of the same name by Tennessee Williams, as Taylor had joined Burton in Mexico for its filming. Taylor affectionately referred to the brooch as her 'iguana' and the couple considered this brooch a celebration of their early love. Taylor later recalled that Burton had wanted another reminder of this time – a new Schlumberger jewel was the only option. Taylor chose the Fleur de Mer brooch (illustrated here) by Schlumberger for Tiffany on a 1965 visit to the Schlumberger Salon at Tiffany's Fifth Avenue flagship showroom. Schlumberger had designed the brooch in 1956 and it was made by the atelier of Henri Picq in Paris. The flower's petals and raised centre are set with sapphires and diamonds. Taylor also referred to this brooch as her 'orchid' and wore it in multiple different styles, including as a hair ornament and pinned to a sash around her waist.

Taylor is renowned, to this day, for her romances, all of which she marked with jewellery. She was engaged 10 times, though only seven of these engagements were sealed with an engagement ring (Eddie Fisher proposed with a bracelet, whilst Richard Burton proposed twice without a ring, though he is known for being the biggest donor to her jewellery collection). Of these seven rings, three were set with sapphires, including a sugarloaf sapphire set in a diamond-set surround from fiancé Michael Wilding, whom she married in 1952.

The Fleur de Mer brooch designed by Schlumberger for Tiffany in 1956 and made with sapphires and diamonds by Henri Picq in Paris, chosen by actress Elizabeth Taylor in 1965 as a gift from her husband Richard Burton. Taylor is shown here wearing the Fleur de Mer brooch in a photograph by Ron Galella, arriving in New York on 19 August 1968 aboard the *Queen Elizabeth II* with Burton and their daughters.

Mauboussin was founded as a traditional Parisian family jeweller in 1827; it is quite remarkable that, more than 100 years later, the business was not only still operating but designing and making contemporary jewels in its own workshops. The firm branched out to the Unites States and in 1936 merged with American company Trabert & Hoeffer, becoming Trabert & Hoeffer-Mauboussin. They set up in New York, and between 1936 and 1953 created jewels that adorned actresses and socialites on and off the screen either side of the Atlantic. In Paris, René Sim Lacaze, a talented designer who also designed for Van Cleef & Arpels, created styles that complemented the fashions of the time. In 1946, the *maison* moved to Place Vendôme, Paris, and for the first time had a shop window.

This demi-parure of a necklace and bracelet is a fabulous example of a statement piece of the 1940s that is also transformable. Each of the two jewels is composed of a *polonaise* chain with hexagonal links, which supports clusters of cones set with circular mixed-cut sapphires and brilliant-cut diamonds, with similar-set tassels. The central motifs on the necklace and bracelet can be detached and worn as brooches. It is signed Mauboussin, with maker's marks for the workshop of Verger Frères.

Front and reverse views of a demi-parure, *c.*1940, by Mauboussin. The necklace and bracelet consist of hexagonal links with detachable motifs of gold cone clusters and sapphire tassels, which can be worn as brooches.

HIGH
SOCIETY

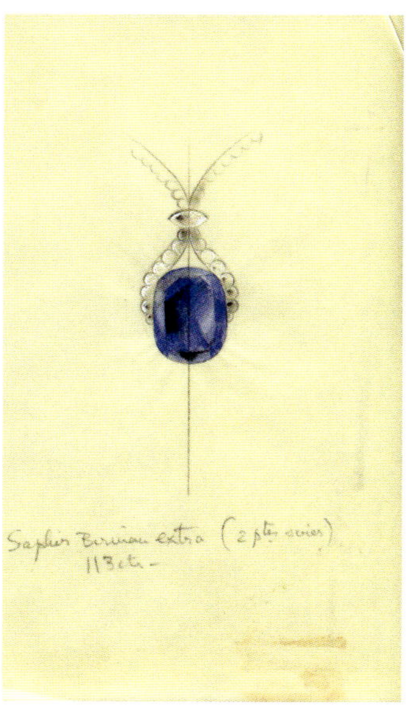

The 114.30-carat Burmese sapphire, known as the Blue Princess, bought by Claude Arpels on one of his gem-hunting 'safari trips' to his wealthy clients in the Indian subcontinent. The design, left, by Van Cleef & Arpels, c.1955, shows the Blue Princess in its original necklace setting with diamond-set scrolls supporting the sapphire from a horizontally set marquise-cut diamond, as worn by a model in a photograph from the period. The necklace was bought by Florence J Gould (1895–1983), photographed here in 1956 arriving in New York from the Mediterranean, a jewellery collector married to a railroad tycoon, known as much for her notorious political connections as for her art patronage.

When looking through catalogues of jewellery collections once owned by famous heiresses, philanthropists and Hollywood icons of the twentieth century, there appear to be key common denominators: quality and superior craftsmanship are the most obvious ones, but also wearability, noticeability and, perhaps above all else, originality. Jewellery is expected to work hard to keep up with its custodians. Collections had to reflect their owner's standing and importance in the circles they moved in and, importantly, needed to complement their individuality and dress. Just as these collectors helped to drive jewellery trends, so the jewels themselves played a significant role in defining their owners. From an imposing single sapphire like the Blue Princess (facing page), once owned by Florence J Gould, to a sapphire-spotted panther sold to the Duchess of Windsor, or a brooch set with carved sapphires, each jewel was more often than not a work ahead of its time and made its presence felt, just like its owner.

Statement jewels and jewellery collections were once mandatory for those wanting to make a mark in high society — a trend that seems to have died off, at least for the time being. Notable figures who owned incredible sapphire jewels include Daisy Fellowes, the Singer sewing machine heiress (featured in this book's New Horizons chapter), whose jewellery collection has been noted as one of the twentieth-century greats; Marjorie Merriweather Post, the Postum Cereal Company heiress, who had a particularly discerning eye for artistry, heritage and craftsmanship; and Elizabeth Taylor, whose collection broke five world-record prices for pearls, diamonds, rubies and Indian jewellery when it was sold at Christie's in December 2011. This posthumous sale realized a total over US$115 million, the highest to date for any private jewellery collection, beating the long-held world record set by the Sotheby's sale of the Duchess of Windsor's collection. That sale, held in Geneva in April 1987, achieved £30 million, six times the expected figure.

Princess Margaret was a great champion of British goldsmiths. Her collection, which was sold at Christie's in 2006, included jewels made by some of the great British jewellers of the 1960s and '70s, including the late John Donald, whose work is avidly collected today. Another great British designer of the period was Andrew Grima, who repeatedly pushed jewellery design to a level that revolutionized the British jewellery industry. Grima won many awards, and some of his pieces were in the private collection of Queen Elizabeth II.

If these jewels could do the writing, they would open up a world of intimacy that we can only begin to imagine. Old secrets are safe, while the jewels continue to accumulate more memories with their new custodians. Jewellery collections once owned by interesting individuals are often sought after for their provenance — owning a jewel with an exotic provenance allows the new owner to feel a part of that legacy (Merriweather Post herself owned diamond earrings that had belonged to Marie Antoinette). But attitudes to legacies can change over time and their appeal dwindle. The exquisite jewels in this chapter are of such rare and exceptional quality that they will always be desirable to discerning new buyers, irrespective of their heritage.

When Wallis Simpson (1896–1986) married the Duke of Windsor (who had briefly been King of England but was forced to abdicate when he chose love over duty), the couple went into exile in Paris. They lived in a 30-room mansion in the Bois de Boulogne, for which the couple paid a minuscule rent to the City of Paris after World War II. There had never been any suggestion that they would ever be allowed to return to live in England. The Duke, who felt that his wife had been denied royal status, wanted to buy her jewelled gifts fit for a queen. The collection was nicknamed 'the alternative Crown Jewels'. The Duchess of Windsor religiously wore her jewels to every social occasion, but, after being so visible, these jewels are now hidden away in private collections around the world. Through photographs of the Duchess and the designs for her jewels, we are still able to imagine the impressive and powerful character of this 'royal' collection.

At the beginning of World War II, and undeterred by the crisis facing France just before its fall to Nazi Germany in 1940, the Duke visited Cartier in Paris carrying with him a necklace and four bracelets belonging to the Duchess that he wanted remodelled. The result is regarded as an iconic jewel: the Flamingo brooch (page 16), made with an articulated standing leg so that the piece would not dig into the Duchess when she moved. Though the Duke had commissioned jewels from Cartier before, this creation marked the beginning of the couple's more spectacular Cartier commissions. The Duke of Windsor had favoured the jewellery house Boucheron when he was with his previous love, Freda Dudley Ward, but this situation soon changed as the Duchess, not wanting any reminders of his past loves, requested he change his allegiance to Cartier.

Notable sapphire-set Cartier jewels in her collection included a necklace, c. 1940, of sapphire beads suspending nine large flower heads of cabochon sapphires; a bracelet, c. 1945, of oval links set alternately with light and dark blue sapphires; a clip made in Paris, 1949, in the style of looped ribbons of oval faceted sapphires joined by a flower rosette of contrasting light and dark blue sapphires (the Duchess is pictured wearing this brooch on page 214); a ring made in 1949 and set with a 46.60-carat light blue sapphire in a surround of royal blue sapphires; a pendant set with a 206.82-carat cushion-shaped sapphire in a border of diamonds, made in Paris in 1951; and a pair of earclips with diamond-set Prince of Wales's feather motifs, each suspending a sapphire drop originally purchased from Harry Winston in 1947 and believed to be redesigned by Cartier.

As a perpetual outsider, the Duchess found a kindred spirit in Jeanne Toussaint, who was never allowed to marry her lover Louis Cartier but who had become, in 1933, the artistic director of Cartier Paris. The two women struck up a friendship that would be of great benefit to Cartier, as the Duke and Duchess subsequently came to be counted among Cartier's most important clients. The Duchess and Toussaint were reputedly very similar in character: they were both resilient and knew what was needed to remain at the top of society; they both had an eye for detail; and, having both come from humble backgrounds, they knew what it was like to be the underdogs and had no intention of ever revisiting that life again – the Duchess is famous for saying, 'You can never be too rich or too thin.' The Duchess of Windsor was always in competition with Daisy Fellowes when it came to being voted the best-dressed woman by fashion magazines, a contest that has never been settled.

Cartier's brilliant designer Charles Jacqueau had worked closely with Louis Cartier for 20 years from 1909 and was one of the pioneers of the Art Deco style but, after Louis's death in 1942 and Jacqueau's return to Paris in 1949, Jacqueau refused to work with Toussaint. He continued to design specific commissions for clients, but times had moved on, and the work of another talented designer, Peter Lemarchand, who had been hired in 1927, now found particular favour with Toussaint. Lemarchand and Toussaint shared a love of animals and birds, and, when Jacques Cartier sent him to India, he was greatly inspired by the wildlife he saw. His close views of tigers and panthers became the catalyst for Cartier's iconic panther jewels. Big cats had fascinated Jacques Cartier since his first visit to India; his great-granddaughter, Francesca Cartier Brickell, related in her book on her family how Jacques, years earlier, while reading Kipling's *Jungle Book* to his son, had paused at the illustration of Bagheera the panther being chased by a bear and later made a note in the margin saying, 'without the bear'. These early inspirations, coupled with Jeanne Toussaint's nickname 'PanPan', coined probably from her time in Africa in 1913 with her then-lover Pierre de Quinsonas, and the big cat skins that decorated her Paris apartment, culminated in the birth of the Cartier *panthère*. It was an obvious progression for Toussaint and Lemarchand to build on the ideas in Jacques's drawings and in Louis's first designs of panther motifs, one of which was on a vanity case for Toussaint in 1917, others on jabot and tie pins.

In 1948 the Duke of Windsor visited Cartier Paris again. This time he brought with him a cabochon emerald, which the Cartier team transformed into a brooch surmounted by the sculptural form of a seated gold panther decorated with spots of black enamel. This jewel sparked an interest in panther jewels and Cartier continued the theme by making for stock another panther brooch. On this brooch (shown on page 215), also bought by the Duchess of Windsor, the panther sits upright on an enormous 152.35-carat blue sapphire, probably from Sri Lanka, pavé-set with brilliant-cut diamonds with sapphire spots and yellow diamond eyes. The Duchess eventually had 12 cat jewels, a collection that stirred a huge appetite for such jewels among high-society women. It is not surprising that these exclusive jewels were so coveted as they represented not only individuality but also female empowerment and paid homage to a strong independent woman with a mind of her own – in French, 'panther' is a feminine noun, so the association between panthers and feminine mystique is one that Cartier has never questioned.

The Duchess of Windsor's 'marriage contract'
Jarretière bracelet by Van Cleef & Arpels,
gifted to her by the Duke of Windsor to wear
on their wedding day in 1937, shown here in
the product card with the original design with
additional sapphires. This flexible bracelet of
baguette- and circular-cut diamonds meets at a
domed clasp set with cushion-shaped sapphires;
it is inscribed 'For our Contract 18-V-37'.
The Duchess wore this bracelet on many
occasions from the late 1930s, above, including
to a ball held in the couple's honour at the
Waldorf Astoria, New York, in 1953 (left).

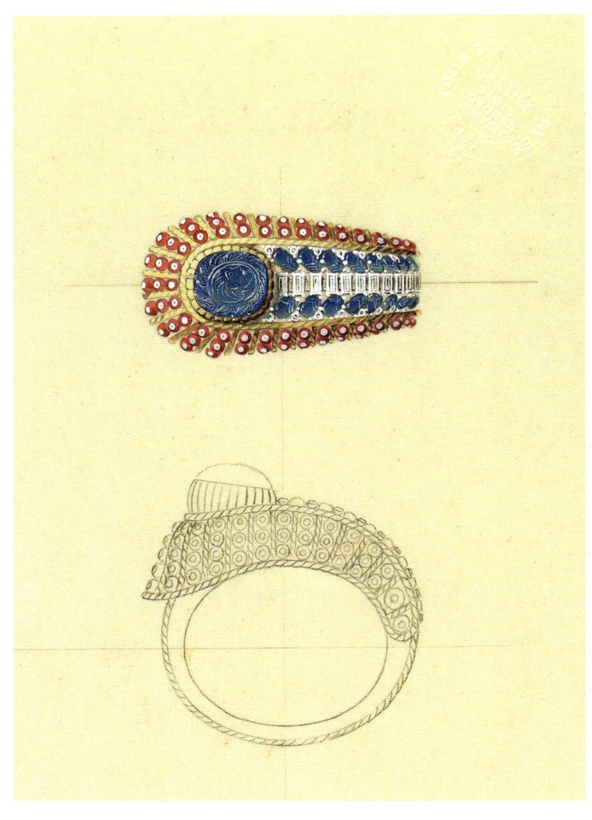

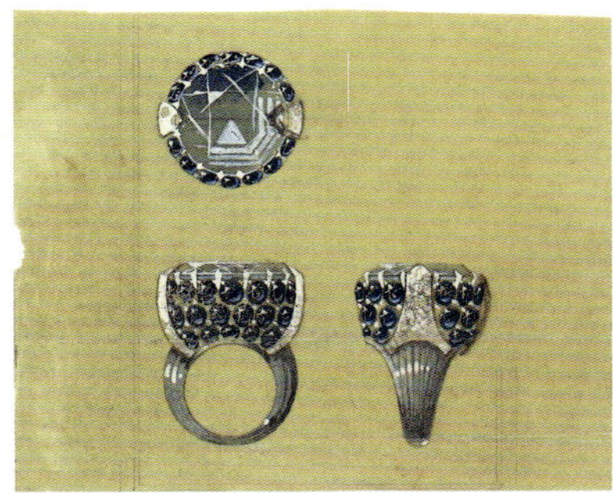

Two Cartier designs for the Duchess of Windsor.
A 1949 design (above) by Cartier Paris for a ring
made with a large, pale blue circular-cut sapphire,
encircled by a gallery of darker blue sapphires and
diamond-set shoulders; and a 1946 design (left) for
a bracelet, the clasp set with a carved sapphire
depicting a woman in profile, to a bracelet of carved
sapphires, ruby beads and diamonds.

The Duchess of Windsor greets Queen Elizabeth II,
the Duke of Edinburgh and Prince Charles at the
Villa Windsor near Paris on the Queen's state visit
to France in 1972. The Duchess of Windsor wears
her impressive clip by Cartier Paris, 1949,
of sapphire-set ribbons and a rosette cluster.

Such was the success of the panther and demand for cat jewels that Cartier has produced a new design or panther motif every year since 1914. Today, there is a team of designers whose exclusive job it is to design panther jewels. Every design is reviewed by the 'creation committee', consisting of the CEO, the director of image and style, the head of the design team and the workshop manager. It is imperative that a panther jewel is easily recognizable as a Cartier piece and that the 'mood', or expression, of the cat is right. There must be a fluidity to a panther jewel, which is conveyed not only through its articulation, but also by how its fur is depicted. There is a special setting technique – called *serti 'pelage'* – that is reserved for the setting of a panther jewel and that reproduces the look of fur and the way it lies. The technique uses tiny grains of precious metal to hold the small gemstones in place, whether they are diamonds, sapphires or black onyx, and the grains are then split and curved to create the eyelashes and to evoke the impression of the animal's coat, emphasizing the character and mood of the animal. The spots are equally important as they help to give the illusion of volume, as well as the feeling of movement; to achieve this, larger spots are placed around the more prominent parts of the body, graduating to smaller spots in proximity to the white belly. Each spot is individually cut by in-house lapidaries, and it can take them up to an hour to achieve the desired shape. Traditionally, the eyes, which determine the mood of the panther, were set with either emeralds or green garnets, but recently other coloured stones have been used. Each jewel takes months to complete.

The Duchess of Windsor's sapphire-spotted panther clip brooch, 1949, the second three-dimensional Cartier panther owned by the Duchess, and its design by Cartier. The panther rests proudly on a 152.35-carat sapphire cabochon. The Duchess wears the panther here, as she leaves Claridge's Hotel with her husband, the Duke of Windsor, in June 1967.

In 1914, at the age of 27, Marjorie Merriweather Post (1887–1973) inherited the Postum Cereal Company worth US$20 million and became one of the wealthiest women in the United States. Attitudes at the time prevented women from officially running companies so, upon her father's death, Marjorie instructed her husband, Edward Bennett Close, to join the company's board, and it was through him that she shared her opinions and advice. Marjorie had to wait until 1936 before becoming the company's first female director.

As Marjorie was the family's only heir, her father, Charles William Post, ensured that his daughter developed an astute business brain from a young age, a strategy that proved invaluable to the further success of the company. While helping to run the company, Marjorie filled her life with social engagements and indulged her interest in collecting eighteenth-century furnishings — which included decorating Burden Mansion, her residence in New York, in the neoclassical style of Louis XVI, which was very fashionable at the time. As a practising Christian Scientist, like her father, the philosophy of giving to others became very much part of her life. Her father had taught her that money can be the root of all evil; instead, it must be used productively, to help people and support education and the arts.

In 1919, Marjorie and Edward Close divorced. Marjorie then married Wall Street financier Edward F Hutton, who helped turn the Postum Cereal Company into General Foods Corporation, which became a huge success. Meanwhile, Marjorie's collecting habits were developing, and her connoisseur's eye helped her acquire fine Sèvres porcelain, French furniture and exquisite objets d'art from Cartier and Fabergé. But following the stock market crash of October 1929, Marjorie remembered her father's maxim that wealth is a privilege best used to help others. Marjorie financed the Marjorie Post Hutton Canteen and began raising funds from hosting luncheons and teas. These events were a big change from the lavish parties she threw in her earlier life.

Her second marriage was not to last, and she divorced in 1935, only to be married again that same year to a Washington lawyer, Joseph E Davies, who soon became the American ambassador to the Soviet Union. This position allowed for a trusted relationship to build between President Franklin D Roosevelt and the Soviet government, and Marjorie and Joseph gained a special access to people, art collections and travel that was only granted to the diplomatic community. This association allowed Marjorie to further her passion for collecting, and she turned her attention to Russian art, an interest she maintained for the next 36 years. Over this time, she amassed the most comprehensive collection of imperial Russian art outside of Russia. As well as retaining her flair hosting lavish and elegant parties and balancing social engagements, Marjorie had become well versed in politics and diplomacy, which prepared her for what awaited the couple when they returned to Washington shortly before the beginning of World War II. Marjorie continued her philanthropic commitments throughout the war years.

Tensions were mounting between her and Davies and, eventually, due to his ill health and falling status, they divorced in 1955. Her fourth husband, businessman Herbert A May, joined the family in June 1958, but that marriage too ended in divorce in August 1964. After this final divorce, Marjorie chose for her name to revert to Marjorie Merriweather Post.

Merriweather Post had enjoyed being based in Washington during Davies's time as ambassador and had bought a house there, Hillwood, which served as her primary residence in the spring and the autumn for the remainder of her life. The great and the good, politicians and businessmen, would never refuse an invitation to a formal dinner at Hillwood. All the lavish functions, magnificent parties and formal occasions meant that her jewels needed to be appropriate for each event, whether impressive or understated. Her astute eye for detail had been finely trained over the many years of collecting artefacts, and she treated jewellery with the same level of scrutiny — her jewels had to be of the highest calibre.

A 1946 portrait of Marjorie Merriweather Post by Frank O Salisbury in which she is depicted wearing her impressive Cartier sapphire and diamond necklace, illustrated on the facing page, paired with sapphire-set earrings, a clip and ring.

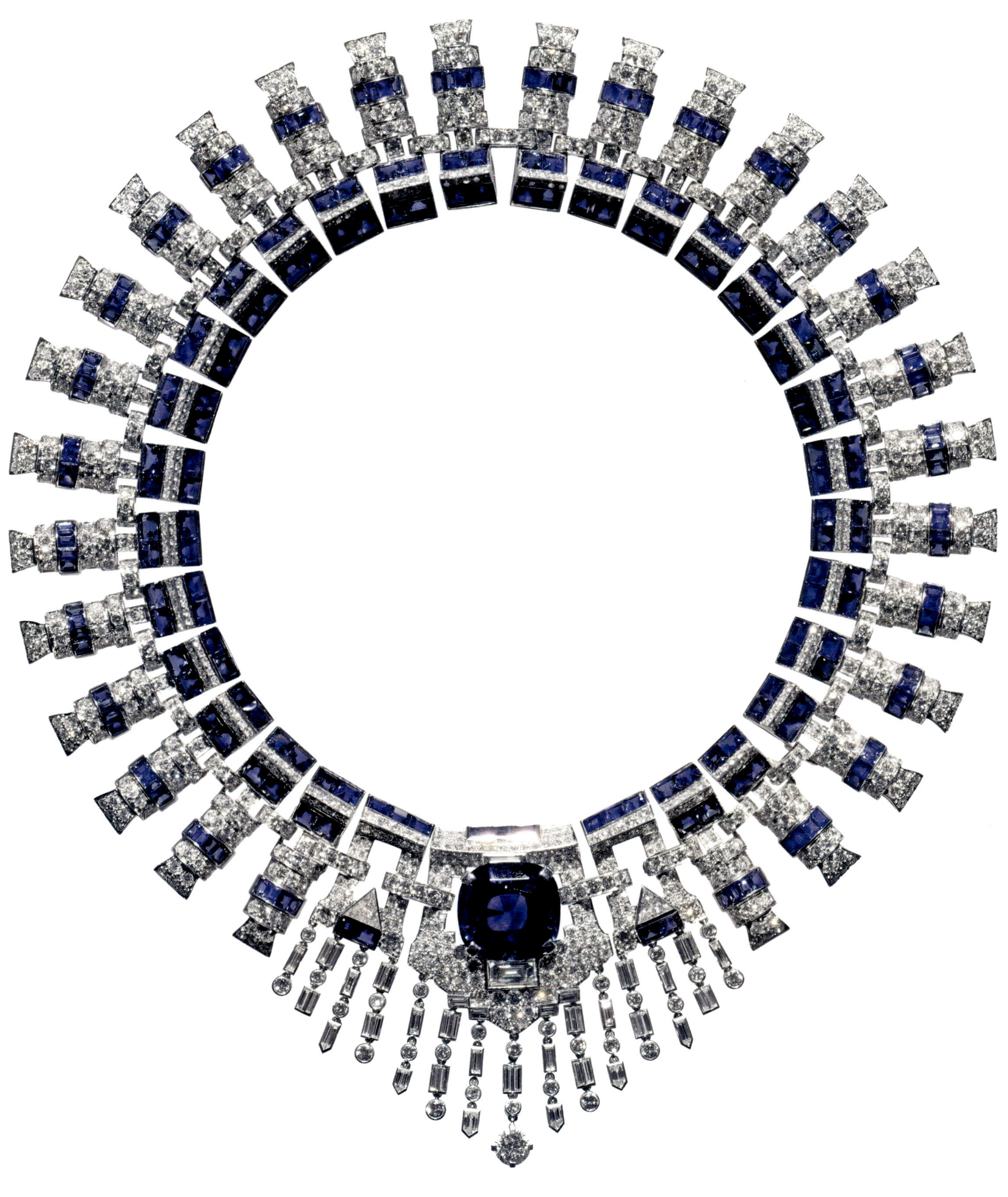

A sapphire and diamond necklace, 1936,
created by Cartier, New York, under the
instruction of Marjorie Merriweather Post from
two bracelets in her collection. The central
motif, set with a cushion-shaped sapphire,
is detachable and can be worn as a brooch.

TRABERT & HOEFFER-MAUBOUSSIN AND JUNE KNIGHT

Trabert & Hoeffer-Mauboussin was a successful American jewellery company. Established in 1926 as Trabert & Hoeffer, the business began as a retail shop that imported gems and jewellery before becoming manufacturing jewellers. Starting a business between the wars was a challenge, but the team at Trabert & Hoeffer-Mauboussin had the creativity, design innovation and clever publicity and marketing not only to survive but to become a house whose jewels were frequently worn by stars of the silver screen.

Founded by two New York jewellery salesmen, Randolph J Trabert and William Howard Hoeffer, this partnership found its early success abruptly interrupted. The market crash of 1929, as well as the sudden death of Randolph Trabert, left Hoeffer rethinking his future strategy. With so many New York families having lost their fortunes, Hoeffer realized that some of the few people who could still afford fine jewellery were the rising stars of Hollywood, so he opened a branch in Los Angeles. Here, he pioneered the practice of lending jewellery to be worn on screen, which resulted in Trabert & Hoeffer being the first jewellery *maison* to receive a credit in a film. Hoeffer aligned his firm with the venerable French jewellery house Mauboussin, which was itself still feeling the effects of economic disaster – a partnership that also helped the French house maintain a presence in the United States while trading in Paris was halted by the war. This mutually beneficial partnership lent Trabert & Hoeffer enormous prestige.

During this time of partnership, which lasted until 1953, Trabert & Hoeffer-Mauboussin manufactured – to great acclaim – jewels that could be used and worn in several ways, such as those in its Reflection series. Hoeffer maintained a lifelong passion for precious gemstones and enjoyed selecting stones specifically to be used in sumptuous, jewelled necklaces that could be taken apart and worn separately as bracelets, pendants, brooches and clips. These so-called Jigsaw jewels, in which the clasps and connecting devices were cleverly hidden, were great examples of micro-engineering. Only about 12 of these innovative pieces were made, one of which was purchased by Broadway and film star June Knight (1913–1987) in 1936. Knight's necklace, set in platinum with baguette- and brilliant-cut diamonds, used carved Indian sapphires that gave the jewel great colour without costing a fortune. Knight could wear the jewel complete as a necklace, or the pendant could be removed to be worn as two clips. The adaptability of its design made this jewel even more desirable because of the financial constraints experienced by many jewellery collectors of the time.

Above: June Knight in 1935, with Tommy Lee, son of Don Lee, auto dealer and broadcasting system operator.

Facing page: June Knight's diamond and carved-sapphire Trabert & Hoeffer-Mauboussin Jigsaw necklace, late 1930s, so called because the pendant can be detached and worn as two clips. Many Jigsaw necklaces could be disassembled and worn as bracelets.

HARRY WINSTON AND
EVELYN ANNENBERG HALL

Some of the most important gemstones in the world have passed through the jewellery house of Harry Winston. Harry Winston (1896–1978) was an astute businessman who did not just know about gems, he loved them and he could excite the same passion in his clients when he showed them a special stone.

Winston entered the jewellery trade when he was 15 years old by working in his father's jewellery shop in Los Angeles. It was a small yet steady business, but Winston had grand ideas. He went to New York, the heart of the North American jewellery industry, and started to make a name for himself. It was not long before he started to buy estate jewellery – jewels sourced directly from previous owners or their estates – and very soon he was buying and selling significant gems, including the famous 337-carat sapphire of Catherine the Great, and resetting them in modern mounts. To this day, he is renowned as having been able to look at a stone and understand immediately its potential, though he often felt sad when he parted with special stones. Winston became jeweller to the Hollywood stars of stage and screen – indeed, in *Gentlemen Prefer Blondes*, 1953, Marilyn Monroe exclaims, 'Talk to me, Harry Winston! Tell me all about it'.

This striking platinum, sapphire and diamond fringe necklace was designed by Harry Winston, Inc, using 36 beautifully matched, graduating cushion-shaped Ceylon sapphires, weighing a total of 195 carats. The jewel was owned by the newspaper heiress, philanthropist and connoisseur of the arts Mrs Evelyn Annenberg Jaffe Hall (1911–2005). As one of seven girls with only one brother, Walter, Evelyn and her sisters were expected to marry well, whilst Walter took over their father's newspaper business and re-established the Annenberg name in magazine publishing (he also became the US ambassador to the United Kingdom). All eight siblings created a new family legacy of philanthropy and patronage – an endeavour that was helped by their position in the fashionable worlds of New York and Palm Springs.

With her second husband, William Jaffe, Evelyn became a great collector of art and artefacts from around the world; the chandeliers, furniture and objets d'art from her New York mansion at 640 Park Avenue were auctioned by Christie's in a special sale in May 2021. Throughout her life she represented many art institutions as a trustee and made numerous gifts to their collections, including works by Edvard Munch, Pablo Picasso, Joan Miró and Pierre Bonnard. Perhaps inspired by their sister Janet Annenberg Hooker's donation of the Hooker emerald brooch (as illustrated in *Emerald*, page 79) to the Smithsonian Museum in 1977, sisters Evelyn and Lita (Mrs Annenberg Hazen) donated their respective sapphire and diamond Harry Winston necklaces to the museum in 1979. Evelyn's sapphire necklace is on display at the Smithsonian National Museum of Natural History and sits alongside the Logan Sapphire and the Bismarck Sapphire Necklace. At her death in 2005, Evelyn divided her fortune between family and medical and art institutions.

Above: Leonore Annenberg (left), former US President Dwight D Eisenhower and Evelyn Annenberg Hall (right) at Walter Annenberg's Sunnylands estate in Rancho Mirage, California, *c.*1968.

Facing page: Evelyn Annenberg Hall's Harry Winston necklace of Ceylon sapphire and diamond clusters, now on display at the Smithsonian's National Museum of Natural History.

'Jewels were a source of pure happiness for her and she loved wearing them, because she could share with others their magic powers of joy and excitement.' – Richard Burton on Elizabeth Taylor, as recalled by Paolo Bulgari, president of Bulgari.

For Elizabeth Taylor (1932–2011), jewels were an extension of her being. Each of her jewels had a personal story or message, and never was anything added to her collection without her knowing about its history. These jewels collectively help tell the story of a woman who not only mesmerized her audience but lit a passion in seven men; it was as thrilling for her suitors to witness her reaction when they presented her with splendid jewellery as it was for her to receive such gifts.

Taylor's collection featured jewels from almost all of the most celebrated twentieth-century jewellery houses, including Van Cleef & Arpels, Cartier, Boucheron, JAR and Schlumberger, but it was Bulgari that seemed to attract her special attention. Her relationship with the Italian jewellery house grew from her first visit in 1962 while she was filming *Cleopatra*. In her book *My Love Affair with Jewellery*, Taylor explains: 'Undeniably, one of the biggest advantages to working on Cleopatra in Rome was Bulgari's nice little shop. I used to visit Gianni Bulgari in the afternoons and we'd sit in what he called the "money room" and swap stories.' It was also while she was on set that she fell in love with her on-screen Mark Antony. Paolo Bulgari, writing in the catalogue accompanying the sale of Taylor's jewellery, remarked: 'Later, Burton would say that "the only word Elizabeth knew in Italian was Bulgari"'.'

Bulgari was founded by a young Greek silversmith, Sotirios Boulgaris (later Italianized to Sotirio Bulgari), in 1884 on Via Sistina in Rome. In 1894 he opened a shop on the fashionable Via dei Condotti. By 1915 Bulgari had established itself as a jeweller's, appearing in that year's *Guida Monaci* under the heading 'Jewels, Diamonds, Pearls and other Precious Gemstones'. In 1908 Sotirio's son, Giorgio, joined the company. Giorgio travelled to Paris and witnessed the interest in the new white metal — platinum — which, along with the 'garland' style, had become highly fashionable. Bulgari made its jewellery both in Rome and Paris in the 1920s and its designs embraced the Art Deco style, for which the colour of choice was white and achieved by using colourless diamonds.

In 1932 Sotirio died, leaving his sons Costantino and Giorgio to run the business. By this time, Bulgari was well known worldwide thanks to its rich, international clientele, including Hollywood film stars, royal families, the aristocracy, politicians, heads of state and titans of industry. All would come to Rome to visit Bulgari. Great names are found in the sales ledgers in the Bulgari archives: jewels had been commissioned or sold to the likes of millionaires Frank Jay Gould, Robert Lehman, Samuel H Kress and Barbara Hutton, as well as Italian dignitary Count Vittorio Cini, former finance minister and governor of Libya Count Giuseppe Volpi, Italian minister of foreign affairs Count Galeazzo Ciano di Cortellazzo and his wife, Edda, who was Mussolini's daughter.

The outbreak of World War II inevitably had a considerable impact on the European jewellery trade. On 3 September 1941 Italy introduced a ban on the purchase or sale of precious metals, pearls, gemstones and any item incorporating these materials; the legislation was amended one month later to allow the use of precious metals in conjunction with the arms industry or items that were under 5 grams in weight that were being used for religious purposes, wedding bands and silver watches. Jewellery could still be made if commissioned or remodelled, as long as the metal and the stones were being reused from the client's own jewels. Towards the late 1940s, Bulgari designed fluid and floral jewels, predominantly in yellow gold, that replaced the earlier geometric designs in platinum and diamonds, as access to these materials had been restricted during and immediately after the war. These new jewels could be worn more casually than previous styles. Cabochon sapphires and rubies and lighter blue and yellow faceted sapphires were set in jewels that could be transformed and worn in more than one way, such as the double clip that could be worn as a single brooch, two brooches, or clipped onto a bangle. Large, gem-set earclips became popular with the shortening of hairstyles and the invention of the clip mechanism.

The post-war years saw the return of lavish jewels, with jewellery houses making extensive use of diamonds. Large, coloured gemstones were also used, but these were always associated with diamonds of multiple different cuts, including marquise, baguette, pear-shaped and circular, that were often alternately set to

Two brooches by Bulgari: The top,
made in 1988, is set with large yellow, blue and
pink sapphires and an emerald in a diamond-set
waved surround; the bottom, a *giardinetto* brooch,
1968, with a cabochon sapphire vase and carved
sapphire and ruby buds and leaves.

Richard Burton and Elizabeth Taylor photographed
in 1975, Taylor wearing her Bulgari sautoir, 1972,
set with a *c.* 62-carat Burmese sapphire.

create jewels that had spikey and pointed outlines; the combination of cuts created jewels with impact. Paris had dictated jewellery trends and designs for decades, but in the 1960s other countries started to introduce their own styles. Whereas in Paris there had been an emphasis on faceted emeralds, rubies and sapphires with diamonds, in Italy, under Giorgio's guidance, Bulgari was using colour imaginatively by combining lesser-known coloured gemstones with the cabochon cut, the house's preferred stone cut. Traditionally, emeralds, rubies and sapphires of fine quality would be faceted to emphasize the stones' clarity, but Bulgari did not differentiate gems by quality and used the cabochon cut for high-quality stones. The marriage of the rounded cabochon shape with soft curved designs became the 'Bulgari style'. In the 1960s, Bulgari created a series of brooches in the form of baskets of flowers — a motif known as *giardinetto* that originated in the eighteenth century — which allowed its designers to explore the pairing of faceted and cabochon cuts with a variety of coloured gemstones.

Bulgari placed emphasis on a jewel's colour and did not choose stones for their intrinsic value. The late 1940s combination of cabochon rubies and sapphires and large faceted Sri Lankan yellow and pale blue sapphires was more fully embraced in the 1950s and '60s. Bulgari's designers not only used cabochon pale blue sapphires, but also combined these pale stones with dark cabochon sapphires, to the extent that the sapphire became the firm's trademark, as seen in jewels in the chapter The Art of Collecting of this book.

In 1966, after Giorgio Bulgari died, the business passed to his sons Gianni, Paolo and Nicola, and with that came the opening of stores in New York, Geneva, Monte Carlo and Paris. Bulgari flourished during the 1950s, '60s and '70s and enjoyed the boom years of Hollywood. Films such as *Roman Holiday*, *La Dolce Vita* and *Cleopatra* helped to ensure that celebrities never left Rome without visiting 10 Via dei Condotti. The 'A list' of Gina Lollobrigida, Anna Magnani, Ingrid Bergman, Monica Vitti, Sophia Loren, Elizabeth Taylor and Capucine would invariably be photographed in Italy wearing a Bulgari jewel — it was almost as if no one dared to be seen without one.

Hardly a photograph of Taylor was taken in which she was not wearing a Bulgari jewel, whether that was on formal or informal occasions. Bulgari was proud of the wearability of its jewels, creating and promoting jewels that could be worn casually — a key trend of the 1970s. In a desperate attempt to salvage his marriage with Taylor, Eddie Fisher bought her jewels from Bulgari, but they were not enough to divert Taylor's attention from Burton; Fisher eventually had Bulgari send the invoices to Taylor. Burton loved buying jewellery with Taylor, and she loved choosing the jewels with him: 'I used to get so excited, I would jump on top of him and practically make love to him in Bulgari', she later exclaimed in an interview with the *New York Times* in 2002. 'One day Richard said "I want to buy you a present. I feel like buying you a present." And I said "Wow! What did we do today that you… That's amazing! Where? Where shall we go?" "Bulgari, of course", he said.'

Burton bought the stunning Bulgari sugarloaf cabochon sapphire sautoir necklace in 1972 for Elizabeth's 40th birthday. Made in 1969, the jewel is set with a superb *c.* 62-carat sapphire from Burma (Myanmar). The stone is a fine example of Bulgari's use of the cabochon cut for top-quality gems; it is very rare to find a sapphire from Burma of this size and quality, and other jewellery houses would have directed their lapidaries to facet it. The articulating necklace is composed of hexagonal and rhomboid links that are pavé-set with diamonds and calibré-cut sapphires. The pendant can be detached from the necklace and worn as a brooch.

Elizabeth Taylor's Bulgari Trombino ring, inset, made
in 1971 and set with a sugarloaf sapphire between
round and baguette-cut diamond shoulders, that she
wore with her sapphire-set Bulgari sautoir,
pictured above and on the facing page.

Elizabeth Taylor's Bulgari sautoir, 1969,
with a *c.* 62-carat Burmese sugarloaf sapphire
and gifted from Richard Burton to Taylor on her
40th birthday in February 1972.

A necklace and bracelet by Bulgari: The double-
row necklace, 1954, of cushion-shaped sapphires,
graduated in size along the two rows, and round and
baguette-cut diamonds, is characteristic of post-war
jewels that drew attention to coloured stones by
using combinations of contrasting diamond shapes;
the bracelet, 1990, articulates with its tessellating
diamond-set panels of blue and yellow sapphires,
interspersed with cabochon amethysts.

Top: A necklace, 1932, of cabochon sapphires set by Bulgari in links of round and baguette-cut diamonds, an early example of the cabochon-set jewellery that became a distinctive feature of the house in the years to come.

Above: The interchangeable pendant setting, made by Bulgari in 2006 for the 321.27-carat Burmese sapphire known as the Grand Kathé, is a more recent example of Bulgari's magnificent cabochon sapphire jewels. This sapphire is named after the Kathé district in Burma (Myanmar), where it was found.

A photograph by Tullio Farabola
of Sophia Loren navigating through cameras
in 1962 wearing matching sapphire and ruby
earrings and necklace by Bulgari and
a large cabochon sapphire ring.

A photograph taken by Gian Paolo
Barbieri in 1966 for *Vogue Italia* of actress
Capucine wearing sapphire-set Bulgari jewels,
including a necklace, facing page,
of cabochon sapphires.

It is every jeweller's dream to have their jewels worn and show-cased by beautiful, celebrated women. Bulgari's striking and colourful jewels attracted A-list celebrities from all over the world. The 1950s and 1960s saw Bulgari's designers create jewels that used the soft, light blues of Ceylon sapphires with the bright and deep reds of rubies, both often cut *en cabochon*, the now-signature Bulgari cut, which gave the jewels a smooth, soft feel, whilst conveying a sense of volume and drama. It proved to be a very popular colour combination. Italian actress Sophia Loren, born in 1934 in Rome – the home of Bulgari – proudly wore Bulgari jewels, including a necklace and earrings of cabochon sapphires and rubies mingled with diamonds. In this photo from 1962, the cabochons of Loren's necklace and ring shine in the light of camera flashes.

The French actress and model Capucine (born Germaine Lefebvre, 1928–1990) was photographed by Italian fashion photographer Gian Paolo Barbieri for *Vogue Italia* in 1966 wearing a striking pink evening dress with a pink feather boa by Lancetti. Against the pink of her dress, the blues of Capucine's eyes and of her sapphire-set necklace, earrings, bracelet and ring take centre stage. Capucine wears the Bulgari necklace illustrated on the facing page in which diamonds have been cleverly used to high-light the cabochons and emphasize the curves of the sapphires and the necklace shape. Sapphire cabochons also hang from her earrings and one is set in her ring. Capucine's sapphire brace-let is set with a combination of dark and light blue cabochon-cut sapphires, a classic Bulgari colour combination during the 1960s. The elegance and femininity of the evening dress complements the contrasting colours of the sapphire parure – if paired well, jewels and outfit can enhance each other.

The Bulgari necklace worn by actress
Capucine for a *Vogue Italia* photo shoot in 1966,
set with perfectly matched oval cabochon
sapphires, each in tear-shaped surrounds
of round brilliant-cut diamonds.

ANDREW GRIMA

'I did not want to be a run-of-the-mill jeweller, I wanted to do my own thing, my own way.' – ANDREW GRIMA

No jewellery collection is complete without a jewel by Andrew Grima (1921–2007), one of the most sought-after jewellers of the latter part of the twentieth century. Having no formal training in design, making or gemmology, Grima came to the industry with fresh, original ideas, without being encumbered by traditional methods of design and production. His instinctive approach to jewellery design ensured his success. The 1950s and early 1960s were a low period for the British jewellery industry – nothing original and exciting was being produced – but post-war British studio jewellers like John Donald, David Thomas, Gerda Flöckinger and Andrew Grima were paving the way for a new era of innovative, design-led jewellery .

Grima was born in Rome in 1921 and moved with his family to London when he was five years old. He studied mechanical engineering at university and wanted to pursue his passion for drawing but, unable to enrol onto an art course, he signed up for a secretarial course. It was on this course that he met his first wife, Hélène, whose adopted father, Franz Haller, had a jewellery business called H J Company. After serving with the 7th Indian Division in Burma during World War II, Grima returned to London and later married Hélène. Grima joined his father-in-law's company primarily to help with the accounts, but a visit from two Brazilian stone dealers changed his career path completely. Mesmerized by their collection of Brazilian gemstones, Grima's imagination was ignited; having managed to persuade his father-in-law to buy the whole collection, Grima then set about designing jewels for these new stones.

It was a life-changing moment for Grima who, after his father-in-law's death in 1951, took over the business and became the company's lead designer. He employed over 24 craftspeople in his workshop in Gray's Inn Road. To this day, many say that this was the most exciting time to be a goldsmith in the Grima workshop, since the creations Grima wanted his artisans to make were always different, pushing technical boundaries. If a goldsmith questioned the viability of constructing a certain element, Grima would often reply that he was the designer and that it was their job as the goldsmith to work out how to make it. He was designing jewellery that no one had ever seen before, changing the emphasis from figurative to organic, bold and abstract designs inspired by the natural world. He allowed the gemstones, both cut and rough, to take centre stage. Grima preferred textured yellow gold, asymmetric designs and colour for his jewels, using diamonds sparingly as elements that served only to accent the other stones and gold.

In 1961, H J Company were selected to cast jewels from the wax models of various well-known artists for the *International Exhibition of Modern Jewellery 1890–1961*, as well as exhibiting six of their own designs, at the Goldsmiths' Company's Hall. Grima followed his success at the exhibition by expanding his reach and travelling to international clientele in the United States and Europe. By the mid-1960s, Grima's jewellery was worn by royalty and high-society women such as Jackie Kennedy Onassis and Ursula Andress; today, it is avidly collected by designers Marc Jacob and Miuccia Prada. Grima was the only jeweller to be awarded the Duke of Edinburgh's Prize for Elegant Design – in 1966 – and Prince Philip gave the winning brooch to the Queen (a jewel featured on

A line-engraved gold ring, 1999, set with
a 25.98-carat cushion-shaped yellow sapphire
in a surround of square-cut diamonds.

A Cuttlefish bangle watch from
Grima's About Time collection for Omega,
1969, with a round, sapphire watch glass
and diamond-set bezel.

A platinum ring, 1976, set with
a marquise-cut green sapphire in a swirled
surround of tapered baguette-cut diamonds.

page 144 of *Ruby*). In the same year, he opened his first shop at 80 Jermyn Street, London, with a workshop in the basement. He won the De Beers International Award, an annual competition that was fiercely contended by international jewellers, a record 11 times; in 1970 Grima received a royal warrant from Queen Elizabeth II.

In 1969 Grima was given the opportunity to design a collection of watches for Omega. The result is arguably still one of the most innovative and original watch collections to be designed to this day. Grima replaced the watch 'glass' with a gemstone so that time was seen through a gem. When the collection, consisting of 55 watches and 31 matching jewels, was launched in 1970 it was only a matter of days before they were all sold.

Shortly after meeting Grima in 1974, Jojo Maughan-Brown, the great-granddaughter of Sir Thomas Cullinan of the famous diamond family, joined the business as a trainee goldsmith and worked at the bench for the next 10 years; together, Andrew and Jojo Grima expanded the business. Grima and his family moved his business to Lugano, Switzerland, where they opened a shop; six years later, in 1992, they moved to Gstaad, Switzerland, where they opened another shop and undertook private commissions. Grima arranged for one of his longstanding craftsmen, Tom Scott, to continue to make the bespoke orders in London; each of these jewels bears Scott's maker's mark alongside the Grima hallmark.

In 1991 the Goldsmiths' Company hosted a retrospective of Grima's work to celebrate his 70th birthday, with over 300 pieces on display. In the words of Dr C E Gordon Smith CB, who was Prime Warden at the time, 'This is an opportunity to see jewellery by Andrew Grima, representing a unique cameo of the work of one of the most original design minds in modernist jewellery.'

Emily Barber, Bonhams jewellery director in the UK, who has been instrumental in bringing the world of Grima jewels to the forefront of the international auction market, recently shared with me her take on Grima's enduring popularity:

> Post-World War II, the UK jewellery market was stagnant and repetitive. Grima wanted to give the wearer a morale boost after the deprivations of war and rationing and to match jewellery design with the modernist spirit that prevailed in art and architecture. Huge and unusual coloured gems — often in uncut crystal form — were mounted in richly textured gold settings that were striking, audacious and full of impact. His designs perfectly captured the free spirit of the 1960s and '70s and quickly invigorated the market. Not only was Grima embraced by actors, pop stars, celebrities, artists and bohemians — the whole Swinging Sixties culture — but he was also embraced by the Establishment, including royalty and aristocracy. These are statement jewels that appeal to the design-conscious as wearable works of art, and the fact that they remain as popular today as they were when they were first made is testament to their excellent design and wearability.

Andrew Grima died in 2007. I vividly remember his memorial service in 2008 for the array of Grima jewels being proudly worn by his family, friends and colleagues; it was a celebration of no ordinary jeweller, for he had changed British jewellery design for ever. Since 2007, Jojo Grima and their daughter Francesca have continued the business, creating a limited collection of jewels each year that are made by goldsmiths who were originally employed by the Grima workshops over 40 years ago.

Actress Ursula Andress wearing a selection of Grima jewels, including earrings set with opals, sapphires, emeralds and diamonds, from the winning collection of the 1966 Duke of Edinburgh's Prize for Elegant Design; and a 1972 articulated bracelet by Grima, with square sapphires and diamonds set in platinum box collets.

JOHN DONALD

John Donald (1928–2023) was one of the most respected goldsmiths working in Britain. His distinctive jewels have caught the attention of all jewellery lovers, including royal collectors. In the early 1950s, John initially trained as a graphic designer until he met Barnett Freeman, a governor at the Royal College of Art, who suggested he try jewellery making; indeed, the metalwork department was looking for artists that could bring a different perspective to metalworking. John spent the next four years at the RCA and remembers them as some of the most rewarding and enjoyable of his life.

As John began his jewellery career, the British jewellery industry was still suffering the lingering effects of World War II. A crippling purchase tax had been imposed on all new luxury goods, including jewellery, since 1940, and the tax reached 125 per cent in 1947, reducing only to 60 per cent by 1955. Purchase tax made buying fine jewellery more expensive for consumers, and the tax was paid initially by the jewellery makers, not retailers, so cashflow was restricted. Jewellers were reticent to make new jewellery in case it did not sell and, consequently, any jewels that were made conformed to past trends and lacked originality. By 1961, the tax had reduced to 27.5 per cent and the industry was ready for a change. In the same year, the Worshipful Company of Goldsmiths held the *International Exhibition of Modern Jewellery 1890–1961* to inspire makers and the public with a display of exceptional jewels and a competition for new designs – the jewellery industry was finally being reinvigorated. People were tired of food rationing and the austerity that had followed the war years and were ready for fresh designs for cars, furniture, ceramics and even jewellery; it was a time of experimentation and the exploration of new ideas and new techniques.

Starting a business in this period was difficult, but John persisted. In 1960 he found a mews house in Bayswater, London, with space for a workshop on the ground floor, which had once been stables for horses. John created a rough-and-ready workshop space and, before too long, hired a young, talented, Australian goldsmith,

Gary Bradley. Together, they turned John's designs and drawings into reality. John had the innovative ideas and Bradley had the practical knowledge, which was more comprehensive than what John had ever learnt at college.

One day, out of the blue, the Earl of Snowdon contacted John and asked if he, his wife, Princess Margaret, and her mother, Queen Elizabeth the Queen Mother, could come to visit his workshop. Princess Margaret loved jewellery and was a great supporter of modern design with a keen appreciation of craftsmanship and new techniques. John was concerned that the workshop would not be up to royal standards and that the uneven cobbled street – too narrow to fit their car – may prove too difficult to walk on. Nevertheless, the royal entourage arrived. That first visit was the beginning of a very fruitful and cherished association that lasted for nearly 25 years, until Princess Margaret's death in 2002. John and the Princess shared a similar sense of humour. On one occasion, there was an event at the Bulgari showroom in London and Princess Margaret asked if she could bring her jeweller along; the response was affirmative, so long as he did not steal any of their designs. John remembered entering the store with Princess Margaret, who announced, to much amusement, that John was her personal jeweller who would not be stealing any jewellery designs.

When Christie's sold Princess Margaret's jewellery in 2002, there were 16 pieces of John's work among the collection.

Top left and right: Two brooches and a pair of earclips by John Donald with carved sapphires and rubies that swing within their textured gold surrounds; the right brooch and earclips, early 1980s, are from the collection of Princess Margaret, who supplied the rubies inset with diamonds.

Centre: John Donald and Princess Margaret at the opening of the 1971 Tecla Pearls exhibition, Bond Street, where John presented Princess Margaret with a brooch of freshwater pearls and diamonds.

THE ART OF COLLECTING

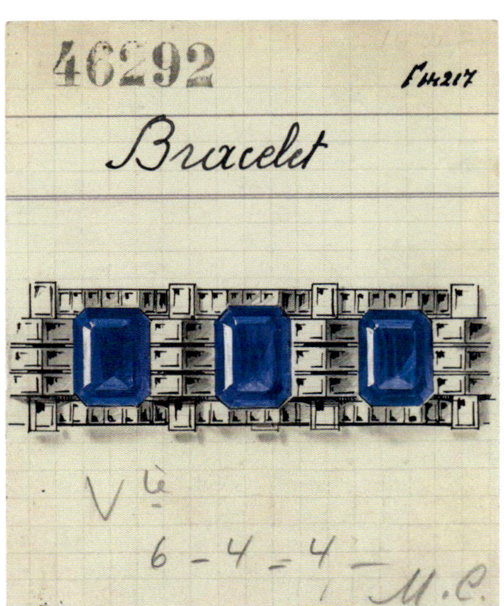

Above: From the New York Collection (page 244),
a spectacular bracelet by Van Cleef & Arpels,
made in 1937, set with seven emerald-cut
sapphires subtly graduated in size and baguette-
cut diamonds in articulated surrounds.
The product card from Van Cleef & Arpels
shows the design for the bracelet.

Facing page: From the LA Collection Privée (page
239), a late nineteenth-century brooch attributed
to Austrian jeweller Ernst Paltscho, set with three
high cabochon Kashmir sapphires weighing 46.86,
10.09 and 9.93 carats, respectively, in surrounds
of cushion-shaped and circular-cut diamonds.

The Art of Collecting

Collecting jewellery is a complex pastime: it involves multiple parameters dictated by many variants. However, in my opinion, most jewellery collectors do not make a conscious decision to collect; their activities seem to be more like a gravitational pull, driven by an obvious passion for their journey as a collector. In contrast, buying items with the sole idea of financial gain creates a totally different type of 'collection' – typically made up of unusual large gemstones that are stored in safes and do not see the light of day – and one that is not covered here.

Then there is the collector, or the custodian of a collection, who often attracts just as much curiosity as the jewels they collect. For this chapter, I spoke to collectors from around the world about their collections and the inspirations driving their selection. They have been extremely generous in sharing with me images of some of their impressive and iconic sapphire-set jewels from their individual collections, which appear over the following pages. Understandably, some wish to remain anonymous, while others are happy to be mentioned; but they all have one common criterion: the jewel or gemstone has to be beautiful.

Later in this chapter, Kazumi Arikawa, collector and president of Albion Art, stresses the paramount importance of beauty in a jewel when he is considering it for his collection; he connects with the positive energy inherent in a beautiful object, and this energy feeds his soul. Through his collection, he shares this energy with others – beauty is a reciprocal exchange that is understood in every culture.

Though beauty is in the eye of the beholder, when it comes to fine jewellery this beauty lies in a combination of factors: the jewel's design, exemplary craftsmanship and rarity, as well as the scarcity and beauty of its gemstones. But how do you decide what to buy? After establishing where a jewel sits in relation to these factors, the final decision comes down to personal preference – some collectors focus mainly on the gems, others on the jewel. Other collectors, still, are influenced by designs or time periods. They might be drawn to Art Deco, for example, because its geometric style and monochromatic colour combinations resonate with them; in fact, Art Deco encompasses diverse designs, which is why the jewels are highly collected and deemed timeless. Often, it is quite simply the gem or jewel 'speaking to you' – it is how it makes you feel.

To start a collection, it is always wise to begin by sourcing signed antique or vintage pieces, as this signature gives the jewel credibility and reassurance to you, the buyer. Not every jewel is signed, but buying an unsigned jewel takes knowledge and courage. In this instance, a trusted, helping hand can provide guidance until a level of knowledge and confidence has been acquired. With this understanding comes the ability to recognize the feel of a jewel: some unsigned jewels have a certain feel that tells you that they may have been manufactured by a luxury jewellery house, but these jewels are rare as most of the jewels from fine jewellery houses are signed. Some view signed pieces as 'liquid

currency' to trade with when a better piece is found, or to save, like money in an account. Documentation is important to substantiate the jewel, along with, if possible, the original fitted box and drawing or sketch. Catherine Cariou, a world-leading jewellery expert and curator, explains later in this chapter the lengths to which one can go when researching a jewel's provenance. Collectors of this calibre will always require an expert's opinion, whether that's from an auction house, archivist or respected dealer, but there are only a handful of specialists that are as knowledgeable about jewellery as Cariou. The jewel must speak to the collector, but this connection must be supported by the jewel's provenance and an expert opinion.

Since auction houses offer a wide range of jewellery and are able to introduce jewels that may not have been on the market for a very long time, many collectors start buying jewels at auctions. With the help of specialists employed by the auction houses, new collectors start to learn and understand jewellery design and creation. Not all buyers want to be educated in the history of jewellery, though; some just want to own something fabulous – there is jewellery for collectors, and jewellery for luxury to be worn.

As appreciation grows, so does the awareness of rarity; this combination often results in buyers enjoying the challenge of the chase. Finding that extra-special jewel or gem that has not been seen – or not appeared on the market – for a very long time is truly thrilling. As one collector said to me, 'Collecting is like fishing: you always remember the ones that got away. Sometimes you make financial mistakes, maybe you paid too much, but, if you still like it, then that is what matters.' Later in their collecting journeys, some collectors sell earlier pieces as their tastes in jewellery grow and change; others like to keep their first purchases as a reminder of how they have developed their collection and how their knowledge has evolved.

A collection needs to demonstrate variety: jewels from different periods and makers. When I am shown collections that have only concentrated on one style or one maker, my heart sinks, since it tends to be indicative more of an obsession than an intellectual discernment for the subject – two very different approaches. If gemstones are included, they should be bought for their unique qualities and not because a gemmological report stipulates a particular country of origin. Laboratory reports are crucial in confirming whether the stone is natural and whether it has undergone any treatments; however, their assessment of a stone's country of origin, though very interesting from a gemmological perspective, should not, in my opinion, be the sole reason for buying a gemstone.

Today, more jewels are being sold online, especially as, at the time of writing, we are living through a global pandemic that has severely curtailed the opportunity to travel

From the LA Collection Privée (page 239),
a clip in the form of a bird resting on a branch
by Cartier, c.1929, with a body of carved
sapphires, circular- and baguette-cut diamond
feathers and an emerald eye.

to different locations to view jewels at fairs and auctions. Yet, the buying and the chase will continue. Though buyers may be reluctant to pay that bit extra to secure a jewel if they have not seen it, known jewels will still generate huge demand.

Collecting habits have changed over time. We have seen in previous chapters how Hollywood stars and socialites made a point of wearing substantial, impactful jewels, and how jewels have been altered in response to current fashions or trends. Auctions were not as prominent and accessible as they are today, so many people would commission pieces from jewellery houses and develop a relationship with their favoured house; today, it is often the reverse: many collectors will buy jewels at auction, as opposed to buying new from a jewellery house. Having said that, the stunning, one-of-a-kind jewels that are created by the fine jewellery houses will become the antiques of the future — a trajectory that all the jewels in this chapter can attest to. People have always been interested in fashion, but today beautifully crafted jewellery with unique design, whether vintage or antique, is starting to be viewed as an art form.

When does a dealer become a collector? For me, there are very few dealers that are passionate about the subject, seeing it instead as a means to make money. A dealer that is also a connoisseur, who has his or her own collection but will buy and sell, is very rare. Kazumi Arikawa, Alisa Moussaeiff and Carlo Zendrini, all included in this chapter, are professional jewellery connoisseurs who believe that you cannot educate a client without being able to show them your passion through the jewels you yourself assemble. People want to buy from an expert — it gives them total reassurance.

In my experience, women collectors do wear their jewellery, but not many jewels are bought to be worn; in fact, quite a few collectors are men. There is very often a close association between collectors' professions and their reasons for collecting jewellery: it may be that a highly skilled cardiologist is able to appreciate the detailed 'engineering' of fine jewellery, or that those who work with the sleek lines of buildings, or the minute mechanics of high-performance cars, are particularly attuned to the design and craftsmanship of exemplary jewels. The opportunity to broaden and nurture the artistic side of one's brain is often recognized by professionals who are required to be creative and inventive in their work; collecting beautiful jewels and gems is a pastime that fits with that ethos.

When so much thought, time and effort has gone into accumulating a collection, the icing on the cake is when one of the high jewellery houses, such as Van Cleef & Arpels, Cartier, Bulgari or Chaumet, asks to borrow a jewel for one of its spectacular exhibitions. This request is an endorsement and the ultimate recognition of a collection's careful curation. As one collector told me, 'Owning a part of history comes with a responsibility

From the LA Collection Privée (page 239),
an iris brooch, made in 1940 by Cartier London
and once owned by heiress Daisy Fellowes.
Set with cushion- and oval-shaped sapphires and
baguette-cut and cushion-shaped diamonds
on a rectangular-cut emerald stalk.

to look after these pieces for future generations. It is also important to loan items to exhibitions so that the importance of understanding good-quality jewellery can be shown. It also gives you a sense of belonging to the collecting community, where you will meet the academics and archivists. There is one thing for sure: you never stop learning.'

Collections are, ultimately, very personal. They can track the journey of one's growth as knowledge increases and the buying evolves. The very nature of private collections is that they are private; not every collector wants people to know what they have. I have therefore been incredibly fortunate to have been put in touch, through Catherine Cariou, with a few private collectors from locations in the United States. Catherine has helped me with both previous books, *Emerald* and *Ruby*, and I again asked for her assistance with *Sapphire*. It felt appropriate to include a chapter on collecting in *Sapphire* because these incredible collections bring together all the elements of jewellery design, history and stewardship that I have discussed in this trilogy. Catherine very generously reached out to some private collectors, who, with her encouragement and support, have agreed to feature some of their sapphire jewellery in this book. It does not surprise me that these great private collectors are from the United States. While the collecting of jewellery in the United Kingdom has, in my opinion, been hindered by the legacy of Victoriana and inherited jewels, American collectors on the contrary are very open to jewellery, and have warmly accepted Art Deco jewels in particular.

Although each collector I spoke to owns between 200 and 700 pieces that together run the gamut of all styles of antique jewellery, they could still remember the stories connected to each jewel. To convey the essence of each of their collections, the sapphire jewels in this chapter have been grouped according to collection. I feel very fortunate to have been in contact with these collectors and grateful that they allowed me into their world. The jewellery world, and particularly its collecting side, is built on trust, and I am delighted and humbled by the trust they have placed in me. It is exciting for me to rediscover in their collections pieces I remember being sold at auction; normally, these jewels vanish from public view after being sold, so it is reassuring to know that top-quality jewellery is appreciated and protected. The unprecedented situation at the time of writing has brought challenges to many, but the solace that collectors have found in their collections has been a very welcome diversion – looking at beauty makes people feel good. The collectors I spoke to were enjoying their collections, allowing them moments of peace and reflection and inspiring them to daydream about what they might find next.

From the LA Collection Privée: an Art Deco bracelet by Cartier Paris (above), *c.*1929, with cabochon sapphires on a central diamond-set plaque joined by two sapphire- and diamond-set loops; and an Art Deco brooch by the workshop of Henri Droguet for Cartier Paris, *c.*1925, set with a rectangular step-cut Kashmir sapphire of 12.64 carats in a diamond-set surround.

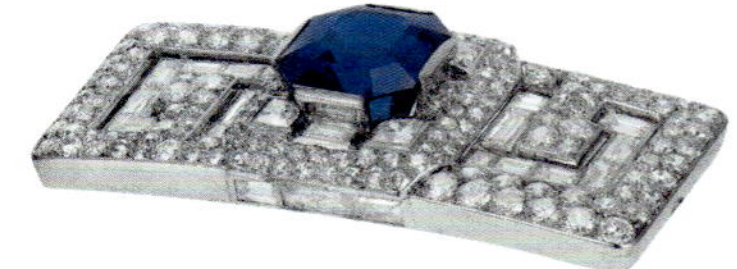

This collector has an enviable collection of high-performance vintage cars, so it is no wonder that his collection of jewellery, known as the LA Collection Privée, is just as spectacular. His move into jewellery collecting only began in the mid-2000s; he enjoyed buying jewellery at auction as presents and very soon realized that there was a difference in the final price between buying at auction and buying in a retail store. He started to look at jewellery in more detail, and his growing interest in fine jewellery was nurtured by auction house specialists. His discerning eye already understood craftsmanship, and the principle of buying jewellery was the same as buying a high-performance car, albeit on a different scale. He quickly noticed that rare jewels did not come onto the market often, but, when they did, they caught his attention. Having a wealth of experience in collecting beautiful cars made by well-known manufacturers, such as Ferrari and Lamborghini, he appreciates the guarantee of quality in jewellery made by the renowned *maisons*; the jewellery he sources is almost all signed. Art Deco is a recurring style in his collection, as he feels that the best jewels from this period are the very finest available. It is the overall look of the jewel that is important to him — it has to be the right jewel for him personally, and he does not think of it as being an investment: 'You must buy it because it brings you pleasure, it's a thing of beauty.'

He never sells a piece of jewellery he has bought as he likes his collection to reflect his evolution as a collector. The collection gives him enormous pleasure and he hopes that, in time, along with his cars, it will bring pleasure to others. He plans to open a gallery for his cars and jewels, which he hopes will be used for charitable events to support children's, animal and environmental charities. Driving this dream is his belief that being surrounded by beautiful objects creates an atmosphere of optimism, which might help people to give generously.

From top: A clip by Cartier New York, 1937, centrally set with a 9.09-carat Burmese sapphire above a fan of calibré-cut sapphires and baguette- and circular-cut diamonds; a bracelet by Cartier London, 1937, of three articulated lines of step-cut sapphires, graduated in size to the centre and spaced by diamond-set geometric motifs; and a Cartier New York Art Deco bracelet, 1937, set w th a cushion-shaped Burmese sapphire of around 22 carats between lines of sapphire beads and diamonds.

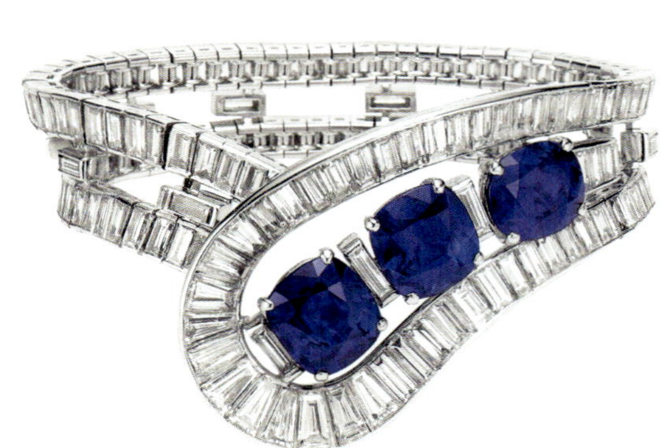

Top: A circular brooch, 1939, by Van Cleef & Arpels,
set with concentric rings of Cambodian sapphires,
weighing approximately 45 carats, and baguette-cut
diamonds to a convex centre.

Above left: A Belle Époque ring by Van Cleef & Arpels,
c.1917, set with a 21.73-carat cabochon Kashmir
sapphire surrounded by a diamond-set
gallery, claws and shoulders.

Above middle: A bracelet of cross-over design by
Van Cleef & Arpels, 1947, set with three impressive
cushion-shaped Kashmir sapphires weighing
approximately 6.56, 5.73 and 5.72 carats, between
a swirled line of baguette-cut diamonds.

Left, right and above right: Three jewels by
Van Cleef & Arpels: a Campanule brooch and earclips,
1959, with Mystery Set sapphire flowers on diamond-
set stems; and a Mystery Set Feuille de Vigne clip,
1963, of sapphires and a diamond-set stem.

Above: A magnificent Van Cleef & Arpels bracelet,
1965, formerly in the collection of art collector Janice
H Levin, set with five sugarloaf sapphires from Ceylon
between articulated diamond-set plaques.

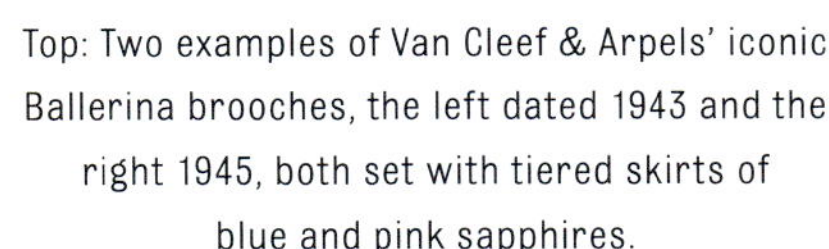

Top: Two examples of Van Cleef & Arpels' iconic Ballerina brooches, the left dated 1943 and the right 1945, both set with tiered skirts of blue and pink sapphires.

Centre: A *genre ancien* necklace by Van Cleef & Arpels, 1956, with a fringe of 11 oval cabochon sapphires in surrounds of gold corded wire and diamonds suspended from a line of sapphires and diamonds in gold corded-wire frames.

Above and left: A Hawaii bracelet, 1938, of tiered sapphire, ruby and diamond flower heads on stems, and a Hawaii brooch, 1938, with cascading sapphire and ruby flower heads. Both by Van Cleef & Arpels.

Right: A pair of Ludo Hexagon lapel clips by Van Cleef & Arpels, 1937, with a buckle motif of Mystery Set sapphires and circular-cut diamonds and a curved band of hexagonal interlocking links, set with sapphires.

Top: A bracelet of geometric design set with five
cabochon Sri Lankan sapphires, including two star
sapphires, by Chaumet, c.1930s.

Above: A bracelet by Mauboussin, c.1930, with a
flexible band set with carved sapphires, rubies and
emeralds in the shape of leaves, interspersed with
sapphire, ruby and emerald beads and diamonds.

Left and right: A Dome ring by Suzanne Belperron,
c.1938, set with a cushion-shaped sapphire of
approximately 5.5 carats surrounded by cabochon
sapphires and pavé-set diamonds; an articulated,
red-enamelled gold bangle by Jacques Lacloche,
made in 1938 by Verger Frères, with carved
sapphires in the shape of leaves and diamonds,
with a cabochon sapphire set within each link.

Below: A brooch, c.1920, centrally set with
an impressive cushion-shaped Kashmir sapphire
of 17.21 carats, flanked by two marquise-shaped
diamonds and two pear-shaped table-cut diamonds.

There are very few jewellery archivists in the world, as very few *maisons* have saved their archives; but, in Paris, the jewellery archivist – along with their archives – has always been held in high regard. As more archives become digitized, the role of the archivist is changing; yet, within the world of fine jewellery, it is the personal knowledge and hands-on involvement with jewellery houses that make archivists and specialists a great asset for collectors. Catherine Cariou is a greatly respected archivist and curator who has spent her career in the Place Vendôme, the heart and home of the headquarters of many of the world's leading jewellery brands. Cariou's knowledge of fine jewellery, especially jewels of French origin, is of the highest calibre, and she is much sought after by jewellery collectors around the world for her guidance and advice before they make a purchase. Cariou helps collectors purchase jewels at auction by viewing the pieces and authenticating their provenance; I therefore asked her if she would kindly write about the process involved in assessing and authenticating a jewel. She chose Elizabeth Taylor's ear pendants and, as we talk about Taylor's incredible collection in the High Society chapter, it felt like the right choice. In her own words, Catherine explains the excitement and the extraordinary sale of Elizabeth Taylor's jewellery at Christie's. Her in-depth knowledge of the ear pendants from the LA Collection Privée demonstrates the added value a specialist can bring to a collection.

On 13 December 2011 Christie's New York held one of the most wonderful auctions that has ever taken place (along with that of the Duchess of Windsor's jewels at Sotheby's in 1987): the fabulous jewellery collection of Elizabeth Taylor. It has justifiably been described as the sale of the century. The great buzz surrounding this collection that was felt by the public, professionals, museums, jewellery lovers and the merely curious was truly exceptional. Born in England and ennobled by the Queen of England, the Oscar-winning actress who had given unforgettable performances in *Cat on a Hot Tin Roof* and *Cleopatra* had an all-consuming passion for jewellery, a passion bordering on addiction. Didn't she once say 'There's more to life than money, there are also furs and jewellery'? During her lifetime, she built up a legendary collection made up of the most sumptuous jewels by the greatest jewellers including Bulgari, Van Cleef & Arpels, Boucheron, JAR, Cartier, Schlumberger and Tiffany.

Suffice to say that those who coveted certain exceptional pieces were left sorely disappointed, since the prices fetched by the jewels were stratospheric. Individuals, dealers, museums, jewellery houses, all eagerly wanted to own a little bit of the legend. The disappointment was all the greater as the

likelihood that some of these jewels would reappear quickly was very low. But that was without counting on providence or destiny, because, five years after the legendary sale, in 2016, once again at Christie's, these earclips resurfaced, as if by magic. Here, now, was an unexpected opportunity to enrich the collection by acquiring this pair of Mystery Set sapphire earrings, signed by Van Cleef & Arpels.

The upper part, in the form of an oval drop, dates from 1984 and was produced by the Le Henaff workshop, which moved with Van Cleef & Arpels from France to the USA in 1939. It is set with a pear-shaped sapphire in a triple surround of circular-cut diamonds and calibré-cut sapphires in a platinum mount. The style, although more classic, is close to the earrings of the 1970s in yellow gold and ornamental stones or rubies and emeralds. Each earclip has a detachable drop with Mystery Set sapphires surmounted by a collet set with circular-cut diamonds, in a platinum and white gold setting. They were produced by the Oscar Heyman workshop in 1985, the only workshop authorized to make the Mystery Set jewels across the Atlantic using the infinitely complex and demanding technique created by Van Cleef & Arpels in 1933 and improved by additional patents. The idea behind this invention was to do away with the classic setting and its too visible claws or grains, which, by hiding some of the stones, deprive them of their shine and lightness. Purity was the ultimate goal.

Unlike jewels displayed in museums, which will never be worn and are very rarely even handled, jewels in private collections have the opportunity to be appreciated in different ways. The generous collector may allow pieces to be worn and handled by family and friends, or lend them to exhibitions for educational purposes and to be appreciated by jewellery lovers and students of the craft. In putting together and caring for his fabulous collection of jewellery, this East Coast American collector understands that jewellery is made to be worn; he ensures that his partner tries on each piece to see how it sits before he commits to adding it to his collection. The brain behind this collection, which was started in the late 1990s, is a cardiologist whose profession requires that he work with extremely fine accuracy; as a result of his training and experience, the attention to detail and the engineering that has gone into making each jewel resonate with him, and he has full admiration for the craftsmanship. 'People are always looking for a diversion that makes them happy, and jewellery makes people happy'.

The collector initially grew up wanting to be an architect, having been enthralled by the architecture in New York, most of which was built under the influence of Art Deco. Even though he never became an architect, his love of the Art Deco style has never waned and is reflected in the jewels he collects. This collection is diverse in its geographic spread and includes the work of a range of jewellers. Many of his pieces are loaned to exhibitions around the world, which goes a long way to validate the collection and endorses the scholarship invested in his selection. This recognition gives him a personal sense of pride, 'It's a good feeling as it confirms you are learning in the right way.' These exhibitions are also a way to connect with the collecting community and with academics working in this field.

The chase and the challenge involved in finding new pieces to add to his collection are a source of great enjoyment, 'We are always attracted to what people do not have, and the value is gauged by what people are wanting.' When I asked specifically about sapphires, his advice is to go by the character of the stone, its rarity and colour; he also suggests looking at how closely stones match one another if the jewel is set with more than one sapphire,

for a good match in itself is a rarity which will be appreciated. Kashmir sapphires really attract his attention because they are so difficult to find; finding a matching pair is very exciting.

Sapphires feature strongly in this collection, and one bracelet in particular – by Van Cleef & Arpels and shown on the opening page of this chapter – is absolutely stupendous. The sapphires are each 'textbook' examples of fine, rare-quality blue sapphires. To find one of these stones is remarkable, but to have seven emerald-cut fine blue sapphires in one bracelet is quite extraordinary; it would have taken Van Cleef & Arpels many years to source these matching stones. The main criteria for their selection would have been outstanding colour matching and clarity; their country of origin would have been of far lesser concern, as the gems would have been judged primarily on their appearance. This bracelet is the epitome of late 1930s modernist design, influenced by the machine age. Its contrasting heights and shapes are emphasized with baguette- and square-cut diamonds, which reflect the geometric designs of Art Deco and invite the eye to travel along the bracelet.

Art Deco designers championed clean lines with minimum fuss, to which Van Cleef & Arpels responded by patenting, in 1933, the revolutionary Mystery Set, a technique that set sapphires, rubies or emeralds alongside each other without any metal visible from the front of the jewel. The Mystery Set technique creates a carpet of uninterrupted colour, while allowing formidable articulation, as seen in another bracelet (below) in this collection. The experience of feeling and holding such a remarkable piece is truly awe-inspiring. When a collection includes a Mystery Set jewel such as this bracelet, I recognize in the collector a genuine sense of stewardship and a very deep understanding of beautiful, rare and unique craftsmanship.

The selection of sapphire-set jewels from this collection that are shown here is but a glimpse into an exemplary world of knowledge, understanding and appreciation for the art and the craft of fine jewellery.

A flexible bracelet, 1962, of beautifully matched
Mystery Set sapphires bordered by circular-
cut diamonds by Van Cleef & Arpels.

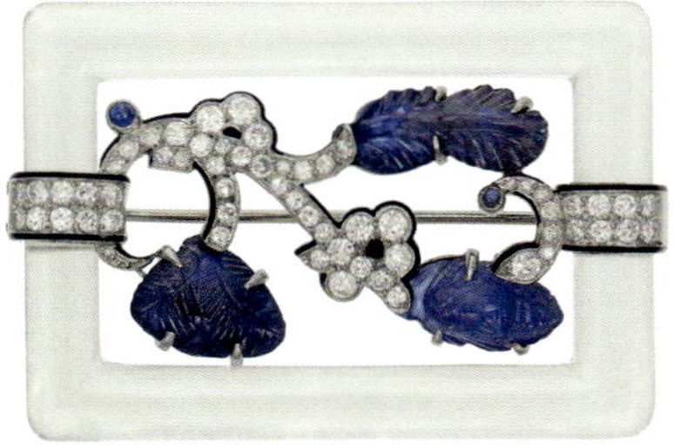

Above left and right: A set of four Floral Bouquet brooches by Cartier London, 1943, 1944 and 1947; each 'bouquet' is set with multi-coloured sapphires and diamonds, with four pavé-set diamond stems.

Left centre: A pair of Cartier ear pendants set with two oval-shaped Kashmir sapphires of 5.03 and 5.36 carats, respectively, suspended within concentric diamond-set surrounds with a diamond-set fringe.

Top: Curled clips of interlocked pale cabochon sapphires and diamonds by Suzanne Belperron and a Van Cleef & Arpels Ballerina brooch, 1942, with a tiered skirt of sapphires, rubies and diamonds, once owned by Barbara Hutton.

Right centre: A pair of Art Deco ear pendants by Cartier, 1927, of graduated geometric shapes set with diamonds and two oval-shaped sapphires and two cabochon sapphires.

Centre: A buckle brooch by Cartier New York, 1928, with a rock-crystal frame enclosing a diamond-set and enamelled scrolled branch and three carved sapphire leaves, one with an insect, and two cabochon sapphires; a Cartier pendant, 1920, in the form of a tall plant with articulated, diamond-set stem and leaf sections with sugarloaf sapphires and millegrain edges, the plant pot suspending a pendant sapphire and diamond.

This collector first started buying vintage jewellery in the mid-1980s because she considered old jewellery to be more interesting than new jewellery: it was far better value and, importantly, it was bought to wear. Collections evolve as knowledge increases, and this large collection has seen pieces swapped and changed over the years, either for better jewels, or indeed when the enjoyment from wearing a particular jewel wanes. The jewels in this collection act like a diary: the collector remembers exactly when she bought a piece, and even the dress she was wearing at the time.

Most of the jewels in the collection are French, because of the *maisons*' high level of craftsmanship, and do not predate the 1920s, as jewels made before this date can be quite fragile and therefore harder to wear. Many of the jewels feature playful motifs and have a fun character. The thrill of chasing down that special jewel is exciting and gets quite addictive, she explains. After years of collecting, her confidence has grown, and she now feels able to buy unsigned pieces for the collection. With experience, you can tell if a piece has been made by a workshop that may have produced jewels for one of the big houses. There is a skill in recognizing 'a hidden gem', and those moments of recognition give collectors a real buzz. Nothing pleases this collector more than being invited to the opening night of an exhibition she has lent to; to be able to dress up and mingle with other collectors while wearing a jewel from her own collection — it is an endorsement of her trained eye and good taste.

'Iconic' is a description very occasionally applied to a jewel that is universally recognized, and one that needs no further explanation within jewellery circles. Jewels that are referred to as iconic receive their title because of their timeless design. They are often instantly reflective of an era, generally postdating the late nineteenth century, and many eras are associated with a particular brand that dominated high jewellery creation at that time. Collectors appreciate jewels for these qualities and choose them to enrich their collections. The sapphire jewels in this collection have not necessarily been chosen for their gems, but because of their unique design, the boldness of their overall appearance and, importantly, their wearability and craftsmanship. These sapphire jewels are only a very small sample of the wide variety of jewels in this collection, but we get the immediate sense that jewels are only chosen for the collection if they make an impact on their custodian. Great collections have a voice and a narrative that reflect the collector; this narrative could be centred round a theme, style, period or size of the pieces, or sometimes round the gems that have been chosen.

I was delighted to discover in this collection the Van Cleef & Arpels pair of Pylône clips made in 1939 and ingeniously inspired by electricity pylons. I am always in awe of designers who can turn a mundane, utilitarian object into a jewel of beauty. During this period, designers had no CAD/CAM technology to use in the planning of these pieces, which makes the exceptional precision in the setting of the sapphires even more remarkable.

Van Cleef & Arpels are known for their innovative transformable jewels, which they created in response to the needs of the modern, independent woman of the 1940s, who wanted to wear jewels that could be adapted for any occasion. This 'bold look' was in vogue due to the popularity of the large on-screen jewels worn by Hollywood actresses. The Passe-Partout necklace in this collection was, and remains, revolutionary in its ability to transform from a necklace into a bracelet, and even into a belt. Its large pastel yellow and pale blue sapphires — the gems of choice to complement this bold design — were set in the bouquets that could be detached and worn as clips. The ingenious linking system in the jewel's incredibly flexible hollow chain was inspired by gas piping and was, therefore, referred to as 'gas pipe linking'. This style was championed by actress Paulette Goddard, wife of Charlie Chaplin. It was, nevertheless, a time of austerity and so jewellery was set with large, inexpensive gems; large pastel-coloured sapphires were more accessible to jewellers than brighter gemstones and, in turn, became very popular.

Jewels with animal and flower motifs are also present in this collection, including the collectable Bulgari articulating snake bracelet, its head set with a Kashmir sapphire, and the Tiffany jewels designed by Jean Schlumberger, whose inspiration from the natural world is beautifully articulated in his designs and his use of gemstones.

A pair of Van Cleef & Arpels Pylône
clips, 1939, with outreaching spikes
of calibré-cut sapphires.

The transformable nature of Van Cleef & Arpels'
Passe-Partout jewels means that these two pale
blue and yellow sapphire clips, 1947, can be worn
on their own or, as illustrated, on this flexible
gold linking as a necklace, bracelet or belt.

A fabulous demi-parure by Bulgari, *c.*1960.
The necklace, bracelet and earrings are all set
with oval cabochon sapphires and trefoil motifs
of cabochon rubies and diamonds on highly
flexible gold snake chains.

The line of coloured gemstones in this Jasmin
necklace by Jean Schlumberger, 1973, includes
blue, pink and yellow sapphires, aquamarines,
citrines and peridots, intertwined with
diamond-set bands and suspending diamond-
set jasmine flowers and buds.

Above and below: A classic Serpenti watch by Bulgari, the curled bracelet set with diamond scales to a diamond-set head with a Kashmir sapphire and cabochon emerald eyes. The snake's mouth opens to reveal the watch face.

Left: A charming gold bird brooch by Pierre Sterlé with wings of green sapphires and diamonds closed over gold tail feathers and a head of granulated gold beads with a diamond-set eye.

Right: A flower brooch by Jean Schlumberger with a tiered diamond-set centre to sapphire and diamond pavé-set petals with gold detailing.

A necklace by Schlumberger for Tiffany designed
as a diamond-set rope with twisted swags of
sapphires and emeralds and a border of marquise-
shaped and brilliant-cut diamonds; and Sea Shells
bracelet, designed for Mrs Peggy Rockefeller by
Jean Schlumberger for Tiffany, *c.*1958, with an
openwork band of pear-shaped sapphires and
circular-cut emeralds around six
diamond-set sculpted shells.

During the early twentieth century, Turin was the richest city in Italy, home to successful entrepreneurs, industrialists and aristocratic Italian families. The art of conversation and cultural exchanges were part of Turinese society; within this atmosphere, Venetian noblewoman Amabile Zendrini opened a salon where men and women could meet and exchange ideas and be surrounded by beautiful artefacts while having afternoon tea. Amabile particularly envisaged the salon as a place for women to meet, as many remained largely within their own homes and were not involved in the workplace. Her salon became a popular destination and gave Amabile the opportunity to introduce the clientele to the vintage jewellery she had on display. Her daughter, Carla Zendrini, used the success of the salon to build a business in vintage jewellery. Carla became arguably one of the first Italian women who truly understood the rarity of vintage jewellery, especially those jewels that had been created by the French jewellery houses.

In 1951, the same year her son, Carlo, was born, Carla Zendrini opened a jewellery shop selling fine second-hand signed jewellery. For the next 30 years, she travelled all over the world sourcing exceptional jewels. This was a period when most people wanted to buy new items after years of rationing, so not many retailers were selling antique and vintage jewellery. Very quickly, Zendrini, who was self-taught, became known for her discerning eye in tracking down the best jewels and gems in the world. Her appreciation and expert knowledge meant she was in high demand from prestigious private families and jewellery collectors alike, who wanted her to find important jewels for them.

In the 1980s, the business was handed to the third generation, and to the first man of the Zendrini family, Carlo Zendrini, who had inherited his mother's passion for historical jewels. Carlo was brought up surrounded by beautiful jewels and was taught how to appreciate and recognize good quality by one of the best jewellery experts: his mother. She also instilled in him a keen business acumen, so Carlo was well equipped to expand the business. He opened shops in strategic locations: in Rome, opposite Bulgari, and in Monte Carlo, next to the Hermitage Hotel.

I have often wondered how someone who is passionate about jewellery can part with jewels they have found, some of which will have taken months or years to find, but Carlo Zendrini provides the answer: you keep some for yourself. Zendrini demonstrates the difference between a trader, or dealer, and a professional connoisseur. Zendrini has over 200 pieces in his private collection, which he has built up over the years and which he displays in showcases in his new location, Monaco. As Carlo explained to me, 'When you are an adviser to someone, you must prove that you know what you are talking about', and what better way than to show your own collection – 'It is a way of transferring my knowledge to other potential collectors.' His collection has been validated by famous jewellery houses such as Cartier, Van Cleef & Arpels and Bulgari, who have each asked to borrow many of their pieces from his collection to show in their high-profile exhibitions around the world. Once on display, you can be sure there will be a label saying that the jewels have been lent by Carlo Zendrini, rather than 'from an anonymous collection'. When you have a lifelong passion for jewellery and gemstones, to find the best jewels in the world is rewarding, but to have that collection endorsed by the jewellery houses is an honour that is as rare as the jewels themselves.

What are Carlo's tips when advising his clients on fine jewellery? 'There needs to be an understanding of jewellery for collections and jewellery for luxury.' If the jewels are signed, they need to reflect the soul of the brand. Sometimes pieces have been specially commissioned from the big jewellery houses and yet lack the 'signature or soul of the brand'. 'There are "normal signed" pieces and there are jewels that are signed that are rare – a big difference. The piece must have beauty and rarity, but also needs to reflect the brand's soul: the essence of what they stand for.' And I agree with him – though it could be argued that all jewels from brands reflect their ethos, there are some years when that is truer than others, as brands can sometimes lose their direction; it is knowing the difference and how to spot it that makes a purchase meaningful and a collection fulfilling. Zendrini particularly appreciates the Art Deco period as one 'of exceptional taste and quality'. Today, he explains, it is easy to find people with money, but much harder to find people with taste and money.

Facing page: A necklace of diamond-set silver bows suspending pendants of cushion-shaped and step-cut sapphires, commissioned from Chaumet in 1894 by a Polish nobleman.

When asked about blue sapphires, Carlo remembers his mother's advice: 'You want to see "the water" inside the sapphire, like the water of Sardinia', a clarity which is full of life. This quality is hard to describe on paper, and cannot be translated into a scientific classification; it is all about the feeling when looking at a sapphire. When it comes to blue sapphires, a fine blue Kashmir stone is the top of the leader board for gem and jewellery connoisseurs, and Zendrini confirms that Old Mine material is highly sought after and can be found in antique jewels. However, he concurs that people are too fixated and swayed by gemmological reports stating a particular origin; instead, what is important to know is 'if the stone has been treated, heated or is natural. Years ago, there were no gemmological reports and people would buy because they fell in love with the stone' – an instinct that Zendrini himself has always followed when buying stones.

All these nuggets of information and expert knowledge, gathered over 40 years, are now being passed down to Carlo's sons, Ferdinando Ferrero Zendrini and Bernardo Ferrero Zendrini. Ferdinando and Bernardo will, in turn, ensure that the Zendrini heritage continues to assist future jewellery buyers and collectors so that the precious world of vintage and antique jewellery carries on being appreciated and admired.

Above left, left and right: A brooch and pair of earrings by Bulgari, late 1950s, with diamond- and sapphire-set flower heads. The flower heads of the brooch are set *en tremblant*.

Above right: A Cartier Paris brooch, *c.*1970, in the form of a stylized flower with diamond-set petals and a central cabochon sapphire, of approximately 25–30 carats, encircled by 13 oval-shaped sapphires.

Facing page: A Van Cleef & Arpels 'Indian-style' necklace, *c.*1962, of flower motifs, each with a white or grey natural pearl surrounded by sapphires, rubies, emeralds and diamonds, suspending a pendant pearl in a sapphire and diamond frame.

COLLECTING GEMSTONES
WITH ALISA MOUSSAIEFF

In the upper echelons of the jewellery establishment, there is one woman who sits alongside some of the great jewellery connoisseurs: Alisa Moussaieff. Alisa and her (now late) husband Shlomo Moussaieff have been fearless in their passion to seek out the best gemstones and pearls in the world. Even though they may not be collectors in the traditional sense, I have included them here because the stones and jewels that have passed through their hands — and have been given the Moussaieff seal of approval — will end up in private collections. When you walk into their London shop in New Bond Street, you are surrounded by incredible gems and jewellery — and there, in the middle of all this splendour, is Mrs Moussaieff herself. Every piece tells a story about how she acquired it, and the same journey, care and knowledge unite her with the other collectors in this chapter.

Mr Moussaieff's great-grandfather began work as a pearl dealer in the 1850s. His grandson Remo established himself as a gemstone trader in the 1920s, selling rare gems and pearls to the famous jewellery houses of Paris. In 1963, Mr Moussaieff opened a discreet jewellery showroom by the Hilton Hotel on Park Lane, London, and the couple ran the business in partnership. I remember, as a young diamond dealer in the 1980s, visiting the exclusive showroom with trepidation to see if I could sell a diamond or two. In 2006, they opened their New Bond Street showroom, along with stores in Hong Kong, Geneva and, during the winter skiing season, in Courchevel.

Mrs Moussaieff speaks with a quiet, confident, authoritative voice. She learned all she knows about gemstones and pearls from her husband and, since his death in 2015, has continued to run the formidable family jewellery business. When I asked how she chooses stones, Mrs Moussaieff asserted her firm belief that stones choose you, rather than you them — sometimes, she will not show a client a particular stone, and this is not through a conscious decision, but rather one of quiet intuition. Some cultures — particularly those in the Middle East and the Indian subcontinent — are very much in tune with gemstones, which are part of their heritage; they have an instinct about stones that cannot be taught, and Mrs Moussaieff understands those sentiments: 'Sapphires are appreciated by all cultures; except in Indian Hindu belief, according to which they embody energy, which is not perceived as being positive in the Indian psyche.'

Before Moussaieff buys a stone, she always comes back to view it at different times of the day — morning, afternoon and evening — so that she can get a real sense of what the stone looks like in different lights. Light will make a stone 'sing' or render it 'lifeless', and you want to make sure you are not buying a stone that reacts in this latter way. She has described top-quality blue sapphires from the mines of Kashmir as 'embodying an otherworldly, slightly milky haziness, which is very subtle but very real'. Though the firm's international clients include the great and the good of society, Moussaieff is as discerning with her clients as she is with her gems: sometimes she will not sell a rare gem unless she feels it will be appreciated:

> During my career spanning around 60 years in the gem world, I have noted several character traits that buyers of gem-quality sapphires have in common. Their personality seems to embody not only unusual depth and diversity; rather, their choice and selection seems to be dictated by metaphysical, spiritual thinking, as well as investment considerations. It seems that gem-quality sapphires provide a window into the 'beyond' and the 'faraway' and partake of their mystery!

Stone dealing is very much dominated by men and I asked if that has ever hindered her in business, but, as soon as I asked the question, I already knew the answer: no, just be the best at what you do.

Left: A ring by Moussaieff, 2018, set with a 13.53-carat oval-shaped pink Sri Lankan sapphire and a gallery of pear-shaped and triangular rose-cut diamonds with rubies set underneath, to diamond-set shoulders.

Right: A spectacular sugarloaf sapphire of 50.52 carats set in a ring with diamond-set claws, gallery and shank, by Moussaieff, 2017.

A 'gem mini-sculpture' by Wallace Chan for
Moussaieff, 2008. A diamond-set chain necklace
suspending a pendant set with an oval-shaped
Sri Lankan sapphire of 119.37 carats in a
sapphire- and diamond-set titanium surround;
the sapphire sits atop a disc of lapis lazuli.
On working with Wallace Chan: 'By applying his
highly focused sensitivity to colour and texture
to the combination of lapis and sapphire he
managed to add a highly artistic aesthetic
dimension to this gem mini sculpture.'

There are very few jewellery connoisseurs that truly appreciate the important contribution jewellery makes to history and understand that jewels offer a narration of past events. Kazumi Arikawa is a world-class jewellery connoisseur; his discerning eye seeks out the very best the jewellery world has to offer.

Arikawa opened his private jewellery showroom, which is part of his Albion Art Jewellery Institute, in 2015 in the Minato-ku district of Tokyo. Here, he receives distinguished guests, jewellery curators and historians from around the world, but, before sharing his collection with them, he first ushers them into the showroom's traditional tearoom, with paper sliding doors and tatami mats. For Arikawa, a tea ceremony demarcates the space, both physical and mental, from the hustle and bustle of the outside world to bring a calmness and tranquillity before viewing the jewels – his exceptional pieces deserve to be viewed with a clear and uncluttered mind.

Though jewellery has long been part of his life – his mother sold contemporary jewellery – it was only later that Arikawa made the decision to work with antique jewellery. The young Arikawa studied for a political science and economics degree at Waseda University, Tokyo, followed by two years training as a Buddhist monk. He then decided to join his mother's contemporary jewellery business. However, it was when he visited the jewellery gallery at the Victoria and Albert Museum in London that he knew his calling was to work with antique jewels and gems.

Arikawa's substantial collection, which consists mainly of jewels of Western manufacture, is an invaluable resource for jewellery historians and one that he continues to add to: while choosing pieces to stay in the collection, he sells others in order to buy additional jewels that will enhance the collection. He purchases these jewels, when they appear on the market, with the help of a few invited, Japanese patrons. Arikawa recognizes his collection's importance and is generous in lending pieces to museums and exhibitions around the world to inspire and spread the beauty of these jewels. His mission is to open a world-class jewellery museum, which would become part of public heritage, in Japan, and a world-first jewellery research centre in Paris. In an interview with Rachel Garrahan of the *New York Times* in January 2020, Arikawa described his jewels as being 'the ultimate world of paradise to human beings', and he believes 'beauty can make the world a better place'.

Throughout this book, significant sapphire jewels from Kazumi Arikawa's collection are included as invaluable examples to illustrate the chapters on historic jewels. When I started my research on coloured gemstones, I was very fortunate to be able to begin a correspondence with Mr Arikawa about his collection, and his answers are recorded here. It was a wonderful opportunity to speak to such a respected dealer about his collection. Since Mr Arikawa has generously assisted with all three books in this series, it feels like a special ending to the trilogy to include in *Sapphire* the collector behind this extraordinary collection.

Above: A platinum millegrain ring, *c.*1910, set
with a cushion-shaped Kashmir sapphire of 3.44
carats, flanked by two brilliant-cut diamonds
and diamond-set shoulders.

Facing page: An impressive necklace, *c.*1885,
of sapphire and diamond saltire-shaped links
between eight oval clusters, each set with
a cushion-shaped sapphire in a diamond-set
double surround, that can be separated and worn
as brooches. The six largest sapphires are from
Kashmir and together weigh 35.55 carats.

Joanna Hardy: What is it that makes you feel you have to own a certain piece of jewellery?
Kazumi Arikawa: Perhaps it is something in my soul that urges me onwards.

Is there a particular period or style you prefer to collect?
There is beauty in objects of any age or style. I'm more interested in the quality of each piece than in specific categories.

You spent two years in a Zen temple; did your experience change your appreciation of beauty?
My Buddhist training taught me about spirituality. It is an understanding that has something in common with concepts of beauty.

But do concepts of beauty not differ from person to person?
Sense of beauty varies according to personal preference, and there is also the issue of one's own individual experiences; nevertheless, I still believe that a universal beauty exists.

Does it matter where the sapphires in your collection have come from? Is origin important, or is it just the beauty of the stone?
The beauty of the stone is what matters. Although the superb beauty of Kashmir sapphires is unique, some Burmese and Sri Lankan sapphires are also amazingly beautiful.

Do you feel your mother's business influenced you? You mentioned feeling a sense of familiarity on your first visit to the Victoria and Albert Museum's jewellery gallery, a visit you have said helped spark your interest in collecting jewellery.
I already had some experience dealing in jewellery when I visited the V&A; it was more a thrilling shock than a sense of familiarity, or of relief.

We are custodians of objects, as they live on after we die. Do you have a feeling of responsibility to the jewellery industry? Do you like to own pieces so that they are safe with you?
I feel a big responsibility. I would like to collect as many high-quality jewellery items as possible, in order to protect them.

How many pieces do you have in your collection?
Its main body consists of some 700 museum-quality pieces. I hope to collect more while keeping the same high quality.

What will happen to your collection? Nothing is eternal, so do you worry that these objects may not last for ever?
I am hoping to create a museum for them. I understand their impermanence, but I would still hope to leave them in some sort of public collection.

You are very generous in lending your pieces to museums and for reproduction in books, so do you hope to touch the hearts of many through showing beauty in objects? Do you sometimes sell pieces?
I like to give people a sense of wonderment when seeing beautiful jewels, and I will always continue to collect. Sometimes I sell pieces, but that is not because I have fallen out of love with them.

You always acquire the best you can buy; what do you think about items increasing in value?
High-quality work is expensive, but prices are reasonable if you think in terms of the work's essential value. The better-quality pieces will always increase in value.

Do you ever wonder what the people were like who owned the pieces?
Sometimes, but that is not what interests me the most. The piece has to have an aesthetic vitality: that to me is everything.

Sometimes I feel history can glorify people unnecessarily; how important is a piece's provenance for you?
Provenance is important, and knowing it conveys a certain allure; however, the beauty of the piece itself is more crucial.

Japanese craftsmanship is superb; do you collect Japanese works of art?
Yes, I do. I've also been collecting Buddhist art for 40 years.

Do you believe that craftspeople can reach perfection?
Craftspeople might strive for perfection to realize their ideas, but, the moment an idea is realized, perfection moves to a higher level. In that sense perfection evolves and goes on for ever.

A medieval gold stirrup ring, made *c.*1400, set with a sapphire that would have acted as a marker of status and as an amulet.

SHOWSTOPPERS

A necklace, 2018, mounted by Etcetera, set with
a 396.89-carat Burmese sapphire on a necklace of
rose-cut sapphires and marquise- and circular-cut
diamonds. This sapphire entered the *Guinness Book
of World Records* for being the largest cabochon-
cut sapphire ever discovered in Myanmar (Burma).
Sapphire dimensions: 42 x 32 x 29 mm.

Showstoppers

There is always great excitement when the biggest and the best of any gemstone is found and brought to market, let alone when considering the possible cachet of owning these showstopping gems. The stones featured in this selection are spectacular gems the world has heard about, but there may of course be others that are still below ground or in a vault.

The majority of large sapphires that have appeared, or reappeared, on the market are Ceylon sapphires, as the Sri Lankan deposits are the only ones known to have produced large crystals. The three exceptionally large faceted sapphires here – the Logan Sapphire, the Blue Giant of the Orient and the Blue Belle of Asia – are masterpieces of nature. It is extremely rare to see large crystals with an even intensity of colour; so often the colour in sapphires displays banding or patches of obvious light and dark areas within the stone. These stones were first cut more than 100 years ago near their origins in Sri Lanka (then Ceylon) by highly skilled lapidaries, who will have taken months to complete the cut. The starting weights of all these gems would have been at least double, or more, before the first cut was made.

Yet size is not everything. Both Kashmiri and Burmese deposits generally produce smaller sapphire crystals that are still highly sought after due to their rarity and individual characteristics – found only in stones from these specific locations. Of course, there are always exceptions, such as the 369.89-carat sapphire on the facing page, which has been cut from a very rare, large Burmese crystal. The Kashmiri deposits have not been mined since 1908, so any Kashmir sapphires on the market will predate this. The sapphire deposits in Myanmar (Burma) are producing fewer gem-quality sapphires, as the existing sites have been mined for over 800 years and are now close to cessation. Sapphires are therefore harder to find.

In recent decades, Western markets have started to appreciate the rarity of other coloured sapphires, especially the padparadscha sapphire following its use in the engagement ring made recently for Princess Eugenie. Demand for other colours has soared, but there are very few such sapphires on the market; so many sapphires will be heated and treated to enhance their colour to accommodate this demand.

I have included the Siren of Serendip, a heated sapphire, because it is an impressive stone that is part of a finely crafted jewel made by some of Britain's best craftspeople. It is the combination of the stone and its necklace that makes it a showstopper. Heating sapphires is, in fact, a skill in itself required of many specialists in the gemstone industry; the wrong combination of heat and time can ruin a crystal.

Some of these sapphires have been sold at public auctions and achieved record prices. I have not cited these prices as I believe that these stones should be judged less by price and location than by the stone's impact on the eye. A price acknowledges the financial appreciation of the gem at a specific time; depreciation in monetary value and increases in desirability will always differ over time.

The 486.52-carat Blue Giant of the
Orient, mined and cut in Ceylon
(Sri Lanka) around 1907.

Reputed to be the world's largest faceted blue sapphire, the Blue Giant of the Orient was reported in the Ceylon *Morning Leader* on 23 August 1907 as being 'a monster sapphire worth £7,000'. It had been discovered in the Ratnapura district and was being offered for sale. The rough stone weighed approximately 600 carats and was purchased by Macan Markar, leading Sri Lankan jewellery and gemstone exporters, now for more than 150 years. After many months of planning and studying the crystal, this family business tasked one of their experienced gem cutters with cutting and polishing it. Weighing approximately 466 carats after cutting (which differs from its stated weight of 486.52 carats when it was later gemmologically assessed and offered at auction by Christie's; the stated 466 carats is likely to

have been a reporting error), the Blue Giant was described as a beautiful, brilliant, medium-blue coloured gem.

After the gem was cut, it was sold to an anonymous American collector. For nearly 100 years, the gem was not heard of, until it suddenly appeared in a Christie's Magnificent Jewels auction in 2004. The sapphire, set in a brooch of platinum with a pavé-set diamond surround, was heralded as the largest faceted sapphire ever to be offered at a public auction. With its display of a good, full-colour saturation, it was an extremely rare opportunity for someone to own such a significant sapphire. Only time will tell whether it will be another 100 years before this stone surfaces again.

The 422.99-carat Logan Sapphire, named after
Rebecca Pollard 'Polly' Guggenheim Logan, who
in 1960 donated the sapphire in its diamond-set
surround to the National Museum of Natural
History in Washington, DC.

At the time of writing, and unless a larger stone is hiding in a safe somewhere, the Logan Sapphire is the third largest faceted sapphire in the world, at 422.99 carats, after the Blue Giant of the Orient (facing page) and Queen Marie of Romania's sapphire of 478.68 carats, profiled in the chapter At Court.

This crystal would have most likely originated from Sri Lanka, in the area of Ratnapura, but exactly when is a mystery. Large stones, if found in the last century, would sometimes have been reported in local newspapers because their surfacing generated such excitement. As there is no known record of a sapphire of a similar weight to the Logan Sapphire, we can surmise that the stone was found and cut before 1900. Gemstones have been mined in Sri Lanka for more than 2,500 years. As a result of their long mining history, Sri Lankan gem traders try to protect gem cutting in their country and are renowned for their attempts to avoid exporting their rough crystals for cutting — a practice that dates back to ancient times. Though some stones may have been recut later in Europe to reflect tastes, it is not known if this was the case with the Logan Sapphire.

The sapphire has a wide table, which allows your eye to easily enter the stone. The stone is said to have good clarity, but recent developments in photomicrography would undoubtedly uncover inclusions that the naked eye and a 10x lens cannot perceive. The colour is a wonderful, rich blue, with a secondary, violet undertone, and it has a remarkable evenness of colour saturation. The cutter has used his expertise to maximize the stone's colour potential — a skill that has been developed and passed down through generations.

The Logan Sapphire, which is set in a brooch surrounded by 20 cushion-shaped diamonds, is named after Rebecca Pollard 'Polly' Guggenheim Logan, a philanthropist and patron of the arts, who in 1960 donated the stone to the National Museum of Natural History, part of the Smithsonian Institution, in Washington, DC. She retained the sapphire in her possession before it went on display at the museum in 1971. The sapphire had previously been owned by Sir Ellice Victor Sassoon, 3rd Baronet of Bombay.

Sri Lankan sapphire deposits have a reputation for yielding very large crystals; they live up to this reputation in the Blue Belle of Asia. This extraordinary blue sapphire, said to have been found in a paddy field in Ratnapura, Ceylon, in 1926, reappeared in a Christie's Geneva auction in 2014 with a finished weight of 392.52 carats.

According to Christie's, the Blue Belle of Asia, as the stone came to be known, was cut and polished between 1926 and 1928. The well-known local exporters of gemstones Macan Markar owned the stone at the time, but then in 1937 sold it to William Richard Morris, 1st Viscount Nuffield.

Lord Nuffield founded Morris Motors Ltd, a British car manufacturer that produced its first car, the Morris Oxford, in 1913 and went on to become a huge success. With no heirs to his fortune, Lord Nuffield began to plan his succession. His youthful ambition was to become a surgeon and, though his parents could not afford the fees for him to train as a surgeon himself, his later business success enabled him to become one of the biggest benefactors to medicine the Commonwealth had ever seen. In his lifetime he gave away more than £28 million and established the Nuffield Foundation, the Nuffield Trust and Nuffield College, providing funding for scientific research and innovative projects to support education and social policies.

It is said that Lord Nuffield bought the Blue Belle of Asia with the intention of offering it to the Queen Mother to celebrate her coronation in May 1937, but that did not seem to happen; what became of the stone for the next 35 years remains a mystery. It was not until the 1970s that the stone appeared in the inventory of a well-known Swiss gem dealer, Theodore Horovitz, only to disappear again from public view for another 50 years. It resurfaced in the Christie's auction in Geneva in 2014 with the announcement that it was 'one of the most prestigious coloured gems to have come to the market for many years, worthy of any leading collection'.

The Blue Belle was set in a multi-strand diamond necklace with diamond-set tassels cascading from the pendant. The jewel generated strong interest and, after competitive bidding, it achieved the highest price ever paid for a sapphire at auction. It was bought by an anonymous private buyer.

Above and facing page: The 392.52-carat cushion-shaped Blue Belle of Asia sapphire from Sri Lanka, set in a necklace with tassels of oval and brilliant-cut diamonds.

In the early twentieth century, miners in Ceylon (modern-day Sri Lanka) reportedly found a rough crystal with an astonishing weight of 2,670 carats. Like some other huge crystals, it was kept under wraps for a number of years. It was then cut as a modified cushion-shaped mixed-cut weighing 422.66 carats, with the dimensions 48.18 x 36.98 x 25.52 millimetres. At some point the stone was heated, a process that can enhance a gem's colour and some-times improve its clarity.

The Houston Museum of Natural Science (HMNS) acquired the sapphire in 2018. For the grand unveiling, the curator wanted to reveal the sapphire in a jewel that complemented such an important gem, rather than exhibiting the stone loose. Joel Bartsch, president and CEO of the museum and curator of its Cullen Hall of Gems and Minerals, had met Hans-Jürgen Henn 40 years before at the popular annual Tuscon Gem and Mineral Show. Bartsch and Henn, the creative director of Henn, a family gemstone business from Germany, struck up a friendship and stayed in close contact. Henn's sons, Ingo and Axel, both followed their father's love of gemstones and the three established Henn of London, a company that specializes in bespoke gem-set jewellery commissions and collection pieces. When HMNS acquired the sapphire, it was logical that Bartsch should commission Henn of London to make a necklace that was fitting for the stone.

After 20 proposals for the necklace, all sketched by Henn of London, three were shortlisted and Bartsch chose the final drawing. The result was a beautifully crafted necklace created to house one of the most important sapphires in history. Far from an easy task, the piece had to be made to the exact specifications of the very deep sapphire. Since the gem was not allowed to leave Houston due to its extremely high value, the craftspeople in the Henn workshops in London had to work with a replica. After 10 months of work, creative director Ingo Henn and their top setter travelled to Houston to finally set the stone. The museum built a workbench on site to enable the Henn of London team to complete this task.

On 3 March 2019, with great fanfare, the Siren of Serendip necklace – whose name references both the dangerous but irresistible creatures of Greek mythology and the ancient Persian name for Ceylon – was unveiled in the Lester and Sue Smith Gem Vault of the Houston museum. When seeing the stone set in the finished necklace, Ingo Henn recalls being able to feel the stone's magnificence. Knowing how many children visited the museum, their noses pressed to the glass showcases, made him realize the power of gems and jewellery to encourage, inspire and initiate thought.

Facing page: The Siren of Serendip necklace, 2019, created by Henn of London for the Houston Museum of Natural Science. This necklace and its 422.66-carat Ceylon sapphire were brought together at the museum by setter Ian Read, pictured above, to a design (top) by Jennifer Bloy.

Though it may have an unusual name for a sapphire, the stone itself is nevertheless spectacular. It has been tumbled and polished, following the natural shape of the rough crystal, which would have originated in Sri Lanka. Today, it weighs 547.71 carats and measures approximately 60 x 30 millimetres. The sapphire was originally presented in 1698 by the Russian Tsar Peter I (1672–1725) to his friend and ally, Augustus II the Strong (1670–1733). Its slightly humorous name – presumably a reference to the shape of Peter I's nose – may have been a later attribution and belies what would have been a very important diplomatic gesture, intended to bring good fortune to the recipient.

Augustus the Strong, Elector of Saxony and King of the Polish-Lithuanian Commonwealth, was a collector of art and objets d'art; his seat in Dresden established the city as a centre for cultural patronage. His palace was lavishly decorated in the baroque style and was home to his treasury, full of secret jewels. The rooms holding his jewels are known to this day as the Green Vault, due to the colour of the interior decor. Supposedly, when Peter I first toured Europe, this spectacular palace was his first stop; after many subsequent visits, Peter built his own treasury and took the Green Vault as his inspiration. During his reign, Augustus the Strong decided to open the treasury – now part of the Dresden State Art Collections – to the public. Peter the Great's Nose is still displayed in the Green Vault, even escaping the 2019 heist that has left many other jewels missing.

A tumbled and polished sapphire crystal of 547.71 carats presented in 1698 to Augustus II the Strong by his ally Tsar Peter I, pictured above in an engraving from 1700 and after whom this stone has gained its name: Peter the Great's Nose.

This is an exceptionally rare 62.02-carat Burmese sapphire because stones of its size and clarity are uncommon from Myanmar (Burma). The stone's clarity means that its rectangular step cut, which is an unforgiving shape, emphasizes the intensity of its rich royal blue colour.

Rockefeller is a name synonymous with the American Dream; John D Rockefeller (1839–1937) is still considered to be the wealthiest American of all time. He amassed his fortune by providing inexpensive oil to the United States, reputedly controlling 90 per cent of all the oil in the US. With his considerable wealth he was able to choose initiatives to support and he became known for his philanthropy, particularly for setting up educational institutions and supporting medical research. His only son, John D Rockefeller Jr (1874–1960), followed in his father's philanthropic footsteps during the Great Depression and is best remembered for building the Rockefeller Center, which opened in 1933.

Rockefeller Jr was an appreciator of the arts and bought much of his jewellery from jeweller Raymond Yard. Yard, along with close friend and world-leading expert on gems Raphael Esmerian (1903–1976), helped advise him on his purchase of this sapphire, reputedly in 1934 from the Nizam of Hyderabad. Rockefeller had the then even larger sapphire recut by Cartier to 66 carats and mounted as a brooch for his wife, Abby Aldrich Rockefeller (1874–1948). After Abby passed away, Raymond Yard redesigned the brooch for Rockefeller's second wife, Martha Baird Rockefeller (1895–1971).

In 1971, the Rockefeller family sold the sapphire to Raphael Esmerian who then sold it on to an Italian private client. The sapphire continued its journey and, after being sold by the Italian family, was bought back by Raphael's son, Ralph Esmerian; Ralph had the stone recut to its final weight of 62.02 carats, mounted in a platinum ring and sold to an American private collector, who already owned exceptionally rare gemstones. That collection, including this sapphire, was eventually sold at a 1988 Sotheby's auction in St Moritz, this gem fetching a world-record price at the time. The buyer was, again, none other than Ralph Esmerian. As he was familiar with the exceptional rarity and provenance of the stone, he knew that he would be able to tempt another gemstone connoisseur to be the gem's next custodian.

The Rockefeller Sapphire, now set in a ring mount made by Tiffany, last went up for auction at Christie's New York in 2001.

The 62.02-carat Rockefeller Sapphire, a Burmese
stone set with cut-cornered triangular-cut
diamonds in a ring mount by Tiffany & Co, once
owned by John D Rockefeller Jr.

These ear pendants are set with exceptional sapphires, weighing 26.66 carats and 20.88 carats, respectively; both display the unique characteristics of Kashmir gems, particularly an intense blue colour that also has a soft, velvety appearance. Sometimes the colour has been compared to the vibrant blue hue of the cornflower. The large size of these stones suggests that they were likely to have been cut from two different crystals and may have even been found during different mining seasons. It can take years to find stones that match in colour and in final shape once cut, which is why this pair is particularly special.

These ear pendants boast a rich provenance. In 1905, they were given to Odile de la Chapelle de Jumilhac de Richelieu as a wedding gift on the occasion of her marriage to Count Gabriel de la Rochefoucauld, Prince de la Rochefoucauld. The couple were friends of the hugely influential writer Marcel Proust and even appeared as characters in his work. The quality of these sapphires suggests that they were from Kashmir's Old Mine and, given the time it would have taken for them to be matched, cut and set, Odile may very well have been their first owner. As mementos of her wedding day, these earrings remained precious for Odile and stayed in her collection. She presumably then passed them to family members to enjoy, since they appeared at auction in 2013 as one of 14 lots comprising a small portion of her jewellery collection, listed as 'Property of a Lady of Title'. The earrings were successfully sold by Sotheby's Geneva for double their auction estimate, a result that reflected their rarity. The design of the ear pendants has changed three times since 1905 as fashions have evolved, with the stones alternating between diagonal and straight settings. In their latest incarnation, following their sale by Sotheby's, the Richelieu Sapphires now reside in beautiful, elegant platinum mounts made by Cartier.

The Richelieu Sapphires,
ear pendants mounted by Cartier, with two
exquisitely matched cushion-shaped Kashmir
sapphires of 26.66 and 20.88 carats, respectively,
given to Odile de la Chapelle de Jumilhac de
Richelieu as a wedding gift in 1905.

This eye-catching 1930s brooch is set with two remarkable cushion-shaped Kashmir sapphires, one weighing 55.19 carats — making it the largest Kashmir sapphire to come to auction to date — and the other 25.97 carats. Both stones are incredibly important in the world of blue sapphires because their size and quality make them extremely rare. Sapphires were only mined in Kashmir during a very short period in the late nineteenth century, and certainly not every stone was of gem quality. Both these sapphires display the typical Kashmir characteristic: a rich, velvety appearance due to very fine clouds of dispersed iron and titanium nanoparticles that scatter the light and give the stones the 'sleepy' or 'hazy' appearance that is unique to sapphires from the region.

These sapphires were gifted to Hariot Hamilton-Temple-Blackwood (1843–1936), Marchioness of Dufferin and Ava, in 1888, the same year the old mine was depleted. Lady Dufferin was married to Frederick Hamilton-Temple-Blackwood (1826–1903), 5th Baron Dufferin and Clandeboye (with multiple more baronetcies to his name), who was 17 years her senior. He served as Under-Secretary of State for India, and later Viceroy between 1884 and 1888, and as Chancellor of the Duchy of Lancaster under Prime Minister William Gladstone, before accepting the position of Governor-General of Canada. Lord and Lady Dufferin spent many years travelling abroad as British ambassadors, while managing to have seven children.

Lady Dufferin was acutely aware of the struggles women faced in India from the lack of medical treatment, and in 1885 she established the National Association for Supplying Female Medical Aid to the Women of India, the work of which included the erection of several hospitals across India. The Lady Dufferin Hospital, a maternity hospital in Karachi, Pakistan, remains in operation to this day.

The sapphires were passed down to Maureen Constance Guinness (1907–1998), heiress to the Guinness brewing fortune, who became Marchioness of Dufferin and Ava through her husband Basil Hamilton-Temple-Blackwood, 4th Marquess of Dufferin and Ava. They married in 1930, and family tradition has it that Maureen had the sapphires mounted by Cartier into this attractive stylized scroll brooch in the 1930s. The brooch remained within the family until recently appearing at auction with Sotheby's Geneva in 2021.

Two Kashmir sapphires of 55.19 and 25.97
carats, respectively, shown loose and set in a
1930s diamond-set brooch by Cartier. Formerly
in the collection of the Marchionesses
of Dufferin and Ava.

With such a limited supply of Kashmir sapphires mined during the later nineteenth and early twentieth centuries, the sale of an exceptional specimen creates quite a stir among the gemstone connoisseur community. Finding a stone that is over 28 carats, and of extraordinary quality, is truly momentous. In his description of this 28.18-carat sapphire, Christopher P Smith, president of the American Gemological Laboratory, stated, 'For the aficionado of Kashmir material, this gemstone represents a superb example of why Kashmir has earned the status for being the source for sapphire that all others are compared [to].'

As a fine Kashmir stone, this sapphire would have been mined in the late nineteenth century. It must then have lived in many other fabulous jewels until it was set in this Oscar Heyman & Brothers ring mount. The design is known as the Ballerina ring because its radiating tapered baguette-cut diamonds undulate around the stone to look like a tutu. George Heyman, youngest of the Heyman brothers, designed the first Ballerina ring in the 1950s and it quickly became a classic Oscar Heyman & Brothers design. The sapphire was sold twice by Sotheby's New York, first in 1983 and then again in 2014, this time in its Ballerina ring mount.

A square emerald-cut sapphire from
Kashmir weighing 28.18 carats; mounted
by Oscar Heyman & Brothers in a ring with
32 tapered baguette-cut diamonds arranged
to resemble a ballerina's tutu.

With the announcement of a royal couple's engagement, the ring is always of great interest. This was especially true in 2018 with news of the engagement of Princess Eugenie of Great Britain and Jack Brooksbank. Blue sapphires have been a popular choice of gemstone for many royal engagements, but the choice of colour this time was different. Princess Eugenie's sapphire displays a pinkish-orange colour and is known as a padparadscha sapphire.

The name comes from a Sanskrit word meaning 'lotus flower', as the colour resembles its petals. The padparadscha colour is one of the rarest colours for sapphires, although for many dealers it has quite wide parameters as there is not a definitive colour range; some padparadschas may seem more pink, with a hint of orange, or alternatively the reverse.

The two jewels shown here are examples of the wide variety of padparadscha sapphires, both being orange with a hint of pink.

The late Art Deco diamond-set clip brooch, *c.*1930, illustrated here on the left, features at its centre an ovoid modified step-cut padparadscha sapphire of 29.87 carats. The padparadscha set in a ring with a surround of pear-shaped diamonds, illustrated here on the right, is a similar size at 28.04 carats. It was sold by Christie's Hong Kong in 2017. Padparadscha sapphires have risen in value because of the growing appreciation and demand for these unusual sapphires.

Padparadschas originate from the historic mines of Sri Lanka and, more recently, from the newly discovered mines in Madagascar. Finding these stones is not easy, as there are so few natural stones available; their scarcity makes the royal couple's choice even more thrilling. Of course, now there is great interest for this coloured sapphire and, as a result, many sapphires are being heated and treated to alter their colour to one closer to a natural padparadscha stone.

Two pinkish-orange sapphires that highlight the colour range encompassed within the term padparadscha. Left: An Art Deco clip, *c.*1930, set with a 29.87-carat Ceylon sapphire in a diamond-set double surround. Right: A ring set with a 28.04-carat Ceylon sapphire in a tiered surround of marquise- and pear-shaped diamonds.

It is not only the blue sapphires that attract attention from gemstone collectors, as this pink sapphire testifies. A beautiful oval mixed-cut pink Ceylon sapphire, this stone weighs 95.45 carats. It was sold at Sotheby's Geneva in 2018 and, despite its significant estimate, the market got a surprise when it sold for more than double. In the auction catalogue, the sapphire was given a whole page to itself, a layout choice that emphasized the stone's importance and set the tone for future sales of pink sapphires. This jewel not only achieved a new world-record price for a pink sapphire, it showed how desirable pink sapphires of exceptional quality have become.

Unlike a pink diamond, which appears the same colour from every angle, pink sapphires display a secondary colour due to their crystal structure. When you view a sapphire from multiple angles, you will see a subtle change in hue: some have purplish tones, others lighter pinks. Cutters become experts at pairing these hues together when they choose the angles for the cut. The hues of this sapphire marry to become a wonderful candy-floss pink. It is rare to find a pink sapphire that is not heated or beryllium-treated to enhance its colour.

An oval pink Ceylon sapphire of 95.45 carats
in a diamond-set gold pendant.

21ST CENTURY

Thorned Branch necklace, 2018, by Shaun Leane,
set with a natural yellow sapphire of 162 carats,
encircled by diamond-set thorns.

<h1 style="text-align:center">21st Century</h1>

This chapter offers an explosion of colour, a 'sapphire heaven' inspired by sapphire's endless colour palette. The 18 jewellers I have chosen here to represent contemporary sapphire jewellery – a mix of big houses and individual designer-makers – all have three elements in common: fabulous craftsmanship, great designs and a discerning eye. They all recognize when they have been lucky enough to find a sapphire that is exceptional and rare.

High-quality sapphires are appreciated universally, whatever their geographical origins. Examples from Madagascar, Sri Lanka (formerly Ceylon), Kashmir, Montana and Burma are all represented in this chapter; it is the beauty of the stone that matters, not where it has come from. I have mentioned some known origins for individual pieces, but this is for reference only and not intended to imply that one stone source is more desirable than another.

Of emerald, sapphire and ruby, sapphire typically grows into the largest crystals; if these large crystals are also of gem quality, then you have stones of instant impact. Like all coloured gemstones, beauty is often in the eye of the beholder. Because of the vast array of sapphire colours, there is a colour for everybody. Coloured sapphires other than blue are still relatively unknown to the public, yet many of the jewellers represented here have embraced the full range of sapphire colours. With a hardness of 9 on the Mohs scale – one step below diamond at number 10, one of the world's hardest known natural materials – sapphires are harder than any other coloured gemstone. So, as a collector, you can buy beautiful pinks, yellows and other pale colours without the worry of scratching them (unless they rub against diamonds).

The selection of jewels in this chapter includes single stones that are incredibly rare because of their size, colour and clarity, as well as smaller stones that have been used to layer colours in a particular design. This range of stones shows the depth and breadth of possibilities that sapphire can present to every designer-maker, to the delight of every sapphire enthusiast and connoisseur.

Fabergé jewels today are greatly influenced by the jewellery house's legacy of jewels with colour and painterly expression, realized in gemstones and enamel. The house celebrated its relaunch in 2009 through a collaboration with the late Parisian jewellery designer Frédéric Zaavy (1964–2011). Colour continued to be integral in Zaavy and Fabergé's jewels – jewels that also drew on the house's fabled Russian past.

Fabergé x Frédéric Zaavy Vagabonde Bleue ring, 2009, set with a 12.46-carat unheated sapphire encircled by additional sapphires, including six larger sapphires totalling 10.04 carats, and diamonds.

Fabergé Emotion White and Rose Gold Pink Sapphire Grande ring, 2012, set with 14.84 carats of pink sapphires, pavé-set in waves of colour to honour the work of Fauvist painters.

In Kemps Corner, South Mumbai, at the end of a small side street, is a small red door that gives away very little of what lies within. The house of BHAGAT is led by the creative vision of Viren Bhagat, a fourth-generation jeweller, who runs the firm alongside his sons Varun and Jay. India has a long and intimate relationship with gemstones and jewellery – they are integral to its heritage and beliefs – and a tradition of combining enamels and gemstones in gold *kundan* settings. BHAGAT has broken with convention by using platinum, and they hunt all over the world for old stones of exceptional quality to grace each jewel.

Each jewel is drawn by hand, and their designs have paved the way for Indian jewellery to be seen in a completely different light. Inspired by traditional Indian forms and motifs, but informed by their minimalistic approach, the designs allow the gemstones to be the centre of attention. BHAGAT only uses the very best pearls, rubies, emeralds, diamonds, spinels and sapphires (all gemstones that have been part of India's heritage), travelling all over the world to find exceptional examples. They draw each design to actual size, as the finished jewel will be compared to the drawing to make sure all the proportions have been adhered to by their craftspeople. Such is the demand for a BHAGAT jewel that many are sold on the basis of the sketch alone and, given that only around 60 pieces are produced a year, there is always a waiting list.

The ring illustrated here is set with an important cabochon blue sapphire: a stunning, exceptional unheated sapphire from Burma weighing 15.63 carats, rare because of its intense colour and clarity. The richness of its colour is heightened by the sugarloaf cut, which is set in platinum with graduated faceted diamond roundels, terminating with two cabochon sapphires. The flower brooch above is set with specially cut sapphire cabochons, with the bare minimum of metal showing, creating a jewel that seems to be as delicate as BHAGAT's drawings.

From top: Paisley brooch, 2018, set with sapphire cabochons and marquise- and pear-shaped diamonds; and a ring, 2019, with an important 15.63-carat unheated sugarloaf sapphire from Burma set above another sapphire and joined by an open ring of diamond discs, terminating with two sapphires.

Christian Hemmerle, the fourth generation of this family of jewellers from Munich, says that when he looks at a sapphire he always remembers his mother's advice: a sapphire should never be too dark; it should still glow at night, since jewels are worn in the evening as well as during the day. Christian likes to choose sapphires for their 'open colour', a freshness that will never retreat at night. Besides selecting gemstones for their unique colour, Hemmerle also prizes stones that have 'life' or 'fire' and the ability to make your heart miss a beat. Sensitivity to these immaterial qualities distinguishes Hemmerle from other jewellers.

The most important aspect of any sapphire to Hemmerle is that its character 'speaks to you'. Hemmerle chooses sapphires of all colours, but a rich, mesmerizing shade of blue is always present. Some of the jewels may be pavé-set, with smaller stones chosen so that they graduate ever so slightly in tone to highlight a hue in the main stone. Matching a pair of sapphires may take years; Hemmerle will keep some exceptional stones just in case one day they may find another stone that complements one of them. The key to finding any beautiful stone is patience, since the pool of high-quality gems is small. Conversely, slight differences in a matched pair might be embraced by a Hemmerle design, especially in their earrings, highlighting sapphire's dichroic tones. This level of detail is almost meant to go unnoticed — a mark of Hemmerle's understated subtlety.

The sapphire selected for the brooch illustrated here looks like it has just dropped from the sky — I have never seen a sapphire so closely resemble a raindrop. The simplicity of this design suggests that the gem is simply resting on the open lattice. Yet to achieve this illusion takes a great feat of precision, a characteristic common to all Hemmerle settings. In another example, a step-cut 16.10-carat sapphire is set so exactly in its ring mount that the mount angles meet the corners of the sapphire to create lines unbroken by the design. Without the aid of computer technology, these entirely handmade settings go far beyond traditional goldsmithing.

Top: A brooch of interlocked bars, 2011, supporting a 29.44-carat unheated Burmese sapphire; flanked by aluminium and white-gold bangles set with green and blue sapphires, both 2017.

Left and right: Two pairs of earrings with shades of blue and pink sapphires, each brilliant-cut stone set so that its pavilion points outwards. The left pair with tassels of rock-crystal beads suspending sapphire briolettes dates from 2017; the right from 2016.

Below right: A ring set with a 16.10-carat unheated Ceylon sapphire, 2010.

Even as a young boy, Mish Tworkowski would make jewellery for his friends. A New Jersey native with Polish roots and an art history degree from Rutgers University, Mish creates his fine jewels in his own New York workshop and has shown them in a secret-garden of a Bond Street studio in downtown Manhattan and now in his elegant, tropical showroom in Palm Beach, Florida. Warm and congenial, Mish instantly puts you at ease, and is eager for you to sense the tactile quality of his pieces. As soon as you pick up one of his creations, your hands detect contrasting textures that result from his ingenious manipulation of material and surface.

For example, the cuff illustrated here, in 18-carat white gold with bezel-set sapphires and stones resembling the growth knots of a tree, simulates bark. Nothing simply shines in Mish's jewels; even his clasps are patterned. Other signature elements include moulded, textured T-bars and loop fastenings on necklaces and bracelets, which can be patterned to mimic the tessellating shapes of a fir cone or laced with gold veins weaving over pavé-set diamonds. I always look for the attention given to clasps, catches and fastenings, because attention there suggests uncompromising quality. This high standard is well understood by Mish, having been surrounded by a constantly changing selection and variety of the best art in the world while working for 15 years at Sotheby's.

Now in business for 30 years, jewellery and botany remain his two great passions. Mish has worked closely with various horticultural institutions and currently serves on the Board of Trustees of the New York Botanical Garden. He travels the world searching for gemstones unique in colour, such as the beautiful grey-blue sapphire set within the textured mount of the Honeywood Cocktail ring or the padparadscha sapphires set in the Ava earclips, with surrounds of apricot- and lavender-coloured sapphires chosen to highlight the colours in the petrified tree-fern drops.

From top: Arden ring, 2012, with a 29.68-carat emerald-cut sapphire and pavé-set diamonds; Honeywood Cocktail ring, 2012, set with a 13.87-carat cushion-shaped sapphire in a diamond-set honeycomb ring; Comet Tail earrings, 2016, of pavé-set sapphire circles; Ava earclips, 2013, set with 3.10- and 3.04-carat padparadscha sapphires, respectively, each surrounded by pavé-set apricot- and lavender-coloured sapphires and suspending drops of petrified tree-fern slices; and Jeweled Bark cuff, 2004, with sapphires, tanzanites, aquamarines and diamonds set in textured white gold.

Mellerio is a family business that has been making jewellery for more than 400 years, 200 years of which have been spent on Paris's famous Rue de la Paix, making it the first jewellers to occupy premises in this jewellery haven. Reputedly the oldest family jewellers in Europe, Mellerio's extensive archive charts the firm's rich heritage and includes a cornucopia of designs, including those that won a gold medal for excellence at the Paris Exposition Universelle in 1867. From jewels for European kings, queens and nobles to trophies for sporting stars, Mellerio has worked with a diverse range of clients. Laure-Isabelle Mellerio has been president and artistic director of the jewellery house since 2017. A gemmologist, she has designed jewels that display a wonderful appreciation of colour.

Sapphires offer the designer a wide palette of colour choices, and Mellerio's colour combinations are showstoppers. Their designs can be sophisticated and classic, as in this piece below that uses carved mother-of-pearl surrounding a central oval cabochon sapphire, and delightfully fun, like the three rings shown here set with padparadscha, purple and blue sapphires accompanied by bright tsavorites and rubies.

Clockwise from left: Pink Cadillac, Purple Rain and Midnight Blue rings from the Color Queen collection, 2020, set with a 2.22-carat unheated Madagascan padparadscha sapphire, a 2.42-carat unheated Madagascan purple sapphire and a 5.16-carat unheated Ceylon blue sapphire, respectively, in gem-encrusted gold surrounds; and the Madre Perla ring from the Isola Bella collection, 2019, featuring a 3.11-carat Ceylon sapphire set in a carved mother-of-pearl surround, accented by diamonds and sapphires.

High up in Japan's Rokko Mountains National Park, outside of Kobe, lies the workshop of Kaoru Kay Akihara, whose jewellery goes by the name of Gimel. Since 2003, Akihara's workshop has nestled here amongst the cherry trees, its predominantly glass walls reflecting the surrounding beauty and serenity into her workplace. Akihara believes that you can only create a beautiful jewel in a harmonious environment.

Akihara is fascinated with the natural world. Her desire to emulate the perfection she finds in nature is reflected in her exquisite attention to detail – nothing is missed – both in her observations and in the manufacturing of her jewels. Such is the expected precision of a Gimel jewel that sometimes her craftspeople need to work under a microscope to accomplish the level of detail required. When you turn a Gimel jewel over, for example, you are often rewarded with the sight of a tiny, jewelled insect: a hidden surprise that only you, the wearer, are privy to.

With her uncompromising eye, Akihara meticulously sorts her selection of gemstones to achieve either a gentle graduation of colour or a match of colours, as in the necklace illustrated here and set with blue sapphires. Sourcing matching sapphires for a necklace like this can take years, particularly as Gimel jewels are only set with the best, unheated gemstones. To maintain a consistently high level of craftsmanship, Akihara's artisans are all graduates from her own workshop. As she herself puts it, 'Using other craftsmen is unthinkable, the quality of the work differs too much'. Shortcuts are not an option: the focus of a Gimel jewel is always quality, however long the jewel takes to make.

Her sense of joie de vivre is captured in these joyful brooches depicting the activities of playing a violin, walking the dog, fishing or riding a horse. Gimel pieces are prime examples of fine jewellery and emblematic of Akihara's desire to leave behind 'something beautiful for the world to see'.

When I asked Akihara what qualities attracted her to sapphires, she replied that she particularly loves the clear blue ones, 'I feel their nobleness and elegance; not only is colour important, but so is transparency, without inclusions, and brightness.' A stone's origin is not important to Akihara; its beauty is what matters.

A delicate Morning Glory flower brooch, 2016, set with sapphires of graduating hue and demantoid garnets. On its reverse rests a sapphire-jewelled bug, inset here, atop the intricate back holes that are shaped around each gemstone.

A Gimel necklace, 2015, of graduated and carefully
matched sapphire and diamond clusters; surrounded
by Gimel's sapphire-set miniature characters
including Violinist, 1996, Conductor, 2003, Fisherman,
1998, Dressage, 2011, and Going for a Walk, 1998,
complete with two diamond dogs, plus two turtles of
blue and purple sapphires, both 2015.

Cartier's stone buyers only source exemplary gemstones. Its modern creations use the very best blue sapphires, Jacques Cartier's favourite stone. Good design never tires, as the necklace reproduced here shows. It echoes Cartier's emerald-, ruby- and sapphire-set 'Tutti Frutti' jewels, but has been recreated using only emeralds and sapphires, including a fabulous triangular carved sapphire weighing 110.29 carats. The green and blue colour combination first used by the *maison* in the early 1900s — inspired by the Indian enamels that Jacques Cartier had seen when he visited India — is also echoed in the ring centrally set with a 32.35-carat hexagon-shaped cabochon sapphire and seen here on page 288.

The Panther is another iconic design from the days of Jeanne Toussaint, whose feline character was associated with the big cat that has become a signature design for Cartier. Ever since the first 3D sculptural image of a panther jewel was commissioned in 1948 by the Duke of Windsor, panther jewels have continued to emerge from Cartier's ateliers every year. Designing a panther jewel is a precious undertaking; only select designers will be given the responsibility of adding a new one to Cartier's unique heritage. The necklace here on page 288 depicting two panthers stretched out to grasp a cabochon sapphire is unusual as the panther's spots, which are usually set with onyx, are set with different-shaped cut sapphire cabochons.

Forming a collection of sapphires can take years of diligence and patience, as gem-quality sapphires are rare. Sometimes, a heritage stone may be passed down through generations and may be recut to maximize the stone's potential; or, very occasionally, mining teams find a beautiful new crystal. The exceptional 29.06-carat blue sapphire in the diamond-set ring, below, has come from the prestigious old mines of Kashmir, which are located at an altitude of 4,000 metres and have not been mined since the early 1900s. It is an outstanding stone of an exceptional size, displaying the velvety and intense blue colour that is characteristic of top-quality sapphires from this region.

Historic stones have fascinating tales to tell — tales that can be pieced together through entries in ledger books, inventories, letters and diaries. One incredible sapphire with a long and eventful journey (which is mentioned in the chapter At Court) is the impressive Romanov Sapphire weighing 197.80 carats. Its first known appearance was in 1883 in St Petersburg, at a fancy-dress ball with the theme of sixteenth- and seventeenth-century Russian nobility hosted by the Grand Duke Vladimir in honour of his brother Emperor Alexander III. This sapphire was sewn onto the gown worn by Maria Feodorovna, who danced the night away. Eventually, the stone was documented by the Bolsheviks but disappeared soon afterwards, only to reappear mysteriously at Cartier New York. Cartier reset the sapphire into a pendant in 1928 and it was sold in 1929 to the opera singer Ganna Walska, who greatly admired its beauty. Having been married to millionaire Harold McCormick, years later she fell in love with her yoga teacher and, looking for a simpler life, in 1970 sold her jewels, including the Romanov Sapphire. The stone vanished again before re-emerging, 21 years later, at a public auction in 1992. Cartier acquired the sapphire in 2014 and reset it in a bracelet, where it is resplendent today (see page 289). This Ceylon sapphire can be replaced by a carved rock crystal as an alternative setting, both options crowned with two triangular-shaped diamonds.

Blue sapphires are the 'stone of awakening' for Buddhists, and yet can also be dedicated to the god of thunder, Indra, in India. Cartier's majestic necklace with cascades of flowing sapphire beads (see page 289), all carefully matched and complementing the colour of the central pear-shaped Ceylon sapphire weighing 22.44 carats, is an embodiment of this divine beauty and power.

The Cartier designers cleverly mix shapes and curves to emulate kinetic art. The four bracelets also on page 289 demonstrate how they give jewellery volume, with the illusion of mobility and energy, using blue sapphires to guide the eye to follow the moving lines.

Facing page and inset: A Cartier necklace, 2017, with its design above. At its centre hangs a 110.29-carat carved Burmese sapphire that can be removed and worn on a chain. The necklace is set with diamonds and additional carved sapphires and emeralds.

Left and right: Two views of the Cartier Bleu-Bleuet ring, 2014, set with an important cushion-shaped sapphire of 29.06 carats from Kashmir and a pair of triangular-shaped diamonds in a split-shouldered band of calibré-cut and brilliant-cut diamonds.

Above and inset: A sapphire-bead necklace, 2018,
with two crouching, sapphire-spotted panthers
flanking a 41.06-carat Ceylon sapphire cabochon,
with diamond and emerald drops; and three views
of a ring, 2017, set with a hexagon-shaped sapphire
cabochon of 32.35 carats and stepped sides of
uncut emerald crystals, encompassed by
a fan of pavé-set diamonds.

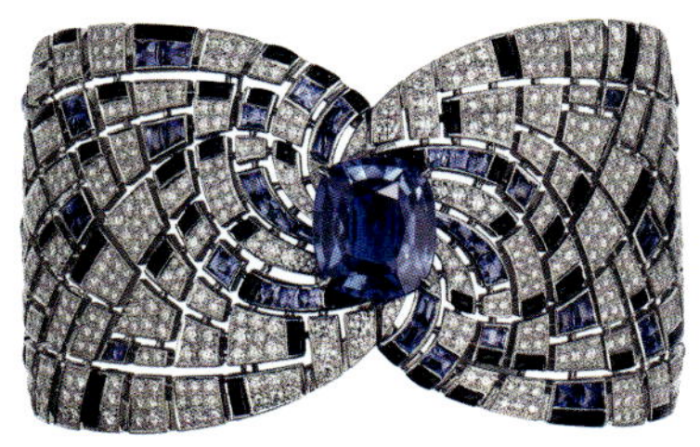

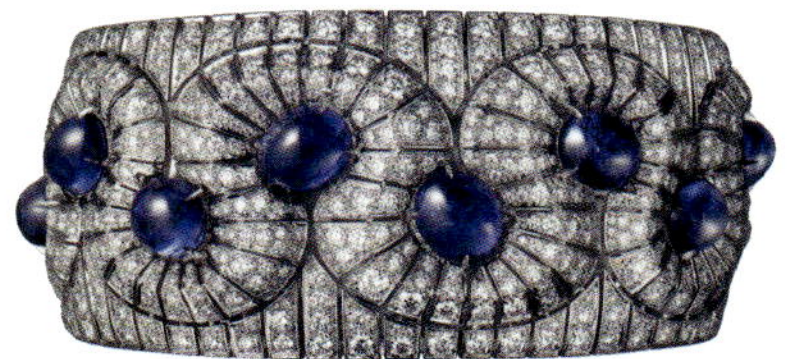

Top: A necklace by Cartier, 2014, with a fringe
of sapphire beads and diamond-set feathers
round a central 22.44-carat pear-shaped sapphire.

Above: Four Cartier bracelets set with sapphires
and diamonds: top left, 2014, with a 69.49-carat
oval-shaped Ceylon sapphire cabochon; top right,
2014, with an 8.24-carat cushion-shaped Ceylon
sapphire; bottom right, 2014, with 12 Burmese
sapphire cabochons, totalling 35.44 carats;
and bottom left, 2015, with design, set with the
Romanov Sapphire, a 197.80-carat cushion-shaped
rose-cut Ceylon sapphire that once belonged
to Maria Feodorovna (see page 118).

Colour is part of Boghossian's heritage and legacy; this jewellery house uses white diamonds to highlight the importance and rarity of the coloured gemstone. Boghossian is steered by the family's sixth generation of gem dealers and jewellers. For more than 100 years, this family has developed an intense understanding and deep-rooted knowledge of gemstones.

The Armenian Boghossian family originated in Mardin, Turkey, once at the centre of the Silk Road and now a UNESCO World Heritage site. Boghossian ancestors journeyed on horseback or by boat to Aleppo, Beirut and Cairo, constantly searching for the best gemstones and pearls to trade with merchants and jewellers. The later generations then moved to Europe, establishing the Boghossian headquarters in Geneva, Switzerland.

The meeting of East and West has continued to play an important role in the designs of Boghossian jewels. The house explores new ideas, pushing goldsmithing technical boundaries while at the same time echoing its traditions and heritage in its designs and choice of gemstones.

The briolette-cut sapphires in this necklace and the sapphires in both pairs of ear pendants are beautifully highlighted by the whiteness of the diamonds, jadeite and the natural pearls, allowing the blue sapphires to glisten. Simplicity and sophistication can be hard to achieve, but the craftspeople at Boghossian are masters at it.

Gem-quality Kashmir sapphires are extremely rare as they have not been mined since the 1920s and, before this time, only very seldom was a top-quality sapphire found. The 22.26-carat Kashmir sapphire set in the ring illustrated top right is therefore an exceptional stone.

Top: Two rings set with unheated Kashmir sapphires; the left, from 2019, an emerald-cut sapphire of 6.13 carats, the right, 2020, a cushion-shaped sapphire of 22.26 carats.

Left: Earrings, 2018, set with kite- and pear-shaped diamonds flanking a cushion-shaped unheated Burmese sapphire, each weighing 11.09 and 9.09 carats, respectively, and backed by carved white jadeite.

Centre: A necklace, 2012, of draped pear-shaped diamonds suspending two unheated Burmese sapphire briolettes of 56.67 and 37.44 carats, respectively.

Right: Earrings, 2017, set with unheated Kashmir sapphires of 8.68 and 8.60 carats, respectively, each paired with a pear-shaped diamond and a drop-shaped natural saltwater pearl in a diamond-set surround.

There is nothing Bina Goenka and her dedicated team of craftspeople cannot achieve in her Mumbai workshop. Her imagination is endless: every piece is a unique combination of materials and technique – all the more extraordinary as each one is made by hand without the aid of digital technology.

Surrounded by the jewellery trade during her early working life, Goenka recognized that traditional Indian jewellery made for wedding dowries had not changed for centuries and was therefore keen to create jewellery that was not constrained by tradition. She hopes that her bespoke pieces will be handed down for future generations to cherish and never broken up. For her the value is as much in the making of the jewel as in the materials used; although this is a concept that has not always been well understood, there is a new generation of discerning buyers who understand the exceptional skills and originality of a Bina Goenka jewel and appreciate that it will have taken months to complete. For traditional Indian craftspeople, these jewels are quite out of the ordinary. Goenka has developed a very close connection to her jewellers in order to convey her vision for each jewel. Her craftspeople are her family and she invests heavily in their training.

Hearing Goenka talk about creations that are still inside her head is infectious; she is constantly dreaming, designing, thinking of gems that she needs to find, and she travels far and wide to find what she is looking for, which can sometimes take years.

A recent trip to see the 2020 exhibition *Kimono: Kyoto to Catwalk* at the Victoria and Albert Museum inspired Goenka to design the Ikebana cuff, illustrated here, based on Japan's national flower, the cherry blossom, or *sakura*. A branch of the cherry tree twists round the wrist revealing cherry blossoms just burst open. Goenka wanted this cuff to be a reflection of spring's new vibrant colours, which is why the flower heads are set with sapphires of all hues. The branch is set with yellow diamonds as if the sun were shining on it. The Ikebana cuff is a celebration of colour and joy.

Three views of Bina Goenka's Ikebana cuff,
2020, depicting a blossoming branch that twists
around the wrist, set with coloured diamonds,
rhodochrosite, emeralds, and flowers of 62.08
carats of sapphires of every colour.

Chaumet's contemporary jewels are a successful marriage of heritage and modernity, taking inspiration from its extensive archives. Its workshop in Place Vendôme has been creating iconic jewels for the past 240 years. Chaumet's founder, Marie-Étienne Nitot, became one of the most sought-after jewellers in Europe and, as court jewellers and official jeweller to Napoleon, the *maison* received a great many prestigious commissions. Head ornaments were key additions to the attire of any woman attending state and important occasions and, since 1780, Chaumet has created over 2,000 unique tiaras. Incredibly, the demand for tiaras has never ceased, as can be seen with the recently created laurel leaf Firmament Apollinien tiara, centrally set with an electric blue sapphire and with cabochon sapphires decorating the band.

Chaumet's contemporary creations are an explosion of colour that utilize a wide variety of coloured sapphires to complement the designs. The Pastorale Anglaise brooch (see page 294) is a stunning example of craft echoing Louis XVI's fashion for bows and antique Chaumet jewels. Setting gemstones on a curve is an extremely difficult process, and this jewel is a testament to Chaumet's highly skilled craftspeople.

Top: Firmament Apollinien tiara from the
La Nature de Chaumet collection, 2016,
of diamond-set leaves and sapphire berries.

Above: Rhapsodie Transatlantique ring from the
Chaumet est une Fête collection, 2017, set with
a padparadscha and a purple sapphire; and
Espiègleries lion brooch from the Les Mondes
de Chaumet collection, 2018, set with blue,
yellow and pink sapphires.

12 Vendôme necklace by Chaumet,
2013, of tanzanite beads and a diamond bow
suspending blue, pink and purple sapphires.
Lueurs d'Orage earrings, left, with pink
sapphires, and Lueurs d'Orage brooch, right,
with purple sapphires, both from the
Les Ciels de Chaumet collection, 2019.

Above and right: Nuages d'Or necklace and Lueurs
d'Orage watch from the Les Ciels de Chaumet
collection, 2019; the necklace of diamond clouds
suspends a yellow sapphire, and the watch
bracelet is set with purple, yellow, orange and
padparadscha sapphires.

Left: Pastorale Anglaise brooch, 2017,
inspired by the bow jewels of Chaumet's past,
set with blue and yellow sapphires,
rubies, emeralds and diamonds.

SILVIA
FURMANOVICH

Silvia Furmanovich's connection to jewellery lies in her memories of her father, a goldsmith who died when she was just 17, working at his bench making jewels. Establishing her own company in 2003 in São Paulo, Furmanovich is a woman whose passion for world cultures, their crafts and techniques is channelled through her jewellery. Refreshingly unconventional, her inquisitive mind never rests; visiting museums and travelling to different countries allows her to explore various crafts, which send her on a journey of research that will eventually form the basis of a new collection. These journeys can be anything between two and three years in the planning.

Furmanovich is passionate about the Amazon rainforest, its flora and fauna, and over the past six years has collaborated closely with artisans deep within the rainforest to develop marquetry techniques for jewellery that reflects the materials and wildlife of the rainforest. This intricate marquetry has become her signature technique. Using the off-cuts from their furniture marquetry work, along with their secret recipe for bringing out the natural colours of the wood without using pigment or dye, Furmanovich's craftspeople have created scenes and patterns – astonishing given the tiny scale. Furmanovich has combined these small works of art with complementary coloured gemstones such as sapphires. The grain of the wood is as unique as the inclusions found in gemstones; nothing is ever the same twice.

Another technique that fascinates this designer is Indian miniature painting. A visit to an exhibition of Rajput painting at the Metropolitan Museum of Art in New York inspired Furmanovich to spend time in Udaipur, India, where she worked with a school that teaches this ancient technique. Miniaturists here use a single camel's eyelash as a brush for the more detailed designs.

Sustainability and responsible sourcing are of paramount importance to Furmanovich. In 2020 she launched a new collection, Amazonia Bamboo, that uses sustainable grass. After learning how to work with bamboo in Japan, where the material is integral to everyday life, Furmanovich returned to Brazil and collaborated with the Instituto Jatobás to develop bamboo cultivation in her homeland. She was the first person to utilize Brazilian bamboo in sustainable products, working alongside craftspeople to teach them Japanese knot-weaving techniques with which to make household products alongside her jewellery range. She conveys her thoughts and ideas to her team using 'mood boards', never drawing or making prototypes, because she knows they understand how to interpret her vison. Furmanovich and her craftspeople then weave and bend the bamboo into curved shapes before dipping these into boiling water containing a pigment. The bamboo is porous and absorbs the permanent dye quite readily. The jewels are delicately complemented by coloured gemstones, as seen in the jewels illustrated that are set with sapphires.

Silvia Furmanovich chooses sapphires for her earrings, above, to complement the natural tones she achieves in her marquetry (middle right, 2019, with sapphire, tanzanite and diamond; and bottom left, 2018, with sapphire and diamond), miniature paintings (top left, 2017, with sapphire and diamond) and woven bamboo jewels (bottom right, 2020, with sapphire and diamond). Also, top right: sapphire- and emerald-set frogs on druzy agate leaves from pear-shaped sapphires, 2019; and middle left: sapphires and diamonds set in fan shapes, 2012.

VAN CLEEF & ARPELS

The craftspeople at Van Cleef & Arpels are masters when it comes to interpreting jewellery designs in colour. Graduating coloured gemstones is an art in itself; choosing the correct hues for a design cannot be rushed, as the wrong colour will easily stand out and destroy the mood of the jewel.

Blue sapphires are used in Van Cleef & Arpels' signature Mystery Set unicorn clip. This brooch is meticulous in its uniformity of sapphire colour, but its stone selection would have taken months to assemble as sapphires can be very hard to match. There is a tendency for natural blue sapphires to have an unevenness of colour saturation, which would be highlighted when used in the Mystery Set technique because the stones are set immediately next to each other.

Madagascar is home to the lemur and also a source for multi-coloured sapphires: two points of inspiration that have been combined in these wonderful Van Cleef & Arpels pieces depicting lemurs. The clip and the ring use a variety of coloured sapphires. All these stones are as nature intended: they have not been heated nor their colour enhanced.

Exemplary in their craftsmanship and design, these jewels have been created by a team of artists whose skilful choice of colourful sapphires conveys a sense of joy and wonder.

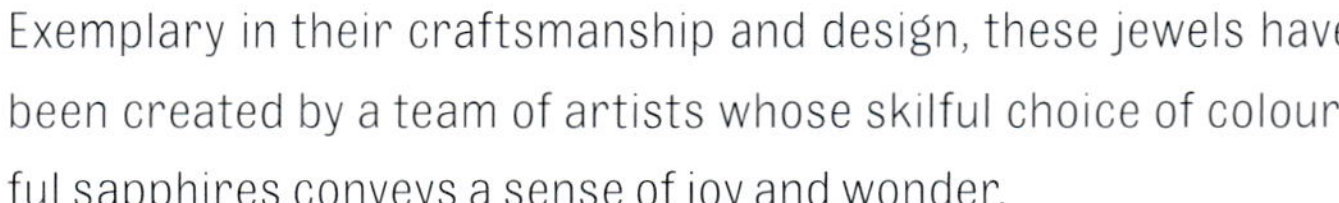

Left, from top: Makis clip and ring, 2010, with diamond-set lemurs holding yellow Sri Lankan sapphires of 20.22 and 10.98 carats, respectively; Licorne clip, 2016, with carefully matched Mystery Set sapphires; and Petit Pavot clip, 2015, set with graduated tones of blue sapphires.

Clockwise from top: Pansy earrings, 2011, each set with a yellow Sri Lankan sapphire, totalling 12.79 carats, and with purple and blue sapphire petals; Libellules clips, 2016, the two dragonflies set with pink, purple and blue sapphires; Pensées clip, 2015, and Mystérieux Butterfly transformable clip, both set with pink and purple sapphires, garnets, tourmaline and spinels; Fennecs clip, 2016, with twpo pink and yellow sapphire-set fennec foxes.

Above centre: Three Van Cleef & Arpels Princesse clips; Princesse Séléné, 2018, Robe Couleur du Soleil, 2014, and Princesse Mer Méditerranée, 2015, all set with sapphires and diamonds in white gold.

From Van Cleef & Arpels, the reversible
Antennae necklace, 2021, from the Sous les
Étoiles collection, set with 154 pink and mauve
Madagascan sapphires, weighing a total of
417.33 carats, collected over two years by
Van Cleef & Arpels gemmologists.
The sapphires are set in articulated
collets to drape like fabric on the neck.

The sapphires in Lauren Adriana's jewels are all of natural colour, which makes the job of finding such stones extremely difficult. Sometimes, Adriana turns to old jewels for inspiration, as seen with the rococo padparadscha sapphire ring. The special cut of this 4.48-carat peach-coloured Madagascan sapphire mirrors the cut of old diamonds; it is a wide pear-shaped stone with a large table and culet, which give it a watery appearance. Many people think that they need to recut old stones to bring out their full potential, yet a gemstone's naïve original cut might be what makes it beautiful and preserves the heritage of its shape and journey. In this jewel, the shape of the ring echoes that of the stone; Adriana has chosen the stone and its cut first and has designed a jewel around it.

The delightful Zig-Zag pair of earrings set with naturally pale emerald-cut sapphires are so clever — elegant and unique in their design and sapphire colour selection. Traditional sapphire dealers will see a pale sapphire and think that they must treat the stones with heat to make them brighter and more saleable; these stones have escaped that fate. Intense colour is always appealing because vibrancy generates an instant reaction, but stones do not need intensity of colour to be desirable. As most sapphires are

in fact treated to make them brighter, finding sapphires with just a blush of colour is rare and very few people work with them as a result. Sourcing this collection of sapphires will have been almost as difficult as finding a fabulous example from Kashmir. I expect that pale-coloured sapphires will rise in desirability because people have not been exposed to them. Adriana's emerald cuts, with minimal faceting, show off the subtle but impactful colour and lustre of these sapphires. Although Adriana has used as little metal as possible in her settings, you can sense just by looking at the earrings that they articulate well. When something looks uncomplicated, you can guarantee that it will have been complicated to make.

Adriana's vision to create jewels that have never been seen before comes from her use of stones to create form, modernity and abstraction. Working with her husband, Nicholas Briggs, Adriana and her team make a limited collection of approximately 40 such jewels each year. Clients are prepared to wait for the availability of such rare pieces because they know the end result will be an exceptional and stunning jewel.

Left: Zig-Zag earrings, 2019, with 160 unheated emerald-cut sapphires, totalling 114 carats.

Right: Shield earrings, 2020, each set with an unheated Ceylon sapphire of 12.31 and 12.69 carats, respectively, single-cut diamonds and unheated Burmese sapphires cut from rough to match the centre stones.

Centre: Rococo ring, 2020, set with a special-cut unheated Madagascan peach sapphire of 4.48 carats and faint pink and white diamonds; Pastel Hoop earrings, 2017, with a colour range of unheated sapphires and pear-shaped tanzanites; and Ribbon ring, 2018, with an oval unheated Ceylon sapphire of 10.5 carats in a spiral of unheated Burmese sapphires and single-cut diamonds.

This Point d'Interrogation ('Question Mark') necklace, here with its peacock feather motif, is the contemporary development of an 1879 design from Boucheron, founded in 1858. The first sapphire, emerald, silver and gold peacock jewel grew from a moment of contemplation for Frédéric Boucheron when he was playing with a peacock feather. The peacock is a beautiful bird whose feathers have powerful symbolic value across many cultures. Boucheron regarded mythical talismans highly, even supposedly gifting his wife a snake necklace to protect her at home when he travelled abroad. Since that time, the *maison* has remained steadfast in its admiration for mythical motifs as a foundational design.

Boucheron and his workshop manager, Paul Legrand, came up with the idea of creating a necklace that could wrap around the neck without using a clasp. The original necklace of 1879 is an incredible feat of engineering: the necklace can twist and move to allow it to fit around the neck, while the peacock feather hangs nonchalantly from the necklace, forming a question mark. The feather demonstrates extraordinary detail, as many of the barbs are able to move individually. The necklace was greeted with much acclaim when it was showcased at the Exposition Universelle held in Paris in 1889. Such has been this necklace's popularity with collectors that, more than 100 years later, Boucheron created a similar design. Though the feather does not detach from this 2020 version, the 'eye' of the feather is centrally set with a faceted sapphire, echoing the nineteenth-century original.

Boucheron was a trend-setter in other ways too: as the only successful French buyer of gems from the 1887 sale of the French crown jewels as well as the first jewellery *maison* to open in Place Vendôme. Frédéric chose the corner building specifically because it is the area of the square to get the last of the sunlight each day; the afternoon sunlight is considered to be the 'selling light' because of its warmth and golden hue that make stones shine. In the twentieth century, Boucheron's boutique drew in many prestigious clients, including the Maharajah of Patiala and his collection of jewels to reset and Mrs Mary-Louise Mackay, a wealthy American socialite who was one of Boucheron's biggest and most frequent collectors.

The reimagined Boucheron Plume de Paon Question Mark necklace, 2020, set with a 10.98-carat Burmese sapphire and diamonds in white gold, above an original Plume de Paon, 1883, set with a sapphire, emeralds and diamonds in silver and gold, with a new diamond necklace.

Shaun Leane's jewels are a union of power and romance, talismans to protect the wearer from harm by harnessing strength. Uncompromising craftsmanship has always been important to Leane, who spent 13 years training as a traditional goldsmith, including seven as an apprentice. Leane set up his own business in 1999, having worked for many years with his close friend Alexander McQueen, making the 'catwalk jewellery' for the fashion designer's collections. Leane's jewels combine tradition with theatricality.

In 2008, Leane was asked to design and make a jewel to celebrate Boucheron's 150th anniversary, which resulted in the fabulous Queen of the Night necklace (facing page). This jewel is centrally set with a purple colour-change sapphire, with the flowers and stalks pavé-set with sapphires, rubies and diamonds. The flower buds open to reveal the ruby stamens. This necklace is a tribute to the Countess of Castiglione, a mistress of both Napoleon III and King Victor Emmanuel of Italy, who was considered to be one of the most beautiful women of the age, but who spent her later years as a recluse lest anyone witness the cruelty of her ageing. She lived in a small apartment and would only venture out at night using a private entrance. When Frédéric Boucheron took over the lease of his shop on the corner of Place Vendôme in 1893, the sitting tenant that he inherited was none other than the 'Countess of Darkness'. Leane chose the plant Queen of the Night as it only blooms at night, while the rare 15.29-carat purple colour-change sapphire was selected to reflect the colour of the Countess's eyes. It can be removed to be worn as a pendant. The necklace resides in the Boucheron archives as a lasting homage to the collaboration between Leane, Boucheron and, of course, the Countess herself.

For Alexander McQueen's 1996 fashion show *Dante*, Leane created a head ornament of a crown of thorns, which later inspired the Thorned Branch necklace of 2018 (illustrated at the beginning of this chapter), set with a stunning natural yellow sapphire weighing 162 carats. The beauty of this gem is protected by the branches and thorns encircling it, motifs that have become Leane's signature style.

While working at the bench in the workshop of English Traditional Jewellery, Leane was constantly exposed to all styles of jewellery, from Victorian to Art Deco, which came in for repair. Those jewels enchanted Leane, who has incorporated some of the designs into his current work. The Victorians were fascinated with the cosmos and celestial motifs, such as stars and crescent moons, which Leane has replicated in this sapphire-set crescent moon pendant. The influence of Art Deco can be seen in the ring, made in 2020, set with a stunning, rich blue pear-shaped sapphire.

The large Crescent Moon pendant, 2015, pavé-set with graduated blue sapphires and diamonds, and Shield ring, 2020, set with a 10.53-carat pear-shaped sapphire in sapphire-set shoulders.

Queen of the Night necklace, 2008, by Shaun Leane
to celebrate Boucheron's 150th anniversary, with
pavé-set sapphire branches and buds that open to
reveal ruby- and diamond-set flowers, protecting
a 15.29-carat purple colour-change sapphire.

The jewellery house of mastermind Laurence Graff is known for its discerning principles and for its association with some of the world's most famous diamonds. So it is no surprise that Graff's buyers, designers and craftspeople implicitly understand and appreciate the concept of rarity when it comes to gemstones, no matter what the variety.

This Graff necklace is designed to show off its exceptional and large blue sapphire, which is unheated and from Sri Lanka. Similarly, the pink sapphires set in the Graff rings illustrated here all show fabulous colours, displaying deep pink to purple pink. Each hue is slightly different, which is the beauty of sapphire — no two stones will ever be alike. It is the colour that 'speaks' to the client. The bangle is set with two sugarloaf sapphires, different from

cabochon cuts as they have four conical sides that come to a soft domed point. The advantage of a sugarloaf cut is that it reveals the true colour of the sapphire because there are no flat facets to reflect and bounce the light through the stone and back to the eye. Colour enthusiasts love a good sugarloaf stone. It is important not to forget that it takes a skilled cutter to marry the dichroic colours of a sapphire, a consideration faced when deciding the best cut for each rough stone. If choosing to facet a sapphire, placing the table facet exactly where the two colours meet ensures that, when you look through the table facet, you see the marriage of colour and the cut maximizes the potential of the gem. For a stone to be set in a Graff jewel, it must be of the highest quality, and this choice of cut would have taken careful thought.

Centre: A diamond necklace, 2019, set with a
58-carat royal blue sapphire from Sri Lanka.

Left: Three sapphire rings set with, clockwise
from top, a 6-carat purple sapphire, 2019,
a 6-carat pink sapphire, 2019, and a 4-carat
unheated pink sapphire, 2018, all from Sri Lanka.

Right: An open bangle, 2019, set with two
unheated Burmese sugarloaf sapphires, each
weighing 20 carats, flanked by 32 additional
sugarloaf sapphires and diamonds.

There are many goldsmiths and lapidaries in the world, but rarely will you find both skills in one person. Mark Nuell, who grew up in the outback of Australia where his father mined sapphires, is something of an exception, being both an accomplished goldsmith and a gem cutter.

Having been taught the skill of lapidary at a young age by Austrian settlers, by age 22 Nuell had already set up his own cutting business. A few years later he left the small mining town of Rubyvale and moved to Sydney, where he enrolled onto a three-year goldsmithing course at a technical college until 1990, when he moved to London. To fund his passage, Nuell sold his cutting machine and did not cut another gemstone for 28 years.

Soon after arriving in London, Nuell showed his jewellery portfolio to a few galleries in London. The esteemed Electrum gallery offered to sell his work, which meant that he had to find a studio quickly. Mark became a regular contributor of jewellery to Electrum and spent his time perfecting his goldsmithing skills. It was only in 2017, after his father died and his mother wanted to give him something in memory of his father, that Nuell returned to gemstone cutting and asked for a cutting machine.

Nuell's early years of finding sapphires with his father and learning how to cut had stayed with him, and the cutting machine felt familiar. Having developed his own style through his goldsmithing, the time now felt right for him to extend his jewellery making by using his own cut stones. Inspired by the German maestro gemstone cutter Bernd Munsteiner, Nuell started to explore cutting unconventional shapes, which led to jewellery designs in which the stones took centre stage.

The jeweller uses Australian sapphires that come in all colours, from pale to dark blues, to yellows, greens and even particoloured; in the past, such stones would have been heated to make the colours a uniform blue by 'burning' out any other colour. Nuell celebrates the Australian sapphire for its natural beauty and cuts the stones to highlight their individuality. While allowing the rough crystal to dictate the shape, Nuell chooses faceting patterns that do not include conventional table and pavilion facets, resulting in wonderful sapphires that are totally unique.

Nuell still returns annually to Rubyvale and buys from the families of miners his father once dealt with; bringing the crystals back to London, Nuell cuts them on his father's wheel. Nuell may have ventured far from his roots, but those early years as a child in the outback, where he experienced freedom and learnt to be comfortable with nature, coupled with the guidance he received from his parents and his own artistic talent, are the ingredients that allow him now to make jewellery with integrity and soul.

From top: A parti-coloured sapphire of 2.70 carats and a teal sapphire of 3.05 carats, both from Rubyvale, Australia, and cut by Mark Nuell in 2020; two rings set with a 2.04-carat yellow sapphire and a 4.72-carat green sapphire, respectively, both from Rubyvale, cut and mounted by Mark Nuell in 2019, now part of the Goldsmiths' Company Collection; three rings with a 1.63-carat green sapphire, a 1.97-carat blue-green sapphire, and a 1.43-carat yellow sapphire, all from Rubyvale and mounted by Mark Nuell in 2018.

A jewellery house ensures its continued success when it creates jewels that are innovative, exciting and different, as Tiffany & Co exemplify. This stunning necklace, which is part of the 2018 Blue Book collection, caught my attention for its asymmetric design of three-dimensional, stylized hydrangea petals nestled amongst a line of 47 emerald-cut sapphires of graduated blue colours; it is a great combination of flora and geometry. The various hues and sizes of the interspersed blue sapphires, including four sapphires from Montana, sit in perfect harmony with the sculptural flowers. This necklace invites the eye to explore it from every angle. It is a fine example of craftsmanship, lapidary and innovative design.

The bracelet, made as part of the 2019 Blue Book collection, follows a similar theme to the necklace. The juxtaposition of strict geometric lines supporting the organic flow of the applied flower motifs is a brilliant combination that ensures this bracelet stands out from the crowd. Together with the necklace, these are jewels that would take pride of place in any jewellery collection.

A necklace, 2018, and bracelet, 2019,
by Tiffany & Co, with sapphire-set flower
motifs scattered over, custom-cut sapphires
and diamonds. The necklace features two oval
sapphires, totalling 17.08 carats, and four smaller
Montana sapphires, interspersed with sapphire-
set hydrangea flowers.

Negative crystal and healed stress fissure,
Ceylon (Sri Lanka).

Spinel octahedra,
Ceylon (Sri Lanka).

Spessartine crystal,
Rock Creek, Montana, USA.

Monazite crystals,
Ilakaka, Madagascar.

Zircon crystals,
Ilakaka, Madagascar.

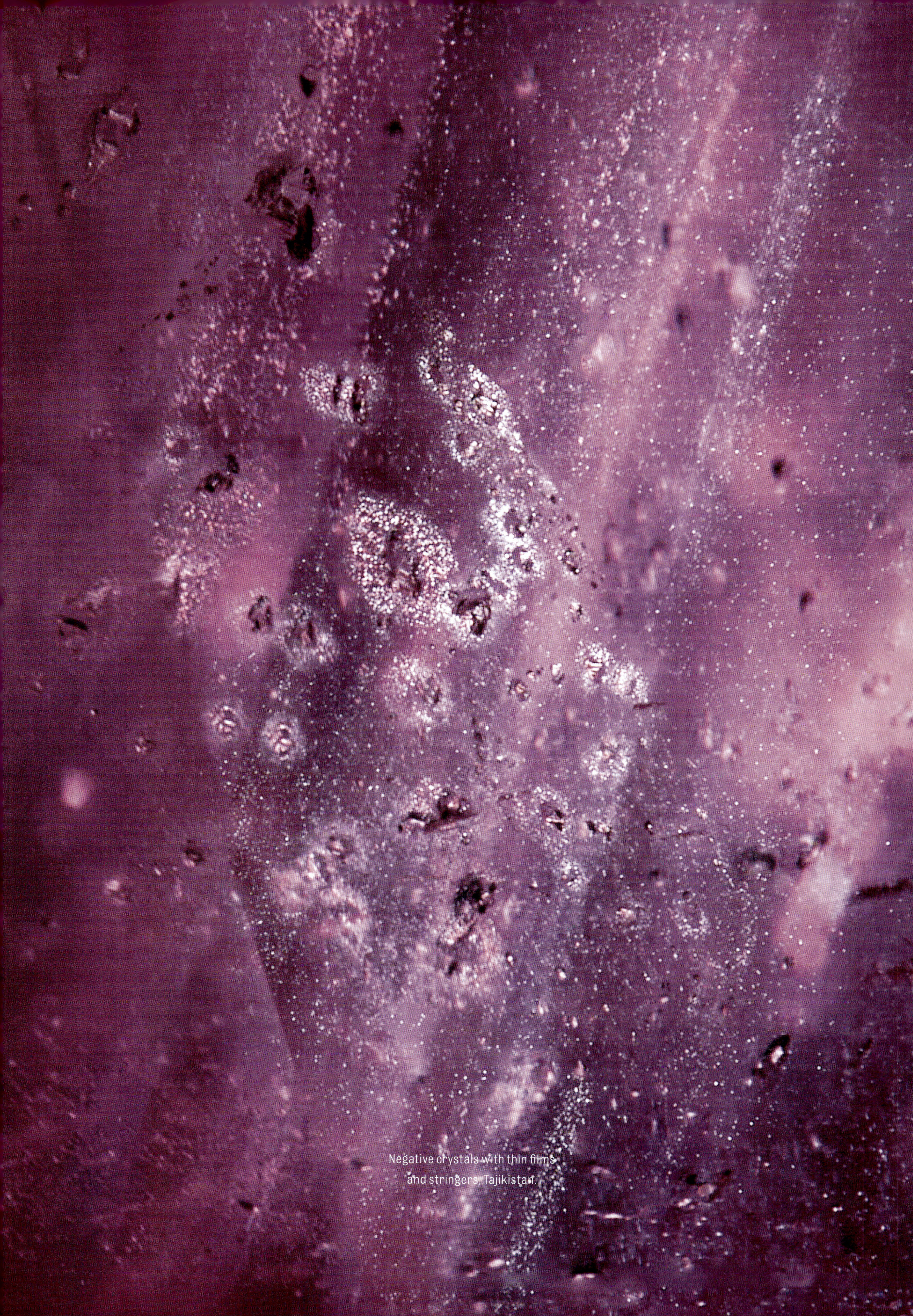

Negative crystals with thin films
and stringers, Tajikistan.

Partially healed fissure,
Rock Creek, Montana, USA.

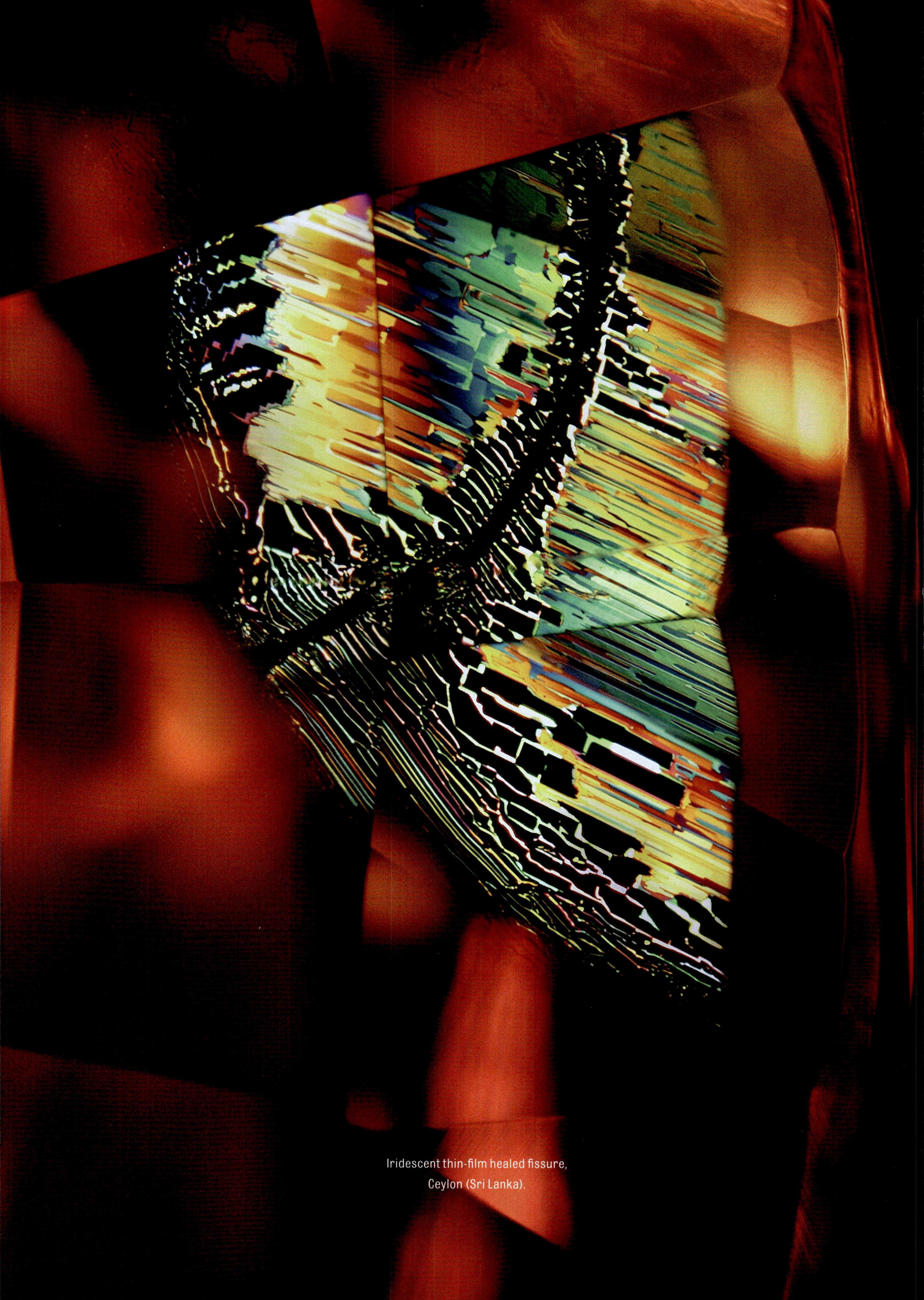

Iridescent thin-film healed fissure,
Ceylon (Sri Lanka).

SAPPHIRE DISCOVERIES

One of four large sapphires carved
as the likenesses of United States presidents,
this 1,318-carat carving depicts the head of
President Abraham Lincoln. It was reportedly
made from a 2,302-carat rough sapphire found
in Australia by Hilda McKinney and bought
by gem dealers the Kazanjian brothers.

Sapphire Discoveries

Sapphires are found in approximately 20 countries, but for this Discoveries chapter I have concentrated on locations that have resonance for me thanks to personal experience or the accounts of friends and colleagues: from speaking with Shirley Nuell, who was a sapphire miner with her husband Mike in the outback of Australia in the 1970s, and remembering my visits to the sapphires mines in Madagascar, Burma and Sri Lanka, to discussing what it is like to be a gem trader with sapphire dealer Santpal Sinchawla and meeting Dr Keith Barron, who owns Rock Creek sapphire mine in Montana. Each location and account is individual, but together they give a global picture of the importance of the sapphire trade and the people whose livelihoods depend on this gemstone.

Since Sri Lanka's historical past has been addressed in the earlier chapter Early Trade, and Montana's sapphire history, in association with Tiffany, is described in Hanoverian Blue, I have focused here on recent observations relating to these two locations. In the case of Kashmir, on the other hand, I write only about the historical context of this important source as nothing has been mined there since the late 1920s. Having recently been offered sapphires from Scotland – much to my amazement – I want to give this much smaller source some attention too.

Mining has always been controversial and involved delicate negotiations between governments and mining corporations. Companies have a duty to build an understanding of the needs of the artisanal and 'informal' miners while supporting the local communities and also, of course, being aware of mining's impact on the environment. A lot is being done to meet the various challenges that mining presents, but the gem community recognizes that there is an awful lot more to achieve.

The first Australian sapphire finds were recorded in 1851 in New South Wales, followed soon afterwards by the discoveries made in 1853 in New England by the Rev W B Clarke, who was a geologist as well as a clergyman. Like most sapphire deposits, they were found by accident rather than directly sourced, often when mining for minerals such as gold or tin. In 1873, near the small town of Anakie in Queensland, locals found zircons, garnets that were thought to be rubies, and some other mystery stones; government surveyor Archibald Richardson later identified these unknown stones as sapphires. By 1881, miners had capitalized on the area and the sapphires were traded through Germany to take their place in jewels and objets d'art for the Russian tsars. The area attracted over 1,000 miners, including many from Russia, but, as World War I approached, mining ceased.

It was not until the 1960s — mainly driven by demand from Asian markets — that mining picked up again, with production reaching a peak during the early 1970s. The area round Anakie, referred to as Anakie Fields, came back into the spotlight as it was famed to be the world's biggest and richest sapphire deposit at that time. Here, in the outback of Queensland, 1,000 kilometres inland from Brisbane, two settlements sprung up, appropriately called Sapphire and Rubyvale. Rubyvale took its name from the red stones that were thought incorrectly to be rubies. Fifty kilometres west grew a settlement called Emerald, named for the rains that fell on the area's black soil and turned everything green. During the 'sapphire rush' of the 1970s, over 100 sapphire plants were in operation; some were large concerns, and Thai stone traders were quick to arrive and take control of the operations, but illegal mining was also a problem.

In its natural state, Australian sapphire has been found in an array of attractive colours, including parti-coloured blue and green, or blue and yellow, orange and colourless; however, Australian sapphires have never fully been appreciated for their natural hues. Instead, stones were usually sent for heat treatment in Thailand, the gem trade's centre for heat treatments, to make their blue colour more uniform. These treated sapphires were often sold as Thai sapphires, but it was commonplace to call the sapphires Sri Lankan, Burmese or Thai, depending on the demand. In the 1980s, 80 per cent of Thailand's export of cut and polished sapphires were from Australia, but, as most were marketed as Thai stones, the only sapphires that were designated as 'Australian' were those of poorer quality, which built an unfortunate reputation for Australian sapphires. One small section of the Anakie gem fields produced a very dark blue sapphire which was marketed as Australian Midnight Blue, although always deemed to be of inferior quality. I remember in the 1980s being led to believe, like many others, that Australian sapphires were always dark and not very attractive, when actually a lot of the dark blue sapphires were mined in China.

In 1985 dealers started to buy lighter-coloured, heavily included (with 'silk' inclusions) sapphires from Sri Lanka, because this material responded well to heat treatment. Around the same time, the Australian trade took a downward turn, not because of reduced availability of the rough, but because of the increased costs of recovery — from machinery to the expensive legal costs involved in navigating land claims and rehabilitation regulations — and the collapse of the Asian market for sapphires in the mid-1990s. Alongside the large-scale mining, Anakie catered for the tourist market by promoting 'fossicking' (recreational sifting) for sapphires. Locals believed that tourists would help spread the word about the area and that local miners would be able to make a living from selling sapphires; however, in an article published in the November 2000 edition of the Australian magazine *Gold, Gem & Treasure*, the writer argued that to attract tourists the area must preserve the wild outback and its remnants of broken mining machinery and the crumbling artefacts of 'clapped-out mines'. It was a great marketing tool that did not add to the financial burden of production, nor to modernization efforts and rehabilitation programmes. As of 2020, however, a couple of new mining concerns have restarted mining recoveries.

An early twentieth-century photograph by
Edwin J Brady of the open pits at Sapphire
Fields in Anakie District, Australia.

Between 1945 and 1972 Australia encouraged immigration to the country to boost its population and economy, promising adventure and hope to those that made the move. As an initiative of the Australian and British governments, families were charged only £10 in processing fees to emigrate to Australia, an opportunity the Nuell family eagerly took up; in May 1963, Mike and Shirley Nuell, along with their two small children, started their journey to Australia. Shirley remembers, 'We had completed the interviews and medicals, attended film showings of the idyllic life that awaited us and generally prepared ourselves for a future without the security and support of our families. The excitement of it all outweighed any lingering doubts that I had. It was what Mike had dreamed of for a long time and he was determined that it would be all that he hoped for.' Ten flight stopovers later, they arrived in Brisbane, but for the next few months life was tough as accommodation for new arrivals was an iron Nissen hut. An apartment, then a house through the Housing Commission followed, until Mike Nuell, who worked for an oil refinery company, was able to save up enough money to build the family a house of their own.

Mike loved the Australian bush and fossicking for garnets, opals and chalcedony, and in 1969, when he heard about the sapphire finds at the Anakie gem fields, he decided to take the family there on a three-week caravan holiday, travelling 1,100 kilometres inland. On the journey the landscape began to change; it became more arid, with less vegetation, and the roads were no longer made of bitumen. Armed with a Miner's Right licence to dig the fields and the contact details of a stone cutter called Roy Spencer, who had spent most of his life in Rubyvale, Mike was eager to explore this new way of life. Roy's stories enthralled Mike, especially the story of how, when he was 12 years old, he found a large rock which he used as a doorstop for 10 years, never thinking it was worth anything until his father inspected it and discovered it was in fact a huge crystal. Roy eventually sold the crystal to American dealer Harry Kazanjian, who cut the stone to produce the largest black star sapphire in the world – the 733-carat Black Star of Queensland, which is featured in the chapter Star Sapphires. These stories fuelled Mike's dreams and his conviction that treasure lay waiting for him to discover, too.

When the family arrived with their caravan in Rubyvale, they set up camp on a hill in an area called Reward. In the morning Mike was up early with picks, shovels and sieves, excited to start exploring the bush, but careful to avoid any claims that were already pegged. Mike found an area he felt was promising and started to dig until he had enough to start washing and sieving. He used a 'leaning screen' to sieve out the larger rocks, leaving the smaller wash ready to be put through a Willoughby sieve, which he dipped up and down in water, as if panning for gold. He flipped the clean gravel onto a sorting table so that he could pick through it. Mike's first day rewarded him with a pile of sapphires that fit into the palm of his hand, and he was now hooked! He left his shovel and picks in the hole to lay claim to it. In the evenings miners and their families sat around camp fires exchanging tales of their finds, and sometimes even showing what they had found. Shirley remembers the tips shared about digging on Reward: 'In the early part of the twentieth century it had been mined by the Withersfield Mining Company. They employed men to dig shoulder to shoulder up the hill. The owners could sell blue sapphires, and anything that showed any other colour had no value and was not picked up. This proved to be fortunate for later generations who were able to find some wonderful yellows, greens and parti-colours. Some of these stones were of a size rarely seen anywhere else.'

The Nuells loved their holiday, but by the end of the three weeks it was time to return to their home on the coast. This trip sowed a seed that grew pretty quickly for Mike, who felt he had found his calling – he had not travelled to the other side of the world to end up doing a mundane job! Six months later, they prepared to leave their newly built house to head back to Rubyvale to start a new life mining for sapphires – it was 16 July 1969, the same day Neil Armstrong landed on the moon.

On Reward Hill there was a hut available for the young family. It had a forty-gallon drum to collect rainwater, a burning stove, a double bed and not much else. Drinking water was seven miles away from a bore along Policeman Creek in Rubyvale. The next day, Mike started digging. Friends they had made from their first trip came to visit, and Mike learnt how to read the land and recognize the old watercourses (which had a higher possibility of revealing good sapphire deposits). Mike also learnt how to clean his stones so that they looked more appealing and revealed any gem material: he would rattle them in a tin box with some valve grinding paste to polish them, and on car journeys he would put the tin in the hub cap and let the car do the work. At this point he was ready to show them to the visiting German gem dealers and cutters who sold to the Russian market.

Early 1970s photographs from the Nuell
family collection: Mike Nuell, right, with Shirley,
their children and a friend round the 'puddler',
ready to wash sapphires on their
land in Rubyvale, Australia.

By the 1970s, more people had arrived in Rubyvale to seek their fortune and Thai buyers visited. Life was still pretty rustic, but the characters were just as colourful as the gems they were searching for. Many wore big hats to protect them from the burning outback sun and grew long beards. Shirley remembers the miner 'Whiskers' Bill, who never seemed to wash himself or his clothes; Harry Whipple, who made a very strong home brew under very unhygienic conditions and who drilled a hole in the side of the caravan where he slept so that he did not need to get up to relieve himself; Alby Naismith, who owned the caravan park and had a museum under his house of rocks that he insisted were in fact petrified dinosaur remains; and Ron Pattle, who always had a roll-up cigarette hanging out of the corner of his mouth and spoke with a very slow drawl. The finds were just as amazing as the people: Hilda McKinney found an enormous blue sapphire on Reward, very close to where the Spencer stone was discovered, that weighed 2,302 carats and was also purchased by the American dealer Kazanjian, for £13 and 15 shillings. In America the sapphire was carved into the likeness of Abraham Lincoln.

For the Nuells' son Mark (see 21st Century chapter), it was an enviable childhood: he and his brother and sister had miles and miles of bush to explore, not to mention a pet emu that followed them everywhere they went. Shirley even looked after orphaned kangaroos whose parents had been run over. Mike was well and truly a miner now, and soon he needed a 'puddler' to be able to process more wash. As Shirley recalls,

> A puddler screened dirt running down a chute into a round receptacle, in this case six feet wide and about one foot deep, with a controlled flow of water. A motor-driven shaft in the middle turned rakes at a steady speed. Sapphires are very heavy and, together with the ironstone found with them, they sank to the bottom whilst all the light material washed out. In this way, one man working alone could greatly increase his output. At the end of the day, gem-bearing wash was scooped out into buckets to be carefully sorted. The family usually gathered around to watch as this was the exciting part. By this time, I was able to help with the sorting and grading of our parcels, and it was usually me who sold them.

They made enough money from sapphires to buy a house that was being pulled down in Rubyvale, and they repositioned it on their land; in those days, the Miner's Right allowed miners to live on the land they were mining.

Mike continued to focus on the area around Reward as he believed the deposit was still rich in sapphires, though mining politics were starting to come into play and there was an influx of miners with large-scale machinery. True to Mike's belief, he and his friend Martin finally unearthed sapphires that were 'good enough to eat'. These finds — the best blue sapphires Shirley had ever seen — allowed the Nuells to buy a new generator, which transformed their living conditions. Thai dealers started to come to Mike and Shirley's house with wads of cash carried in airline bags, and they became friendly with two regular buyers, Mr Chan and Mr Chew. As more

gem-quality sapphires were unearthed, more people arrived, and demand and prices rose. Mike kept most of the green, yellow and parti-coloured sapphires and sold all the blues to Mr Chan and Mr Chew. Some exceptional blue sapphires found in Scrub Lead were labelled as 'Sri Lankan sapphires', because the buyers could command a higher price for fine Sri Lankan sapphires. 'When the Thais came to the gem fields to buy in the mid '70s, it was so exciting', Shirley recalls, 'It was all cash money. Sometimes you would have to queue all night outside their huts so that you could be the first in line to offer your stones for sale'. Shirley would watch intently as the dealers looked through the parcels. They always picked one stone from the parcel without explaining their choice, and Shirley quickly understood this to be a 'silky blue' stone that they could heat so as to remove the silk and be left with a high-grade rough ready for cutting into a special stone. It was difficult to know how to price your rough because every miner would accept a different offer depending on what overheads they had incurred that week or month; they never asked a set price for a particular stone quality, and that made it very difficult for other miners like Shirley to know the right price. Shirley remembers that if the sapphires were really good she would get a maximum of Aus$20 per carat. One of the Thai buyers always wanted to have a look at the three stones that were in Mike's collection: two yellow sapphires and a colour-change sapphire, which was referred to as an alexandrite sapphire; in 1974 he made a serious offer of $10,000 (which the Nuells could live on for two years) for the three stones — an offer they could not refuse.

Shirley and Mike worked together for 30 years in the gem fields, and got to be very knowledgeable about mining for sapphires. Some miners hired bulldozers to dig up the soil quickly, but it was a costly business, so miners often hired a diviner to find sapphire wash. Shirley herself became successful at finding water by divining (dowsing). With two wires loosely held in each hand, she walked around the allocated area and waited to see if the two wires would spin round and cross over each other — the sign that water was detected. By locating an ancient watercourse riverbed, which may be up to 40 feet below ground, Shirley learnt to divine for sapphire wash. When the wires crossed, Shirley would count the number of seconds that they dipped down, which corresponded to the number of feet the wash was below ground. This phenomenon is probably due to the magnetic fields from the iron-rich and magnesium-rich basalt wash. Shirley's party trick was to be blindfolded and spun around while sapphires were hidden in the ground for her to locate through divining only. Shirley did not divine for many people as the responsibility was too great when costs were involved for hiring machinery.

Shirley remembers her time in the sapphire gem fields with great fondness; it was certainly a different way of life. In 2002, as other sources such as Madagascar were taking over production streams, they left the mining behind them and opened a restaurant in Rubyvale called Buddies — although they always kept a keen eye on what was happening in the gem fields.

BURMA
(MYANMAR)

Burma is renowned for its rubies, and indeed its ruby deposit is discussed at great length in the second book of this trilogy, *Ruby*, but its sapphires are also some of the best in the world. I was very fortunate to go to Burma in 2016, and while there I was able to visit the sapphire mines and admire the beautiful countryside.

The Burmese Stone Tract spans two main valleys: the Kyatpyan Valley and the Mogok Valley. The Mogok Stone Tract, located in the Kathé district of Upper Burma, is about 40 kilometres west to east and 20 kilometres north to south; it is here that the famous ruby mines are situated. Sapphires have been found all over the district, though they seem to be more scarce, especially fine examples, in comparison to rubies; however, some very fine sapphires have been found in the valley, such is the random nature of the deposits. Mogok, the centre of Burma's ruby trade, is now a thriving town situated about seven hours' drive north of Mandalay, if travelling on the old road. Good-quality sapphires are found west of the village of Kathé (indeed, this is where Bulgari's 321.27-carat sapphire cabochon the Grand Kathé, featured in the chapter High Society, was found). The majority of really good sapphires appear along a deep valley stretching from south of Sinkwa to Khabaing, and then over the hills to Bernardmyo, which is also the location of peridot deposits. When I was in Mogok, travelling with my guide, the world-renowned gemmologist Vincent Pardieu, we had to get up very early in the morning to make the journey to the sapphire mines in Bernardmyo, which was a good three hours away along a single track that wound up into the hills. On the way we passed some incredible golden stupas that had been erected precariously on top of rocky outcrops to bring good luck and in homage to the deities.

The town of Bernardmyo is an abandoned British military cantonment. Whereas rubies were the dominant find down in the Mogok Valley, here and in most of the western and northern areas of the Mogok Stone Tract sapphire was the main gem, and miners used to find very fine sapphires of large sizes and of good clarity and colour amongst the greater quantities of poor-quality rough. Finds of gem-quality sapphires in this area are now very rare. Sapphires are usually found in secondary deposits, having broken away over time from the rock in which they formed because of weathering; this means that they are mined in an open-pit operation that uses water to wash the ground and sift through the gem-bearing gravel known as byon. (Mogok's rubies, by contrast, are generally found in primary marble deposits, which are broken by a hammer or blasted with dynamite and then sifted.) There are still a few open-pit sapphire mines in operation, though no mining happens in the rainy season between June and October. When I was there, I went to see one of the mines. Between its two sapphire operations was an overgrown walled area that enclosed a cemetery for British soldiers who had died in the late nineteenth century, when malaria was apparently a huge factor in deaths. The sapphire miners, being respectful of the dead and not wanting to upset the spirits, would not touch this area and had in fact mined around the cemetery. I paid my respects and read the tombstones, dated between 1891 and 1898, of soldiers from the Devonshire, Essex, Hampshire, Border and Yorkshire regiments; it suddenly made me realize how tough it must have been, and how far from home these young men had travelled and died.

Gemstones from Burma have been mined for over 1,000 years, though the Burmese rejected blue sapphires as being 'unripe' rubies before they recognized British interest in the blue stone, and trading has survived because of the variety of gems and the range of qualities that the area yields. Burmese sapphires can have a wonderful, rich royal blue colour and are sometimes formed in very large crystals. There is no doubt that a fine Burmese sapphire is quite mesmerizing, just like the beautiful country from which it originates.

A gem merchant inspects a sapphire crystal at her table in the gem markets of Kyatpyin during Joanna Hardy's trip to Burma (Myanmar), in a photograph by Joanna's guide, gemmologist Vincent Pardieu. On her face the merchant wears *thanaka*, a pale paste applied to reflect the sun. The Mogok Stone Tract in the Kathé district has yielded some very fine sapphires, including Bulgari's 321.27-carat Grand Kathé sapphire cabochon of 2006 (see page 227).

Kashmir sapphires were originally found on the remote, north-east face of the Kudi Valley, near the village of Sumjam (Soomjam) in the Padar district of Kashmir, in the Himalayas, at an altitude of approximately 4,500 metres. Known as the 'Old Mine', the deposit was first revealed by a landslide in 1880 and mined in 1881, becoming depleted by 1887. During this brief window, impressive crystals were found, some apparently as large as 12 centimetres long and 7 centimetres wide. More sapphires were later found 250 metres below the 'Old Mine' on the valley floor but were of a lower grade. Three mining areas were eventually identified, now referred to as Old Mine, New Mine and Valley. The geology is such that there must be pockets of sapphires still to be found, but the challenge of the terrain, coupled with political instability, makes it unlikely that further discoveries will be made soon. The deposit's remote location and its often low-grade material has meant mining has been very basic and sporadic in more modern times as it has not been considered profitable. At this altitude, snow made the deposit unworkable for most of the year, apart from a short window of one to three months when the snow receded.

A late nineteenth-century photo-etching of
a photograph by T H D La Touche (1855–1938)
of an expedition group approaching the rock-face
mines of the Kashmir sapphire deposit. La Touche
recorded the first description and images of
the deposit, published in 1890 by the
Geological Survey of India.

There are many stories about how Kashmir sapphires first appeared on the market. According to one, traders in late 1881 or early 1882 who were travelling by mule from Afghanistan to Delhi came across the landslide and discovered the blue crystals, which they traded for salt when they arrived in Delhi. Eventually the stones made their way to Calcutta, where they were cut and sold — allegedly for a price equivalent to around $400,000 at the time, which would be considerably higher today. Word quickly got out, and, upon hearing that the deposit was in his princely state, the Maharajah of Jammu and Kashmir sent guards to protect this new discovery. Theft was a constant problem, and, with revenues decreasing, in 1887 the Maharajah asked the British government for assistance in developing the area. As a result, geologist Thomas Henry Digges La Touche — who had been appointed to the Geological Survey of India in 1881 and was promoted to superintendent in 1894 — conducted the first thorough survey of the area and even tried to recreate a landslide to see if he could find and expose any more localized areas of sapphire. In 1906, when the deposit was greatly depleted and mining had stalled, the Maharajah leased the mines to private concerns. The 'New Mine' was found in 1907 only 200 metres southeast of the 'Old Mine', but it did not produce much, and mining ceased until 1924. The Maharajah of Jammu and Kashmir accumulated a substantial hoard of sapphires over a 40-year period, which are kept safely in the Kashmir State Treasury; one can only wonder what sapphires lay hidden from view both in the location and in the treasury.

Since 1927, very little, if any sapphire has been mined, and the mines lie dormant but under guard. The only way to reach them is by foot or, more recently, by helicopter; such is the ongoing interest in the mines that drones have apparently been used in an attempt to get a better look at the topography, but as of yet no further mining activity has been allowed.

It is remarkable to think that this very short six-year period when sapphires were found in abundance here has made Kashmir sapphires legendary; they have become the benchmark against which the colour and quality of other sapphires have been measured ever since. A marked characteristic of gem-quality blue sapphires from Kashmir is their velvety blue colour; the stones have a haziness that does not degrade their clarity, such is the fineness of the microscopic needles in their criss-cross patterns. I was always told that inclusions in a Kashmir sapphire resemble the wispy clouds seen in the Himalayan skies. Kashmir sapphires can also display distinct growth zoning (which appears as milky colour banding), although so too can sapphires from other locations. Sapphires from Madagascar and Sri Lanka can also have a velvety appearance, so these characteristics are not conclusive evidence of origin, just an indication. Sapphires from Kashmir can take your breath away but, as with all deposits, only a very small number of stones are truly exceptional. The Kashmir provenance of a sapphire attracts much attention from a financial perspective, but this can be deceptive as stones from other locations might be of an equally high quality. The label 'Kashmir sapphire' has become a brand, so caution should be exercised when buying a stone purely on the strength of an origin report, which can be misleading.

Every trip I make to a different country is filled with excitement and trepidation, but there was something special about going to Madagascar, a country I knew very little about and never thought I would get to visit. It was early March 2020 and there was a lot of talk about a virus reaching our shores, so I was pleased to be leaving the UK thinking it would be contained by the time I returned. Little did I know.

Madagascar is unique. As the planet's fourth biggest island, its diversity of flora and fauna is unmatched; it is home to over 1,000 species of orchids, of which 80 per cent are endemic, and to the lemur, with over 100 recognized species that are not found anywhere else on the planet. Madagascar's cultural identity is represented through its 18 officially recognized different indigenous groups, plus many more. It is also home to exciting, recently discovered sapphire deposits. Sapphires were first recognized in sporadic finds in the far south and far north of Madagascar. In the north, the iron-rich alkali basalts produce blue sapphires, with a green secondary hue, and green and yellow sapphires; but it is the south of Madagascar, and its metamorphic-origin deposit, that has yielded the most exciting sapphires of all colours. It was not widely understood that these stones were sapphires when they were first recorded in the 1990s, but in 1997 that all changed. According to some accounts, the stones were noted to be sapphires by French geologist Joël Delorme, who was sorting through a parcel of stones thought to be garnets and tourmalines from an area close to the then sleepy town of Ilakaka. That first encounter changed everything.

In Madagascar, I was hosted by Marc Noverraz, a Swiss gem trader who had been living in Madagascar trading and cutting sapphires for over 20 years; through his business, Colorline, he also gives tours of mining operations. Marc not only spoke French, but also Malagasy, which is the national language of Madagascar, and understood some local dialects, which, as I was to later witness, helped in conversing with local communities.

I was also very kindly hosted by a Malagasy businessman, who spoke to me about his involvement in the early years of the late 1990s sapphire rush but wishes to remain anonymous. He trained as a civil aviation engineer in France and pursued his gem-mining hobby on visits to Madagascar, looking for garnets, tourmalines and the occasional aquamarine. In 1998 he visited Ilakaka and met with a friend who had heard about a new source of sapphires; not knowing anything about sapphires, he went with his friend to find the location, which did not prove to be easy as the information from locals was, understandably, not very forthcoming and sometimes purposefully incorrect. After many hours of talks with people near the area of Ranohira, they set off along dirt tracks to eventually arrive at Ambarazy, where they caught sight of small plastic shelters – they knew they had found the right place. They saw blue sapphires but also yellow, white and even pink ones, which they had never heard of before. No one knew what to make of these colours and no one was purchasing them; it was rumoured that a Malagasy trader bought kilos of the different colours by using the standard market measure of a condensed milk tin, and another bought a half-kilo bag of 'pink stuff' for $20. Everything quickly changed when a buyer known as 'Big Boss' was heard offering 'crazy prices' for blues from his Peugeot 405 that acted as his office and another dealer was seen selecting the pink stones; that was when everyone quickly realized that there was more to these stones. Word of the new deposit very quickly spread, and the hundreds of prospectors turned to thousands, all hoping to make their fortune. The two friends, now gem traders, ended up employing 434 diggers. With this volume of miners, naturally the yield depleted in Ambarazy, but other sites mushroomed around the rivers of Ilakaka, which contained a rich source of gem-bearing gravel – 'Money was flowing like water everywhere.' By the end of 1999, 60,000 miners were digging in the rivers and Ilakaka became a thriving town, though smuggling grew to be a problem. Understandably there has been tension when lands have become overwhelmed with miners. Local communities have felt threatened enough to take measures to protect themselves, as well as what lies underneath the soil. These are sensitive issues today, and Marc is very aware of the importance of reaching an agreeable and responsible balance between landowners and miners so that all can benefit from mining.

In 2003 miners moved to a new deposit near Sakameloka, which had been the location of gem-quality sapphires in 1999. The removal of 8,000 cubic metres of 'easy pickings' of relatively high-grade sapphires exhausted this deposit very quickly. By 2007, the two traders ran the only mechanized mining operation in Ilakaka. They produced approximately 700 cubic metres of gravel a day, seven days a week, but with gemstone mining, unlike commoditized minerals such as copper or iron ore, it is never possible to guarantee reliability or stability as gemstone mineralization is almost always sporadic and unpredictable, yet is still subject to market forces of supply and demand. By the end of 2007 oil had reached $100 a barrel, and in the wake of the crash of Lehman Brothers in September 2008 low- to medium-quality sapphire prices plummeted. Sapphires once at $8 per gram were now only worth $1 per gram. The two friends closed their operations permanently.

Joanna Hardy in March 2020 at a river visited by miners in search of water-worn sapphire pebbles near the village of Lufkadabu, Madagascar.

Since 2008 there have been further 'sapphire rushes' with similar levels of frantic activity. A rush in 2016 near the village of Bemainty brought more than 45,000 people to mine the area and produced world-class sapphires that attracted publicity both in the gemmological world and the luxury retail market. Apparently, more fine blue sapphire crystals over 100 carats were found within six months than from the area in the preceding 15 years, creating an understandable buying frenzy from all over the world.

When I arrived in March 2020, there did not seem to be a fever pitch of sapphire excitement; on the contrary, it all seemed pretty subdued. Flying into Antananarivo airport, the surroundings looked quite green and rural for a capital city. The next day we left the city's narrow, winding roads lined with stalls, and its signs to beware the bubonic plague, and drove for 10 hours through stunning countryside dotted with Madagascar's famous statuesque baobab trees to Ilakaka. As soon as we arrived in the late afternoon, Marc drove us slightly out of town to an area called Sakuviro. Here, the road is lined on either side by a row of huts, one of which was Marc's, where he opened the window hatch for miners to show him what they had found that day – all sorts of gemstones, not just sapphires. Marc placed any stones that caught his eye in a dish of water to 'open them up' so he could see the colour better; by magnifying the crystal, the water revealed if the stone had been dyed as the colour would appear concentrated on the outside of the crystal. Apparently, even leaving a sapphire crystal in the sunshine on aluminium foil can be enough to change its colour. If a particular stone caught his eye Marc would quickly look at the crystal down through the c-axis (down its length) because the colour you see in that direction will be true to the colour after cutting. There is no time to loupe a stone, because you want to see everything available before someone else buys the good ones. This time, there was nothing very exciting.

The mining district of Ilakaka is located in the Isalo massif between the cities of Sakaraha and Ilakaka. These deposits are trapped within alternating layers of sandstone and red clay stone, and the sapphires they yield are very similar to the sapphires of Sri Lanka and Tanzania. The sapphires around Ilakaka occur in many colours, in fact all the colours of the rainbow, and are found alongside zircon, chrysoberyl, alexandrite, topaz, garnet, spinel, andalusite and tourmaline. There are three types of mining in this area: 'informal mining', open-pit mining and mechanized mining.

The first type, 'informal mining', used to be associated with illegal mining under Madagascan law, though, as Marc informed me, it cannot be deemed illegal since no one has been given a licence from the government anyway. I went to see miners engaging in this practice by the Benna River at Sakameloka, an area that had once triggered a sapphire rush. Madagascar has mainly secondary deposits of sapphires that have broken away from their host rock and now constitute gravel layers along the rivers and flood plains. Some areas have five to six gravel layers, so it is best to methodically wash each layer in order not to miss anything. These miners use the 'windlass system', which involves digging along the

riverbeds and in paleo channels based on their knowledge of the local area and hearsay. They dig to a depth of up to 20 metres, but sometimes 55 metres, with very rudimentary means of getting air to the miner in the shaft: they run around capturing air in a big see-through plastic tube, which they feed down the shaft and then squeeze so the miner receives oxygen. Buckets are hauled up and then taken down to the nearby river system to remove by hand any gems found. In the river running through Ilakaka I saw predominantly women washing and panning, and hopefully recovering gems, which the gem buyers will come to collect and for which they will pay the miners on the spot. These miners are funded by the gem buyers: the Thais, Sri Lankans and Malagasy based in Ilakaka, Manombe and Sakaveero, who get to see the goods first. Apparently, in the interests of the local community, the government feel it is beneficial to have manual mining because it keeps the community engaged and employed. When I was there, not a lot was being found; according to Marc, all the easy mining had been done and they now needed to dig below the water table, which requires an injection of money and expertise.

The second method, which is slightly more organized but still largely informal, is open-pit mining. This mining is still done manually by digging a cluster of tunnels that lead to a rich gravel bed and then merging the tunnels to form an open pit. Produce from the gravel bed is then washed and panned at a handmade water sump to recover gems and then sold to the expectant buyers in the nearby villages. Marc was particularly passionate about the 'informal miners' and felt that more organization was needed for them to be fairly treated and given prospects, but this route naturally requires more funding and probably outside investment. When production is low, the miners move from place to place until a significant deposit is found, and then there will be a rush to mine that site. Over the last couple of years, however, there has been very little good production. Mining is like gambling: you live in hope and always think that it may just be your lucky day.

The third method, mechanized mining, is very limited in Madagascar. We went to see the Suzannah mining operation on the Ilakaka River, which employed 150 people. We were shown round by Guillaume Ah Thion, whose family jointly work the mine with

A tray of sapphire crystals (centre) found by the Suzannah mining operation on the Ilakaka River in Madagascar shows the beautiful range of colours of Madagascan sapphires. Joanna Hardy selected one sapphire pebble, of 2.15 carats, to purchase and have faceted locally in Ilakaka. The cutters made a feature of this crystal's straight colour zoning, a common characteristic in sapphires. Back in the UK, the finished 1.04-carat stone was set in an ear stud by goldsmith Roger Elliott.

Madagascar is the home of the lemur, shown above in
Joanna Hardy's own photograph from Isalo National Park.
Van Cleef & Arpels has caught the playful nature of the lemur
in its jewelled Lémuriens clips, 2016, set with diamonds and
black spinels in the form of two lemurs jumping on branches
of blue and mauve sapphires, emeralds and pearls.

a Thai company. It is an alluvial deposit of gravel bedrock reaching 8 metres deep. The topsoil has been removed to reveal the surrounding soft sand of a uniform pale yellow colour, which contrasted with the bright turquoise water. The depths of mining are governed by the water table: when mining hits the water table, they move to another area. Mechanized mining and manual mining are happy to work near each other as they have a mutually beneficial arrangement: the manual miners dig a 1 metre-diameter hole until they reach the gravel bed, which they then follow by digging more holes side by side, and if they find a stone after washing and sieving the gravel they will offer it to the mining company. In the company's main house lay a five-day production of sapphires in trays – it was as if I had walked into a sweet shop with a rainbow of fruit gums to play with! All the stones had been found in the layers of gravel, and I was told that sometimes you can mine and sieve the gravel continuously for five days and find nothing, and at other times you may strike lucky and find a pocket of sapphires in a small quantity of gravel. This was a good area for sapphires, but the deposits are spread out and sometimes take a while to find.

It was here that I chose my sapphire crystal. After much hesitation – there was so much to choose from – I eventually selected a blue crystal weighing 2.15 carats, which I then took back to Marc's workshop and which, in a few hours, his lapidaries had cut to a beautiful oval weighing 1.04 carats.

A few days later, we decided to head to an area that Marc had not been to for over seven years: an off-road journey that was going to test his car, his driving skills and his memory. We needed to get there and back in daylight, so a very early start was key, plus plenty of petrol and water as the temperature was more than 40 degrees Celsius. We headed first to Sakalama via Ilakaka Be ('Big Ilakaka', where the first sapphires were found), but we had to pass a checkpoint and I was asked for my passport, which I hadn't brought with me. Marc read the situation very well and, being the most unfazed person I have ever met, just slowly got out of the car and spent time chatting with the checkpoint officials and having a coffee by a local stall, while my two companions and I sat in the car patiently.

Interestingly, when we returned by that route much later that day, there was no evidence of any stall or checkpoint – everything had vanished as if it had been a stage set. A couple of hours later we made a pitstop to look at a distinctive mound known as a 'sugarloaf mountain', whose geology indicated that it could be a good source of sapphires.

After three more hours of driving, we eventually arrived at the village of Lufkadabu, which consisted of clay dwellings and very small lanes. Obviously, our arrival caused a stir of wary excitement, so it was very important that we found the president of the village to make sure everyone was comfortable with our presence. We were invited into a hut where all the senior men of the village sat with the three of us; I was the only woman. To the right of the president sat a young man of stature with quite a large gun hanging over his shoulder. It was so hot and humid inside that hut, but I tried to ignore the discomfort while Marc was negotiating with the president to allow us to visit the sapphire deposit near the village. Thirty minutes later there was movement, and we were ushered out of the hut and told to follow the president. It was like the story of the Pied Piper, with a string of excited children also following us through the village until they were told to leave us, as we continued to follow a river. It was midday, the sun was at its strongest, and the heat, even with the wind, was incredibly intense; I could feel I was dehydrating very quickly. I asked Marc how long the walk was going to be, and he said about one and a half hours; oh, my goodness, I thought, as I sank my straw hat into the river to then place over my head. The president took the walk in his stride, while I followed behind, trying not to let the side down. At last, after walking through pristine countryside and having to pinch myself that I was walking through the wilds of Madagascar, I was very pleased to have arrived at our destination: another river. The few miners we came across (there were hardly any sapphires being found these days) had stopped for lunch, and the spot was magical. Marc went to have a swim in the river, which despite being clean was very murky with the silt such that there would be no way of telling if a crocodile was lurking underneath – no wonder we needed armed protection!

After cooling down, we set off back to the village and were met again by the children, all wanting to have their pictures taken. I thought we were going to get in the car and head back, as we had at least a three-hour journey ahead of us and needed to make sure we returned in daylight, but instead we were invited into the president's home for a drink and, importantly but unbeknown to me, to negotiate a fee for taking us to the river.

My last day in Madagascar was spent in the Isalo National Park, where we spotted many different lemurs, which was an absolute thrill – they are such magical creatures. The park is mainly looked after by the dedicated local community, which is a challenge in itself as they get very little support. Madagascar is a simply stunning country and clearly capable of producing a wealth of revenue, from its gemstones to tourism, through responsible management.

There are three deposits of sapphire in Montana, USA: Yogo Gulch, which is famous for its natural blue colour, Rock Creek and the Missouri River. The history of sapphires from Montana can be found in the chapter Hanoverian Blue. In addition to this American history, there was a Victorian jeweller by the name of Edwin Streeter who found the world of prospecting very enticing, and who, as well as being involved in mining rubies in Burma and elsewhere, turned his sights on Montana.

In September 1891, Streeter formed a company called the Sapphire and Ruby Company of Montana Ltd to acquire 8,000 acres of land in Montana. He and seven other directors, including Lord Chelmsford as chairman, released a prospectus that stated that there were 'gems lying about in profusion simply waiting to be picked up', but the prospectus was considered too 'flowery'; Streeter had already run into difficulties in Burma, so this Montana project was much maligned by mining press reporters. Streeter had been criticized for not visiting the mines, so he went in December 1891 and cabled back to say that an independent mineralogist was happy with the value placed upon the project. To build confidence in the venture, that same month he held an exhibition in his shop at 18 New Bond Street to showcase sapphires from Montana. *The Morning Post* said that the sapphires displayed a variety of shades and 'were an excellent sample, pure in colour but of varying shades and those made up in a variety of settings excited universal admiration'. Streeter was quoted as saying that he thought Montana was 'likely to become a formidable rival to the gem-producing countries of the Old World'. But all was not well; the founder shareholders, made up of titled directors that had been chosen for their status and rank to lure investors, were supposed to have guaranteed to underwrite the ordinary shares, but had in fact failed to do so. Eventually, the company was taken over in 1898 by Eldorado Gold and Gem Co of Montana Ltd. Edwin Streeter miraculously survived this failed attempt at being a prospector, but he did not learn his lesson.

Sapphires were first discovered in the headwaters of Rock Creek in the summer of 1892, while prospectors were looking for gold. After the identification of these stones, the 'Sapphire King' Mr Emil Meyer went on to successfully sell several hundred thousand carats to Swiss watch companies, before selling out with his partners to the American Gem Mining Syndicate (AGMS). Between 1903 and 1936 AGMS continued to mine millions of carats of sapphires from the alluvial deposit. This production was mainly used for watch bearings, with the gem-quality sapphires used for jewellery. By the mid-1920s, Verneuil furnace synthetic sapphires started to replace natural sapphires for watch bearings as they were a cheaper alternative, with the result that, by the 1940s, most commercial mining had ceased at Rock Creek. The improvement of heat treatments in the 1980s and early 1990s, however, brought renewed interest in the Rock Creek sapphires, which could be heated to improve their colour. Since then, mining in the area has operated on two scales: a smaller scale to cater for the tourist market and a larger scale for commercial production.

In 2020, I met with Dr Keith Barron, who now owns Potentate Mining LLC at Rock Creek. He told me that he fell into sapphire mining because of his love for skiing, as it was while he was skiing in the area in 2012 that he passed a for-sale sign for a few acres of gold- and sapphire-producing ground. He offered half the amount, and, to his surprise, his offer was accepted. Since then, he has acquired over 3,500 acres of sapphire-producing ground in the Rock Creek district and has reinvigorated commercial sapphire mining in Montana.

In the past, mining had been concentrated along the riverbeds, but Potentate has identified new sapphire deposits on the slopes above the valley, and production started in 2014. In a process known as colluvial mining, the sapphires are found in loose rock and soil underneath the topsoil; this rock has been transported only a short distance down the slope by natural movement. The layer of rock and soil can range from 30 centimetres to 4.5 metres in depth. Potentate is mindful of the environment and does not use any chemicals in its processes; after they have mined an area, they backfill it before replacing the topsoil. Rock Creek is also famous for its 'blue ribbon' trout fishing and attracts anglers from all around the world, so preserving the water quality is of paramount importance to the mining efforts today.

The Ponderosa Sapphire, the USA's largest known blue sapphire, found in 2018 at Rock Creek, Montana, by Potentate Mining LLC as a 61.14-carat rough crystal. Shown here before heating, the crystal has the natural yellow core typical of sapphires found at Rock Creek.

The mine is run by 15 people, so it is a low-key operation, which means Dr Barron and his team have more control over how production is handled. At the end of each day, the task of cleaning the screen from the jig by hand to see if any sapphires emerge is always surrounded by great excitement and anticipation. Sometimes, just to add to the excitement, gold nuggets will appear, which help with the mining costs.

Dr Barron did not disclose the production figures, but he said that the area has a longevity of at least 25 years for sapphire mining. The colours range from pink, orange, violet and green to the very popular teal colour that Rock Creek is famous for. The sizes of the rough range from 2 millimetres to 8 millimetres, though the bigger sizes only represent two per cent of the production. The gravel is separated into four different-size fractions, before being further sorted by hand for material suitable for heat treatment. Eighty per cent of their production is heat-treated in Montana and then sold rough to clients around the globe. Rock Creek's production is mainly of pale greenish and bluish colours, of which 80 per cent is gem-quality.

Rock Creek is the richest deposit in Montana today, with the upper Missouri River gravel bars still in production too. Tourism is important in the area's economy, and some tourists come every year to search for their own perfect Montana sapphire. The mining operators are in full agreement about their commitment to maintaining and protecting the environment, and they each take responsibility for restoring the land after an area's mining has finished.

It was only very recently that I learned that sapphires had been found near Loch Roag on Lewis and Harris, the main island in the Outer Hebrides; its oldest rocks were formed over 3,000 million years ago and are some of the oldest metamorphic rocks in Europe. Various gem materials, such as agate, amethyst, Cairngorm quartz and garnet, have been found all over Scotland and sometimes even gem-quality topaz, beryl, tourmaline and zircon. These Scottish stones, especially quartz and agate, became very popular when set in jewellery in the twentieth century, mainly thanks to Queen Victoria, who loved visiting her Scottish residence Balmoral Castle and promoting all things Scottish. Finding sapphires, however, is a very recent development.

Most of the world's gem deposits are found by accident, and this was the case in September 1994 when two amateur gemmologists, Ian and Linda Combe, found a heavily fractured 242-carat sapphire and a 39.50-carat sapphire crystal fragment. A sapphire of gem quality had been found in the early 1980s, which weighed 9.6 carats when cut, and the land was declared a Site of Special Scientific Interest in 1984 to protect the area from exploitation. Very few sapphires therefore exist from this region.

Two chromolithograph images from renowned Victorian jeweller and mining prospector Edwin Streeter's *Precious Stones and Gems* publication, 1892, written soon after Streeter's company first found sapphires at the Missouri River deposits. The left, titled 'Rough Montana Sapphires and Rubies'; above, 'Sapphire in the Matrix', depicting sapphire crystals as they would have grown in their host rock.

Gems were hiding in the soil long before man was walking the planet, and man is thought to have inhabited the island of Sri Lanka (formerly Ceylon) from about 500,000 BCE. From 3,000 BCE, stories of life and gems on this island have been intertwined; in Sanskrit it is referred to as Ratna Dweepa, meaning the island of jewels, and indeed over 40 different gem species can be found here. For thousands of years Sri Lanka has produced astonishing blue sapphires of spectacular sizes, and it continues to be an important source of sapphires today, including the occasional large blue crystal and other colours like the padparadscha, the prized pinkish-orange sapphire, and wonderful yellow. Sri Lanka has been the world's primary source of padparadscha and yellow sapphires. It is astonishing that after thousands of years of mining sapphires are still being found here, and that is thanks to the fact that mechanized mining has always been forbidden on the island.

Corundum is found primarily in the south of the island, which is now home to the bustling town of Ratnapura, the city of gems. I visited Sri Lanka with my mother in 2014, and together we went to some of the small mining concerns near Ratnapura. The terrain was not challenging, as we walked through a field of long grass before arriving at the mining area. The sapphires are found in gem-bearing gravel called *illam* at a depth of three to 20 metres; the depth determines how many *illam* layers there are. Alluvial mining along riverbeds and streams is also a popular form of extraction. The men we visited were mining a small area that had

a shaft about 15 metres deep in the middle that was supported by wooden timbers to stop the earth from caving in. A miner descended the side of the shaft by ladder and a bucket was winched down for him to fill with earth. This bucket was pulled back up to be later sieved and panned in a shallow bamboo open basket. When digging commences, water is pumped out at the same time — a pump being the only form of machinery allowed — so that the shaft does not fill up with water.

The mining area we visited had not yielded anything that week; it is a very slow process and there is no assurance that you will be rewarded at the end of a hard week's work. Landowners and miners work together so that they can be equally compensated when something is found. Little has changed in mining operations and mine construction for thousands of years, and this way of working has long provided continued employment for local communities. I sometimes think the island must be like a Swiss cheese with holes everywhere, but miners must refill their shafts once their period is up for renewal, usually every five years, before moving on to another area.

Sri Lanka is still a very important producer of sapphires today. On a visit to Bangkok, Thailand, I met with Santpal Sinchawla, the managing director of Sant Enterprises Co Ltd, whose company has a long association with Sri Lanka. Santpal has been a gemstone trader for more than 40 years. He comes from a long line of gemstone traders; his parents were born in Burma and were involved with the local gem industry and owned mines. In the late 1890s, overseas buyers, mainly from Western countries, bought gems from his great-grandfather and grandfather in Mogok, Burma. The family remained in the industry and moved to Thailand in 1963, initially to deal sapphires from Pailin in western Cambodia and Thailand which were in abundance during the 1960s and 1970s. His father first went to Sri Lanka in the 1980s as he was being asked for a wider selection of colours than the slightly darker basaltic Thai and Cambodian sapphires he traded in. Blue sapphires from Sri Lanka, on the whole, tend to be paler than Burmese stones, but it is not easy to tell them apart from Burmese, Kashmir or Madagascan sapphires. Santpal's father began visiting Sri Lanka once a year to buy stones, which he would get recut back in Thailand. Every stone that leaves Sri Lanka has to be cut as it is illegal to export uncut stones.

I asked Santpal what has most changed the business in the last 40 years, and he said it was global wealth. Before, there was no market for sapphires in China or India, so buyers came mainly from Western countries and only a few people had a lot of money. At this time, only blue sapphires were in demand; the other colours were literally being thrown away. In the last 20 years the corundum world has completely changed. Sri Lanka produced fancy colours, but the discovery of Madagascar's many sapphire colours made other colours more popular. Rarity and higher demand have since driven up sapphire prices. Sri Lankan sapphires first started to appear in Thailand in the 1980s when Thailand was perfecting its heat-treatment process. A variety of corundum commonly found in Sri Lanka is called 'geuda': it lacks colour but with heat treatment can turn into good-quality blue sapphire.

During the recent pandemic, Santpal has continued to trade, but getting goods has been difficult with only one flight per week to Sri Lanka from Bangkok, and mining operations have ceased in many areas. Santpal and his family have built their business on trust and integrity, which is especially important at the moment when there are no physical gem fairs to buy and sell stones from. Santpal has found that he is not competing so much on price, as people are happy to pay a bit more for goods from suppliers they can trust. The gemstone industry is about the relationships that have been built up over many years with others all around the world; it is a network of miners, traders, burners and cutters, for everyone relies on each other before a gem even reaches a piece of jewellery, after which the journey continues with another chain of relationships.

Above: Three examples of the bipyramid crystal habit of sapphire in pink, yellow and blue, paired with pink, yellow and blue faceted sapphires. These stones come from the collection of prominent sapphire dealer Sant Enterprises Co Ltd.

Facing page: A multi-coloured selection of rough and faceted sapphires from the collection of Sant Enterprises Co Ltd. The large rough crystals were mined in Sri Lanka, the smaller crystals in Madagascar.

SUGGESTED READING

Atterbury, P., and Benjamin, J. *Arts and Crafts to Art Deco: The Jewellery and Silver of H.G. Murphy.* Woodbridge, Suffolk: Antique Collectors' Club, 2005

Bagnoli, M., Klein, H. A., Mann, C. G., and Robinson, J., eds. *Treasures of Heaven: Saints, Relics and Devotion in Medieval Europe.* London: British Museum Press, 2010

Balakrishnan, U. R., Peshekhonova, L., Scarisbrick, D., and Vecherina, O. *India: Jewels that Enchanted the World.* Ed. by Shcherbina, E. London: Indo-Russian Jewellery Foundation, 2014

Balakrishnan, U. R. *Jewels of the Nizams.* New Delhi: Department of Culture, Government of India in association with India Book House, 2001

Bernstein, B. *If These Jewels Could Talk: The Legends Behind Celebrity Gems.* Woodbridge, Suffolk: ACC Art Books, 2015

Brus, R. *Crown Jewellery and Regalia of the World.* Amsterdam: Pepin Press, 2011

Cailles, F. *René Boivin: Jeweller.* London: Quartet Books, 1994

Campbell, M. *Medieval Jewellery in Europe 1100–1500.* London: V&A, 2009

Cartier Brickell, F. *The Cartiers: The Untold Story of the Family Behind the Jewellery Empire.* New York, NY: Ballantine Books, 2019

Content, D. J. *Ruby, Sapphire & Spinel: An Archaeological, Textual and Cultural Study.* Parts I & II. Turnhout, Belgium: Brepols Publishers, 2016

Corbett, P. *Verdura: The Life and Work of a Master Jeweler.* London: Thames & Hudson, 2002

Falk, F. *Schmuck/Jewellery 1840–1940: Highlights, Schmuckmuseum Pforzheim.* Stuttgart: Arnoldsche Art Publishers, 2004

Farges, F., Panczer, G., Benbalagh, N., and Riondet, G. 'The Grand Sapphire of Louis XIV and the Ruspoli Sapphire', *Gems & Gemology*, Vol. 51, No. 4, Winter 2015, pp. 392–409

Field, L. *The Queen's Jewels: The Personal Collection of Elizabeth II.* London: Weidenfeld and Nicolson, 1987

Forsyth, H. *The Cheapside Hoard: London's Lost Jewels.* London: Philip Wilson Publishers, 2013

Gere, C., and Culme, J., with Summers, W. *Garrard: The Crown Jewellers for 150 Years.* London: Quartet Books, 1993

Gere, C., and Rudoe, J. *Jewellery in the Age of Queen Victoria: A Mirror to the World.* London: British Museum, 2010

Halford-Watkins, J. F. *The Book of Ruby & Sapphire: Being a Description of the Minerals of the Corundum Group Used for Gem Purposes.* Ed. by Hughes, R. W. Boulder, CO: R.W.H. Publishing, 2012

Hardy, Joanna, and Self, Jonathan. *Emerald: Twenty-one Centuries of Jewelled Opulence and Power.* London: Thames & Hudson in association with Violette Editions, 2013

Hardy, Joanna. *Ruby: The King of Gems.* London: Thames & Hudson in association with Violette Editions, 2017

Hindman, S., with Fatone, I. and Mantova, A. L. *Toward an Art History of Medieval Rings: A Private Collection.* London: Paul Holberton Publishing for Les Enluminures, 2007

Hughes, R. W. *Ruby & Sapphire: A Gemologist's Guide.* Bangkok: Lotus, 2017

Keay, A. *The Crown Jewels.* London: Thames & Hudson Ltd, 2011

Krzemnicki, M. S., Butini, F., Butini, E., and De Carolis, E. 'Gemmological Analysis of a Roman Sapphire Intaglio and Its Possible Origin', *The Journal of Gemmology*, Vol. 36, No. 8, 2019, pp. 710–25

Kunz, G. F. *The Curious Lore of Precious Stones.* New York, NY: Dover Publications, 1971

Loring, J. *Tiffany Jewels.* New York, NY: Harry N. Abrams, 1999

Lüle, Ç., and Smith, C. P. 'The Buddha Blue', *Women's Jewelry Association News*, September 2012

Markowitz, Y. J. *The Jewels of Trabert & Hoeffer-Mauboussin: A History of American Style and Innovation.* Boston, MA: MFA Publications, 2014

Marsden, J., ed. *Victoria & Albert: Art & Love.* London: Royal Collection, 2010

Martin, S. 'Precious Gems of the Silk Routes', *Gems & Jewellery*, Vol. 28, No. 4, Winter 2019, pp. 24–27

Meylan, V. *The Secret Archives of Boucheron.* Woodbridge, Suffolk: Antique Collectors' Club, 2011

Meylan, V. *Van Cleef & Arpels: Treasures and Legends.* Woodbridge, Suffolk: Antique Collectors' Club, 2011

Munn, G. C. *Tiaras: A History of Splendour.* Woodbridge, Suffolk: Antique Collectors' Club, 2001

Ogden, J. *Jewellery of the Ancient World.* New York, NY: Rizzoli, 1982

Palke, A. C., Saeseaw, S., Renfro, N. D., Sun, Z., and McClure, S. F. 'Geographic Origin Determination of Blue Sapphire', *Gems & Gemology*, Vol. 55, No. 4, Winter 2019, pp. 536–79

Panczer, G., Riondet, G., Forest, L., Krzemnicki, M. S., Carole, D., and Faure, F. 'The Talisman of Charlemagne: New Historical and Gemmological Discoveries', *Gems & Gemology*, Vol. 55, No. 1, Spring 2019, pp. 30–46

Papi, S. *Jewels of the Romanovs: Family and Court.* New York, NY: Thames & Hudson, 2010

Papi, S., and Rhodes, A. *Famous Jewelry Collectors.* New York, NY: Thames & Hudson, 1999

Papi, S., and Rhodes, A. *20th Century Jewelry & the Icons of Style.* London: Thames & Hudson, 2013

Raulet, S. *Art Deco Jewelry.* London: Thames & Hudson, 2002

Raulet, S., and Olivier, B. *Suzanne Belperron.* Lausanne: La Bibliothèque des Arts, 2011

Ritchie, H. *Designers & Jewellery 1850–1940: Jewellery and Metalwork from the Fitzwilliam Museum.* London: Philip Wilson Publishers, 2018

[Romanoff], Prince D., with Snyder, L. B. *Once Upon a Diamond: A Family Tradition of Royal Jewels.* New York, NY: Rizzoli, 2019

Rudoe, J. *Cartier: 1900–1939.* London: British Museum Press, 1997

Russell, W. H. *The Prince of Wales' Tour: A Diary in India; with Some Account of the Visits of His Royal Highness to the Courts of Greece, Egypt, Spain and Portugal.* London: Sampson Low, Marston, Searle & Rivington, 1877.

Schmetzer, K. and Gilg, H. A. 'The Late 14th-Century Royal Crown of Blanche of Lancaster – History and Gem Materials', *The Journal of Gemmology*, Vol. 37, No. 1, 2020, pp. 26–64

Somers Cocks, A. *Princely Magnificence: Court Jewels of the Renaissance, 1500–1630.* London: V&A, 1981

Streeter, P. *Streeter of Bond Street: A Victorian Jeweller.* Harlow: The Matching Press, 1993

Taylor, E. *My Love Affair with Jewelry.* London: Thames & Hudson, 2002

Themelis, T. *Mogôk: Valley of Rubies & Sapphires.* Los Angeles, CA: A & T Publishing, 2000

Triossi, A. *Between Eternity and History: Bulgari from 1884 to 2009, 125 Years of Italian Jewels.* Milan and Rome: Skira/Rizzoli, 2009

Triossi, A. and Chapman, M. *The Art of Bulgari: La Dolce Vita and Beyond, 1950–1990.* Munich: Prestel, 2013

Van Gelder, B. *Traditional Indian Jewellery.* Parts I & II. Woodbridge, Suffolk: ACC Art Books, 2018.

Henn of London 259 all;

Hillwood Estate, Museum & Gardens / Photographed by Edward Owen 216–17;

Courtesy of Hennell Limited 139c, b;

Burg Hohenzollern / © Roland Beck 93;

Houston Museum of Natural Science 268;

Hubei Provincial Museum, Wuhan, China 40 all;

Stift Klosterneuburg © Zechany 94t;

KHM-Museumsverband 82 all;

Lalique Museum, Hakone, Japan 141t, r;

Neil Lane Collection 183tr, 219;

Courtesy Mansfield Library, Archives & Special Collections, University of Montana 148 all;

Maricar Gems 187 all;

Mellerio 283 all;

Mish Fine Jewelry 282 all;

Moussaieff 256–7 all;

Archivo Museo Arqueológico Nacional, Madrid (Inv C594) 80;

Museum of London 19b, 75br all, 102 all, 103tl, tc, tr;

Museum für Kunst und Gewerbe, Hamburg 141l;

Museum of Natural History, Paris 71b;

Napoleon Museum, Thurgau 48, 50r;

© The Trustees of the Natural History Museum, London 16b;

The New York Times (James Oliver © Arthur Brower), courtesy Eyevine 181b;

Mark Nuell 303 all, 317 all;

Dr Jack Ogden 35b / Osmund B 36t, c;

Osenat, Paris 97, 239b;

Peles National Museum 116t;

Private Collection 118l;

Mary Pickford Foundation 184b;

Potentate Mining LLC 324;

Royal Collection Trust / All Rights Reserved 90t, 91, 99l, 124b, 126t, r, 127 all, 129bl, br, 131t, l, r, 135tr;

Sant Enterprises 326–7;

Scala, Florence / The Metropolitan Museum of Art, Art Resource 39b, 73br / 82 / Bildagentur für Kunst, Kultur und Geschichte, Berlin 100r, tl, bl / 108t / RMN – Grand Palais 109 all, 110t;

Schmuckmuseum Pforzheim / Günther Meyer 105, 175 all;

Shaun Leane 278, 300–1;

Shutterstock / AP (Don Brinn) 22l / Ac-AP 22r / Condé Nast (Horst P Horst) 167b / Condé Nast (Carl Oscar August Erickson) 196l / AP (Anthony Camerano) 210br / AP 210t;

Courtesy of the Smithsonian Institution / Chip Clark 167, 184t, 186, 265;

Sotheby's 104b, 111 all, 134, 139tr, 176 all, 192, 208 all, 235, 237, 239c, 240c, tr, bl, 241tc, bl, 242t, tc, 274, 249, 273–4, 276;

The State Hermitage Museum, St Petersburg / Photograph © The State Hermitage Museum / photographs by Vladimir Terebenin, Alexander Koksharov, Aleksey Pakhomov, Leonard Kheifets, Yuri Molodkovets 72tr, 74tc, tr, 94c, 101l, 117t;

Stockpholio / R K Lakshmi 61c;

© Studio Sébert – Photographes 241br;

Swiss Gemmological Institute (SSEF) / Michael S Krzemnicki 71tr;

Symbolic & Chase 139tl, 174 all, 272;

Simon Teakle Fine Jewelry 203 all;

© Thuringian State Office for Heritage Management and Archaeology / B Stefan 46b;

© Tiffany & Co Archives 2021 144–7 all, 149, 150br, 155 all, 157, 206, 207b, 304 all;

Topfoto / Heritage-Images 18c;

V&A Images / Given by Dame Joan Evans 55 all, 56c / 61l, r, 62c, b, 65tl, tr, 72l, 74br all / Bequeathed by Lady Alma-Tadema 101r / 103b / Purchased through the generosity of William & Judith, Douglas and James Bollinger as a gift to the Nation and the Commonwealth 128, 129t / Cory Bequest 138 all / 142br, 210br;

Photograph by Richard Valencia 332;

Van Cleef & Arpels 24, 170–3 all; 198 all, 199, 200 all except top, 201 all, 202tl, 213t, 234br, 247, 296–7, 323l, r / Patrick Gries © Van Cleef & Arpels 246;

Van Gelder Jewellery 62l;

Courtesy Verdura 20t, 204t, c, 205 all;

Roger-Viollet © Boris Lipnitzki 204b;

Walters Art Museum 150t, c;

Wellcome Collection (public domain) 45;

Wikipedia Commons and/or public domain 43, 131c, 194b, 320;

Harry Winston 117b, 220;

Woolley and Wallis 143;

© York Museums Trust (Yorkshire Museum, YORYM: 1991.43) 52 all;

Zendrini 253–4, 255 all;

Every effort has been made to trace all copyright holders and we apologize in advance for any unintentional omissions. We would be pleased to insert the appropriate acknowledgement in any subsequent edition of this publication.

ACKNOWLEDGEMENTS

In the early part of 2019, sitting in a crowded tube train at one of London's busiest train stations, I became aware of the person that had just taken the seat next to me. Though I would not normally look up from my newspaper, something made me turn to this person and I found Robert Violette, the editor and creator of this trilogy of books on coloured gemstones, which began in 2013 with *Emerald: Twenty-one Centuries of Jewelled Opulence and Power*, and continued in 2017 with *Ruby: The King of Gems*. We had not seen each other for more than two years, since the publication of *Ruby*. Some would say that it was coincidence, but I believe it was serendipity, as right there and then *Sapphire* was born, a natural progression from our close collaboration on the previous two volumes.

Writing *Sapphire: A Celebration of Colour* has been an enormous journey – one that I have enjoyed immensely – but it would not have happened without the support, encouragement and generosity of a team of people from around the world. My first debt of gratitude is to Sean Gilbertson, CEO of Gemfields, for his support and generosity and for having the conviction to complete this trilogy while still in the early stages of establishing a sapphire mine. I would also like to thank the team at Gemfields: Emily Dungey, Sophie Ebbetts and Helena Choudhury.

There are two people without whom this book would never have materialized. Robert Violette's and Georgina Izzard's tireless support, editorial skills and monumental patience have ensured *Sapphire*'s completion. Georgina has been my assistant for over four years, and I count my lucky stars that I have her guidance, total support and above all else her friendship. I will never be able to thank her enough for all her amazing encouragement, loyalty and enviable organizational skills. A big thank you also goes to copy editor Sarah Kane, who has worked with Robert, my very understanding editor, and also to Frith Kerr and her team at Studio Frith for their beautiful and innovative design. Together, we have produced three books we can be very proud of.

Researching these books has usually involved quite a bit of travel, but the world was struck by the Covid-19 pandemic in March 2020, which put a stop to international plans. Luckily, I was able to visit Madagascar at the beginning of that year to see the sapphire deposits. I would not have been able to achieve this trip without the expert guidance of Marc Noverraz, Algy Strutt and Anirudh Sharma. In 2016, I travelled to Burma (Myanmar) and visited the sapphire deposits, which would not have been possible without the direction and company of field gemmologist Vincent Pardieu.

Boundless gratitude from me and the editorial team goes to the jewellery houses that have shared their archive material and expert knowledge, especially, from Chaumet, Hélène Yvert and Clémence De Gasperi; and from Van Cleef & Arpels, Solène Taquet, Lise Macdonald, Sarah Corke, Sophie Marquant, Julia Dubreuil and Pascalie Leroy. Additionally, I would like to thank, at Belperron and Verdura, Caroline Perkowski, Caroline Packer, Ward Landrigan and Nico Landrigan; at Boucheron, Emma Beckett and her team and Anaëlle Rouquerol-Gauchard; at Bulgari, Monica Brannetti and Carlotta Sapia; at Garrard, Nadia Carlin, Claire Scott and Sara Prentice; at Graff, Hannah Kayani and Alexandra Paillot; at Mellerio, Diane-Sophie Lanselle; at Tiffany, Cristina Vignone and Annamarie Sandecki; and at Harry Winston, Fernanda Reynoso and Orla Colgan. I also extend my thanks to, at Grima, Jojo Grima and Francesca Grima; at Hennell, Chandra Narang and Hafeez Anjarwalla; at Henn of London, Ingo Henn; and John Donald and Russell Elliot.

Grateful thanks are also due to many historic institutions worldwide, and their staff, including, at the State Hermitage Museum in St Petersburg, Anastasia Mikliaeva and Elena Obuhovich; at the Lalique Museum in Tokyo, Sakiyo Hayashida; at the Victoria and Albert Museum, Olivia Stroud; and at the Goldsmiths' Company in London, Dr Dora Thornton and Charlie Spurrier. Numerous other museums and public collections have been critically helpful in providing important sapphire pieces for this publication, including the British Museum, Royal Collection Trust, Museum of London, American Museum of Natural History, Kunsthistorisches Museum Wien, Metropolitan Museum of Art, Pforzheim Museum, Hillwood Museum, Hofburg Treasury, Smithsonian Institution, Residenz Munich and others.

Many of the jewels illustrated have been sold through auctions and I would like to thank, for their help in sourcing images and gaining permissions, at Bonhams, Emily Barber; at Christie's, Rahul Kadakia, Keith Penton, Alexandra Kindermann and Marie-Cécile Cisamolo; and, at Sotheby's, Edward Hall-Smith and Rachel Barber.

Within the industry there are many highly acclaimed experts that have graciously and generously helped me: Dr Jack Ogden, Derek Content and Dr Usha Balakrishnan; from the Museum of London, Hazel Forsyth; Martin Travis and Sophie Jackson from Symbolic & Chase; Dr Michael Krzemnicki from SSEF; Catherine Taylor from Sandra Cronan; Noëlle Viguurs-van Gelder from Van Gelder Jewellery; Santpal Sinchawla; Vincent Pardieu, again; Salih Maricar of Maricar Gems; Dr Keith Barron of Potentate Mining at Rock Creek; Alan Hart from Gem-A; Pia Tonna from Fuli Gemstones; Charles Bexfield of Adit Fine Minerals; Franck Notari of GGTL; Francesca Cartier Brickell; John Benjamin; Beth Bernstein; Elizabeth Gage; Roger Elliott for setting my Madagascan sapphire; and my daughter, Chloe Emerson, for help with research.

The Art of Collecting chapter would not have been possible without the expert guidance and enormous support of Catherine Cariou, along with the generosity of the collectors who have shared their jewels with me, some of whom remain anonymous. I am also indebted to Mr Arikawa and Keiko Horii from Albion Art; Mrs Alisa Moussaieff, David Warren and Violet Fraser; and Carlo Zendrini and Ferdinando Ferrero.

To Christopher Smith and Bilal Mahmood at American Gemological Laboratories, thank you for the fabulous inclusion images you have so generously donated and for your close collaboration on their selection; they make all three of these books come alive. I would like to thank the numerous contemporary jewellers included in this book for sharing their creativity.

To my husband Craig and my two children, Scott and Chloe, thank you for your understanding and patience while putting up with me locked away for months on end in my office for the third time.

Sapphire is the result of professional relationships and friendships that I have been lucky to acquire over my long career and by which I am so very humbled. Writing this book during a pandemic has highlighted their importance to me all the more, and I remain deeply appreciative of and grateful for their vital contribution to my work.

JOANNA HARDY

THE GEMMOLOGICAL WORLD OF SAPPHIRE
GEORGINA IZZARD

Sapphire is a variety of the gem species corundum. Corundum is the crystalline form of aluminium oxide (Al_2O_3) that grows in single crystals large enough to be cut for use in jewellery. Corundum in its 'pure' state of aluminium oxide is a colourless mineral, but it is common for trace amounts of chemical impurities to mix with the aluminium oxide during the crystal's growing process; these trace elements, including iron, titanium, chromium and vanadium, absorb and transmit different wavelengths of visible light, and it is these different interactions with light that cause us to see the stones as different colours. The term 'sapphire' refers to corundum with trace amounts of iron and titanium, which make the stone blue. Another popular variety of corundum is ruby, coloured by chromium. Corundum occurs naturally in all colours and these stones are known as sapphires too, but we distinguish them from blue sapphires by including their colour in their name: pink sapphire, green sapphire, yellow sapphire, and so on. The orangey-pink variety of sapphire is known as padparadscha. There is no scientific demarcation between a pink sapphire and a ruby – classifying colour nuances is subjective.

Corundum is a durable mineral. It has a hardness of 9 on the Mohs hardness scale (a 1–10 scale that orders minerals based on their ability to scratch each other), and its hardness means the surface of sapphire responds well to polishing. Most transparent sapphires are selected for faceting, as faceted cuts show off the stone's clarity and create brilliance from the reflection of light from internal and external facet faces. Many translucent and opaque sapphires are cut as cabochons, with domed surfaces, to make the most of the stone's colour. For sapphires that have many needle-like rutile inclusions, the cabochon cut directs light to reveal a star-like intersection of bands of reflected light – an optical effect known as asterism. The direction of the needle-like inclusions of these star sapphires, also found in smaller quantities in other sapphires, follows the hexagonal symmetry of the sapphire's crystal structure.

Corundum belongs to the trigonal crystal system, a subdivision of the hexagonal system. Understanding sapphire's crystal system makes it easier to identify rough stones and appreciate how light interacts with the stone, which influences how the stone is cut. Sapphires typically occur in a hexagonal tabular or bipyramid – two hexagonal pyramids placed end to end – crystal habit (the shape the stone grows in). As these crystals grow, so their chemical composition can alter, and this change is evident in hexagonal colour zoning through the crystal.

Crystals of the trigonal system, like corundum, are anisotropic. Anisotropic stones split a ray of light that passes through them into two rays; the two rays vibrate perpendicularly to each other and have minor differences in their wavelengths. These very slight variations are enough to make transparent and translucent sapphires display two different colours, an effect known as dichroism. When we look at a sapphire from one direction, we see more of one of the rays than the other, and the stone may appear bluer; when we then rotate the sapphire and look at it from another direction, we see more of the other ray, which may appear greener. Cutting sapphires involves considering which direction is likely to produce the best marriage of the two colours – a consideration that is particularly important for achieving the full potential of the subtle orangey-pink of padparadscha sapphires. Sapphire can also show a colour change that is distinct from its dichroism. Most of these colour-change sapphires appear blue/violet in daylight and fluorescent light and reddish purple in incandescent light; this is because a trace amount of vanadium in their chemical composition affects how the crystals absorb and reflect particular wavelengths of light.

Blue sapphires are typically from metamorphic or basalt-related magmatic geological origins. Basaltic sapphires are those that were brought to the Earth's surface as foreign crystals in volcanic eruptions of alkali-rich basaltic magma. Metamorphic sapphires were created as the tectonic plates of the Earth's crust collided to form mountain ranges; the pressure of this collision was so great that rocks recrystallized together as new metamorphic rocks. Their different growing conditions give the sapphires different characteristics, particularly in colour, crystal size and inclusion types.

However, sapphires from different deposits around the world can have similar characteristics because these areas were once closer together. Hundreds of millions of years ago, Earth's landmass was concentrated in supercontinents, areas of land much larger than our continents today. One supercontinent, Gondwana, was at its largest around 600–530 million years ago and covered approximately one fifth of the Earth's crust. Gondwana centred over the South Pole and included the areas we now refer to as Africa, Madagascar, Sri Lanka, the Indian subcontinent, Australia, Antarctica and South America. The formation and separation of Gondwana provided the geological conditions necessary for the creation of sapphires, particularly during mountain building and eruptive events, and is the reason we find groups of similar sapphires across the world. Metamorphic sapphires have been found in Sri Lanka, Burma (Myanmar), Kashmir (today on the borders of India, Pakistan and China) and Madagascar (which was once connected to India and Sri Lanka); these sapphires are often lighter in colour and occur in a range of colours. Basaltic sapphires have been found in Australia, Thailand, Cambodia, Nigeria and Ethiopia; these sapphires are typically dark greenish-blue stones. Other sapphire deposits, including those in Montana (USA), Canada, France, Scandinavia, Indonesia and Scotland, are more scattered and have been less extensive and predictable in yield than the more substantial deposits already introduced. Within sapphires, mineral inclusions often reflect the origin of the parent rock, and these variations help us to assess from which deposit a stone may have been mined. Typical sapphire inclusions are rutile needles often called 'silk', healed fissures and crystals of mica, zircon and calcite. Inclusions can be both informative and beautiful, as demonstrated in this book's full-page micrographic images. These remarkable photographs of inclusions and colour zoning, by Christopher P Smith and Bilal Mahmood of American Gemological Laboratories, help us understand and appreciate each sapphire's unique characteristics.

Most sapphires are found in secondary deposits, having broken away from their host rock through magma eruptions or weathering and been set down by now-extinct rivers in gravel beds; we find these sapphires as water-worn pebbles. These secondary deposits are mined by digging pits and washing and sifting the extracted gravel; this process is known as an open-pit operation. The depleted Kashmir mines are an example of sapphire found in a primary deposit of marble.

Linking geological origins to geography can lead to an overemphasis on origin reports and their monetization – the idea that stones from particular deposits are more valuable. Sapphires from Madagascar, a deposit only recently discovered, display similar inclusions to those from Sri Lanka; Madagascan stones are therefore sometimes sold as Sri Lankan sapphires, since an attribution to this more-established deposit is thought to drive up prices. Many sapphires are treated, typically by being heated to specific temperatures; this affects their crystal structure, which can thereby alter their colour. As all deposits yield both high- and low-quality stones, it is more important to check for treatments and choose a sapphire that appeals to you than it is to ascertain its origin.

A pair of repoussé gold earclips, 2014,
by Leo de Vroomen, each set with a low oval
cabochon of Burmese pale grey-blue sapphire
that shows angular hexagonal growth patterns,
reflecting the shape of sapphire's crystal system.

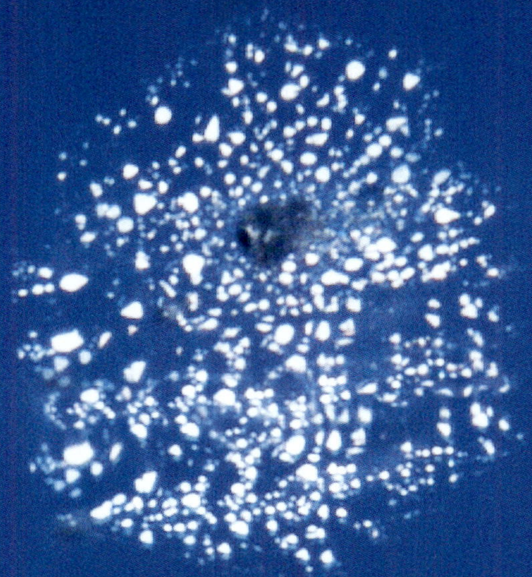

Basal-oriented thin film,
Kashmir.

Growth and colour zoning,
Andranondambo, Madagascar.

Milky zonal clouds,
Kashmir.

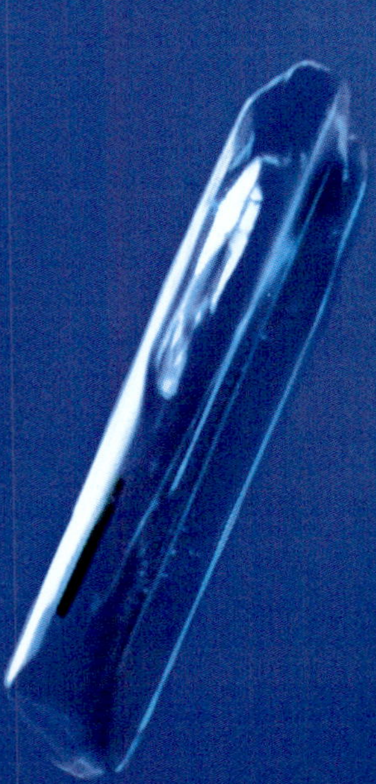

Pargasite crystal,
Kashmir.

Snowflake inclusions,
Kashmir.

V—V dipyramidal growth zoning,
Kashmir.

Highly resorbed plagioclase crystals, Kashmir.

Milky zonal clouds,
Madagascar.